This book comes with access to more content online.

Quiz yourself, track your progress,
and improve your grade!

Register your book or ebook at
www.dummies.com/go/getaccess.

Select your product, and then follow the prompts
to validate your purchase.

You'll receive an email with your PIN and instructions.

Basic Math & Pre-Algebra

ALL-IN-ONE

by Mark Zegarelli

Basic Math & Pre-Algebra All-in-One For Dummies®

Published by: **John Wiley & Sons, Inc.**, 111 River Street, Hoboken, NJ 07030-5774, www.wiley.com

Copyright © 2022 by John Wiley & Sons, Inc., Hoboken, New Jersey

Published simultaneously in Canada

For general information on our other products and services, please contact our Customer Care Department within the U.S. at 877-762-2974, outside the U.S. at 317-572-3993, or fax 317-572-4002. For technical support, please visit https://hub.wiley.com/community/support/dummies.

Wiley publishes in a variety of print and electronic formats and by print-on-demand. Some material included with standard print versions of this book may not be included in e-books or in print-on-demand. If this book refers to media such as a CD or DVD that is not included in the version you purchased, you may download this material at http://booksupport.wiley.com. For more information about Wiley products, visit www.wiley.com.

Library of Congress Control Number: 2022932610

ISBN 978-1-119-86708-1 (pbk); ISBN 978-1-119-86748-7 (ebk); ISBN 978-1-119-86726-5 (ebk)

SKY10033495_032922

Contents at a Glance

Table of Contents

UNIT 6: REACHING THE SUMMIT: ADVANCED PRE-ALGEBRA TOPICS 327

Introduction

Does math really need to be so hard? Nope.

I say this speaking as a guy who has struggled with math as much as, if not more than, you have. Believe me. And a big part of the struggle often has more to do with the lack of clarity in how math is explained than with the actual math.

This is too bad, because the whole idea behind math is supposed to be *clarity.* In a world where so many things are unclear, $2 + 2$ will always equal 4.

My second-greatest joy in teaching math is when a light breaks across a student's face as they suddenly understand something new. My greatest joy, though, is what often follows: a skepticism that it couldn't possibly be this easy.

When you approach math right, it's almost always easier than you think. And a lot of the stuff that hung you up when you first saw it probably isn't all that scary after all. Many students feel they got lost somewhere along the way on the road between learning to count to ten and their first day in an algebra class — and this may be true whether they're 14 or 104. If this is you, don't worry. You're not alone, and help is right here!

About This Book

This book brings together the four components you need to make sense of math:

» Clear explanations of each topic

» Example questions with step-by-step answers

» Plenty of practice problems (with more available online!)

» Chapter quizzes to test your knowledge at the end of most chapters

Although you can certainly work through this book from beginning to end, you don't have to. Feel free to jump directly to whatever chapter has the type of problems you want to practice. When you've worked through enough problems in a section to your satisfaction, feel free to jump to a different section. If you find the problems in a section too difficult, flip back to an earlier section or chapter to practice the skills you need — just follow the cross-references.

Foolish Assumptions

If you're planning to read this book, you likely fall into one of these categories:

>> A student who wants a solid understanding of the basics of math for a class or test you're taking

>> An adult who wants to improve skills in arithmetic, fractions, decimals, percentages, weights and measures, geometry, algebra, and so on for when you have to use math in the real world

>> Someone who wants a refresher so you can help another person understand math

My only assumption about your skill level is that you can add, subtract, multiply, and divide. So, to find out whether you're ready for this book, take this simple test:

$$4 + 7 = \underline{\quad} \qquad 13 - 5 = \underline{\quad} \qquad 9 \times 3 = \underline{\quad} \qquad 35 \div 7 = \underline{\quad}$$

If you can answer these four questions correctly (the answers are 11, 8, 27, and 5), you're ready to begin.

Icons Used in This Book

You'll see the following five icons throughout the book:

Each example is a math question based on the discussion and explanation, followed by a solution. Work through these examples, and then refer to them to help you solve the practice problems that follow them, as well as the quiz questions at the end of the chapter.

This icon points out important information that you need to focus on. Make sure you understand this information fully before moving on. You can skim through these icons when reading a chapter to make sure you remember the highlights.

This icon points out hints that can help speed you along when answering a question. You should find them useful when working on practice problems.

This icon flags common mistakes that students make if they're not careful. Take note and proceed with caution!

When you see this icon, it's time to put on your thinking cap and work out a few practice problems on your own. The answers and detailed solutions are available so you can feel confident about your progress.

Beyond the Book

In addition to the book you're reading right now, be sure to check out the free Cheat Sheet on Dummies.com. This handy Cheat Sheet covers some common "math demons" that students often stumble over. To access it, simply go to Dummies.com and type **Basic Math & Pre-Algebra All in One Cheat Sheet** in the Search box.

You'll also have access to online quizzes related to each chapter, starting with Chapter 3. These quizzes provide a whole new set of problems for practice and confidence-building. To access the quizzes, follow these simple steps:

1. **Register your book or ebook at Dummies.com to get your PIN.** Go to www.dummies.com/go/getaccess.

2. **Select your product from the drop-down list on that page.**

3. **Follow the prompts to validate your product, and then check your email for a confirmation message that includes your PIN and instructions for logging in.**

If you do not receive this email within two hours, please check your spam folder before contacting us through our Technical Support website at http://support.wiley.com or by phone at 877-762-2974.

Now you're ready to go! You can come back to the practice material as often as you want — simply log on with the username and password you created during your initial login. No need to enter the access code a second time.

Your registration is good for one year from the day you activate your PIN.

Where to Go from Here

You can use this book in a variety of ways. If you're reading without immediate time pressure from a test or homework assignment, start at the beginning and keep going, chapter by chapter, to the end. If you do this, you'll be surprised by how much of the math you may have been dreading will be almost easy. Additionally, setting up some solid groundwork is a great way to prepare for what follows later in the book.

If your time is limited — especially if you're taking a math course and you're looking for help with your homework or an upcoming test — skip directly to the topic you're studying. Wherever you open the book, you can find a clear explanation of the topic at hand, as well as a variety of hints and tricks. Read through the examples and try to do them yourself, or use them as templates to help you with assigned problems.

1

Getting Started with Basic Math & Pre-Algebra

In This Unit . . .

Chapter **1**

Playing the Numbers Game

One useful characteristic of numbers is that they're *conceptual,* which means that, in an important sense, they're all in your head. (This fact probably won't get you out of having to know about them, though — nice try!)

For example, you can picture three of anything: three cats, three baseballs, three tigers, three planets. But just try to picture the concept of three all by itself, and you find it's impossible. Oh, sure, you can picture the numeral 3, but *threeness* itself — much like love or beauty or honor — is beyond direct understanding. But when you understand the *concept* of three (or four, or a million), you have access to an incredibly powerful system for understanding the world: mathematics.

In this chapter, I give you a brief history of how numbers likely came into being. I discuss a few common *number sequences* and show you how these connect with simple math *operations* like addition, subtraction, multiplication, and division.

After that, I describe how some of these ideas come together with a simple yet powerful tool: the *number line.* I discuss how numbers are arranged on the number line, and I also show you how to use the number line as a calculator for simple arithmetic. Finally, I describe how the *counting numbers* (1, 2, 3, . . .) sparked the invention of more unusual types of numbers, such as *negative numbers, fractions,* and *irrational numbers.* I also show you how these *sets of numbers* are *nested* — that is, how one set of numbers fits inside another, which fits inside another.

Inventing Numbers

Historians believe that the first written number systems came into being at the same time as agriculture and commerce. Before that, people in prehistoric, hunter-gatherer societies were pretty much content to identify bunches of things as "a lot" or "a little." They may have had concepts of small numbers, probably less than five or ten, but lacked a coherent way to think about, for example, the number 42.

Throughout the ages, the Babylonians, Egyptians, Greeks, Hindus, Romans, Mayans, Arabs, and Chinese (to name just a few) all developed their own systems of writing numbers.

Although Roman numerals gained wide currency as the Roman Empire expanded throughout Europe and parts of Asia and Africa, the more advanced system that was invented in India and adapted by the Arabs turned out to be more useful. Our own number system, the Hindu-Arabic numbers (also called decimal numbers), is mainly derived from these earlier number systems.

Understanding Number Sequences

Although humans invented numbers for counting commodities, as I explain in the preceding section, they soon put them to use in a wide range of applications. Numbers were useful for measuring distances, counting money, amassing armies, levying taxes, building pyramids, and lots more.

But beyond their many uses for understanding the external world, numbers have an internal order all their own. So numbers are not only an *invention*, but equally a *discovery*: a landscape reflecting fundamental truths about nature, and how humans think about it, that seems to exist independently, with its own structure, mysteries, and even perils.

One path into this new and often strange world is the *number sequence:* an arrangement of numbers according to a rule. In the following sections, I introduce you to a variety of number sequences that are useful for making sense of numbers.

Evening the odds

One of the first facts you probably heard about numbers is that all of them are either even or odd. For example, you can split an even number of marbles *evenly* into two equal piles. But when you try to divide an odd number of marbles the same way, you always have one *odd*, left-over marble. Here are the first few even numbers:

$$2 \quad 4 \quad 6 \quad 8 \quad 10 \quad 12 \quad 14 \quad 16 \dots$$

You can easily keep the sequence of even numbers going as long as you like. Starting with the number 2, keep adding 2 to get the next number.

Similarly, here are the first few odd numbers:

$$1 \quad 3 \quad 5 \quad 7 \quad 9 \quad 11 \quad 13 \quad 15 \ldots$$

The sequence of odd numbers is just as simple to generate. Starting with the number 1, keep adding 2 to get the next number.

Patterns of even or odd numbers are the simplest number patterns around, which is why kids often figure out the difference between even and odd numbers soon after learning to count.

Counting by threes, fours, fives, and so on

When you get used to the concept of counting by numbers greater than 1, you can run with it. For example, here's what counting by threes, fours, and fives looks like:

Threes:	3	6	9	12	15	18	21	24 ...
Fours:	4	8	12	16	20	24	28	32 ...
Fives:	5	10	15	20	25	30	35	40 ...

Counting by a given number is a good way to begin learning the multiplication table for that number, especially for the numbers you're kind of sketchy on. (In general, people seem to have the most trouble multiplying by 7, but 8 and 9 are also unpopular.)

These types of sequences are also useful for understanding factors and multiples, which you get a look at in Chapter 9.

Getting square with square numbers

When you study math, sooner or later, you probably want to use visual aids to help you see what numbers are telling you. (Later in this book, I show you how one picture can be worth a thousand numbers when I discuss geometry in Chapter 19 and graphing in Chapter 25.)

The tastiest visual aids you'll ever find are those little square cheese-flavored crackers. (You probably have a box sitting somewhere in the pantry. If not, saltine crackers or any other square food works just as well.) Shake a bunch out of a box and place the little squares together to make bigger squares. Figure 1-1 shows the first few.

FIGURE 1-1:
Square numbers.

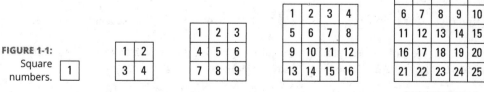

© John Wiley & Sons, Inc.

Voilà! The square numbers:

$$1 \quad 4 \quad 9 \quad 16 \quad 25 \quad 36 \quad 49 \quad 64 \ldots$$

TIP

You get a *square number* by multiplying a number by itself, so knowing the square numbers is another handy way to remember part of the multiplication table. Although you probably remember without help that $2 \times 2 = 4$, you may be sketchy on some of the higher numbers, such as $7 \times 7 = 49$. Knowing the square numbers gives you another way to etch that multiplication table forever into your brain.

Square numbers are also a great first step on the way to understanding exponents, which I introduce later in this chapter and explain in more detail in Chapter 5.

Composing yourself with composite numbers

Some numbers can be placed in rectangular patterns. Mathematicians probably should call numbers like these "rectangular numbers," but instead they chose the term *composite numbers*. For example, 12 is a composite number because you can place 12 objects in rectangles of two different shapes, as in Figure 1-2.

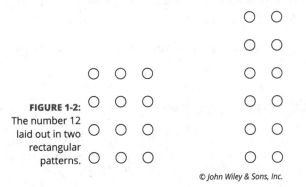

FIGURE 1-2:
The number 12 laid out in two rectangular patterns.

© John Wiley & Sons, Inc.

As with square numbers, arranging numbers in visual patterns like this tells you something about how multiplication works. In this case, by counting the sides of both rectangles, you find out the following:

$$3 \times 4 = 12$$
$$2 \times 6 = 12$$

Similarly, other numbers such as 8 and 15 can also be arranged in rectangles, as in Figure 1-3.

As you can see, both these numbers are quite happy being placed in boxes with at least two rows and two columns. And these visual patterns show this:

$$2 \times 4 = 8$$
$$3 \times 5 = 15$$

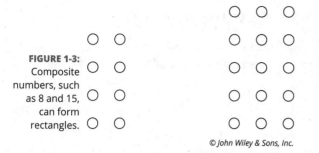

FIGURE 1-3: Composite numbers, such as 8 and 15, can form rectangles.

The word *composite* means that these numbers are *composed of* smaller numbers. For example, the number 15 is composed of 3 and 5 — that is, when you multiply these two smaller numbers, you get 15. Here are all the composite numbers from 1 to 16:

4   6   8   9   10   12   14   15   16

Notice that all the square numbers (see the section, "Getting square with square numbers") also count as composite numbers because you can arrange them in boxes with at least two rows and two columns. Additionally, a lot of other non-square numbers are also composite numbers.

Stepping out of the box with prime numbers

Some numbers are stubborn. Like certain people you may know, these numbers — called *prime numbers* — resist being placed in any sort of a box. Look at how Figure 1-4 depicts the number 13, for example.

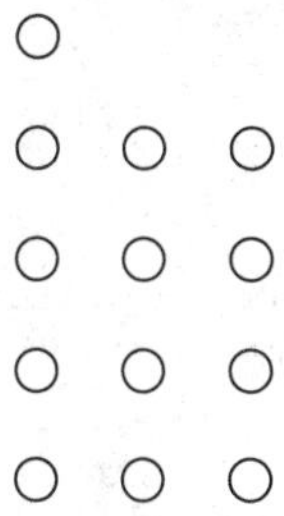

FIGURE 1-4: Unlucky 13, a prime example of a number that refuses to fit in a box.

Try as you may, you just can't make a rectangle out of 13 objects. (That fact may be one reason why the number 13 got a bad reputation as unlucky.) Here are all the prime numbers less than 20:

2   3   5   7   11   13   17   19

As you can see, the list of prime numbers fills the gaps left by the composite numbers (see the preceding section). Therefore, every counting number is either prime or composite. The only exception is the number 1, which is neither prime nor composite. In Chapter 8, I give you a lot more information about prime numbers and show you how to *decompose* a number — that is, break down a composite number into its prime factors.

Multiplying quickly with exponents

Here's an old question whose answer may surprise you: Suppose you took a job that paid you just 1 penny the first day, 2 pennies the second day, 4 pennies the third day, and so on, doubling the amount every day, like this:

$$1 \quad 2 \quad 4 \quad 8 \quad 16 \quad 32 \quad 64 \quad 128 \quad 256 \quad 512 \ldots$$

As you can see, in the first ten days of work, you would've earned a little more than $10 (actually, $10.23 — but who's counting?). How much would you earn in 30 days? Your answer may well be, "I wouldn't take a lousy job like that in the first place." At first glance, this looks like a good answer, but here's a glimpse at your second ten days' earnings:

$$\ldots 1{,}024 \quad 2{,}048 \quad 4{,}096 \quad 8{,}192 \quad 16{,}384 \quad 32{,}768 \quad 65{,}536 \quad 131{,}072 \quad 262{,}144 \quad 524{,}288 \ldots$$

By the end of the second 10 days, when you add it all up, your total earnings would be over $10,000. And by the end of 30 days, your earnings would top out around $10,000,000! How does this happen? Through the magic of exponents (also called *powers*). Each new number in the sequence is obtained by multiplying the previous number by 2:

$$2^1 = 2$$
$$2^2 = 2 \times 2 = 4$$
$$2^3 = 2 \times 2 \times 2 = 8$$
$$2^4 = 2 \times 2 \times 2 \times 2 = 16$$

As you can see, the notation 2^4 means *multiply 2 by itself 4 times.*

You can use exponents on numbers other than 2. Here's another sequence you may be familiar with:

$$1 \quad 10 \quad 100 \quad 1{,}000 \quad 10{,}000 \quad 100{,}000 \quad 1{,}000{,}000 \ldots$$

In this sequence, every number is 10 times greater than the number before it. You can also generate these numbers using exponents:

$$10^1 = 10$$
$$10^2 = 10 \times 10 = 100$$
$$10^3 = 10 \times 10 \times 10 = 1{,}000$$
$$10^4 = 10 \times 10 \times 10 \times 10 = 10{,}000$$

REMEMBER

This sequence is important for defining *place value*, the basis of the decimal number system, which I discuss in Chapter 3. It also shows up when I discuss decimals in Chapter 13 and scientific notation in Chapter 17. You find out more about exponents in Chapter 5.

Four Important Sets of Numbers

In the preceding section, you see how a variety of number sequences extend infinitely. In this section, I provide a quick tour of how numbers fit together as a set of nested systems, one inside the other.

TIP

When I talk about a set of numbers, I'm really just talking about a group of numbers. You can use the number line to deal with four important sets of numbers.

>> **Counting numbers (also called natural numbers):** The set of numbers beginning 1, 2, 3, 4 and going on infinitely

>> **Integers:** The set of counting numbers, zero, and negative counting numbers

>> **Rational numbers:** The set of integers and fractions

>> **Real numbers:** The set of rational and irrational numbers

The sets of counting numbers, integers, rational, and real numbers are nested, one inside another. This nesting of one set inside another is similar to the way that a city (for example, Boston) is inside a state (Massachusetts), which is inside a country (the United States), which is inside a continent (North America). The set of counting numbers is inside the set of integers, which is inside the set of rational numbers, which is inside the set of real numbers.

Counting on the counting numbers

The set of *counting numbers* is the set of numbers you first count with, starting with 1. Because they seem to arise naturally from observing the world, they're also called the *natural numbers:*

     1   2   3   4   5   6   7   8   9...

The counting numbers are infinite, which means they go on forever.

REMEMBER

When you add two counting numbers, the answer is always another counting number. Similarly, when you multiply two counting numbers, the answer is always a counting number. Another way of saying this is that the set of counting numbers is *closed* under both addition and multiplication.

Introducing integers

The set of *integers* arises when you try to subtract a larger number from a smaller one. For example, $4 - 6 = -2$. The set of integers includes the following:

>> The counting numbers

>> Zero

>> The negative counting numbers

Here's a partial list of the integers:

$$\ldots -4 \quad -3 \quad -2 \quad -1 \quad 0 \quad 1 \quad 2 \quad 3 \quad 4 \ldots$$

Like the counting numbers, the integers are closed under addition and multiplication. Similarly, when you subtract one integer from another, the answer is always an integer. That is, the integers are also closed under subtraction.

Staying rational

Here's the set of *rational numbers*:

>> Integers (which include the counting numbers, zero, and the negative counting numbers)

>> Fractions

Like the integers, the rational numbers are closed under addition, subtraction, and multiplication. Furthermore, when you divide one rational number by another, the answer is always a rational number. Another way to say this is that the rational numbers are closed under division.

Getting real

Even if you filled in all the rational numbers, you'd still have points left unlabeled on the number line. These points are the irrational numbers.

An *irrational number* is a number that's neither a whole number nor a fraction. In fact, an irrational number can only be approximated as a *non-repeating decimal.* In other words, no matter how many decimal places you write down, you can always write down more; furthermore, the digits in this decimal never become repetitive or fall into any pattern. (For more on repeating decimals, see Chapter 13.)

The most famous irrational number is π (you find out more about π when I discuss the geometry of circles in Chapter 19):

$$\pi = 3.1415926535897932384626433832795028841971693993751 0\ldots$$

Together, the rational and irrational numbers make up the *real numbers,* which comprise every point on the number line. In this book, I don't spend too much time on irrational numbers, but just remember that they're there for future reference.

multiplication, and division)

» **Adding larger numbers with and without carrying**

» **Subtracting larger numbers with and without borrowing**

» **Multiplying with one-digit and multiple-digit multipliers**

» **Knowing how to do long division step by step**

Chapter **2**

The Big Four Operations

The operations of addition, subtraction, multiplication, and division are at the heart of arithmetic. I call them the Big Four operations.

In this chapter, the focus is exclusively on these four operations. To begin, I discuss each of the Big Four operations in turn. Then, I show you how to add, subtract, multiply, and divide larger numbers.

Even if this information isn't new to you, make sure that you know it before moving on to the chapters that follow. In Chapter 4, I discuss negative numbers, which depend heavily on the operation of subtraction. And then in Chapter 5, you discover a variety of more advanced Big Four topics that I expand upon in later chapters of the book.

The Big Four Operations

When most folks think of math, the first thing that comes to mind is four little (or not-so-little) words: addition, subtraction, multiplication, and division. I call these operations the *Big Four* all through the book.

In this chapter, I introduce you (or reintroduce you) to these little gems. Although I assume you're already familiar with the Big Four, this chapter reviews these operations, taking you from what you may have missed to what you need to succeed as you move onward and upward in math.

Adding things up: Addition

Addition is the first operation you find out about, and it's almost everybody's favorite. It's simple, friendly, and straightforward. No matter how much you worry about math, you've probably never lost a minute of sleep over addition. Addition is all about bringing things together, which is a positive goal. For example, suppose you and I are standing in line to buy tickets for a movie. I have $25 and you have only $5. I could lord it over you and make you feel crummy that I can go to the movies and you can't. Or instead, you and I can join forces, adding together my $25 and your $5 to make $30. Now, not only can we both see the movie, but we may even be able to buy some popcorn, too.

Addition uses only one sign — the plus sign (+): Your equation may read or $2 + 3 = 5$, or $12 + 2 = 14$, or $27 + 44 = 71$, but the plus sign always means the same thing.

When you add two numbers together, those two numbers are called *addends,* and the result is called the *sum.* So in the first example, the addends are 2 and 3, and the sum is 5.

REMEMBER

Take it away: Subtraction

Subtraction is usually the second operation you discover, and it's not much harder than addition. Still, there's something negative about subtraction — it's all about who has more and who has less. Suppose you and I have been running on treadmills at the gym. I'm happy because I ran 3 miles, but then you start bragging that you ran 10 miles. You subtract and tell me that I should be very impressed that you ran 7 miles farther than I did. (But with an attitude like that, don't be surprised if you come back from the showers to find your running shoes filled with liquid soap!)

As with addition, subtraction has only one sign: the minus sign (–). You end up with equations such as $4 - 1 = 3$, and $14 - 13 = 1$, and $93 - 74 = 19$.

When you subtract one number from another, the result is called the *difference*. This term makes sense when you think about it: When you subtract, you find the difference between a higher number and a lower one.

REMEMBER

One of the first facts you probably heard about subtraction is that you can't take away more than you start with. In that case, the second number can't be larger than the first. And if the two numbers are the same, the result is always 0. For example, $3 - 3 = 0$; $11 - 11 = 0$; and $1,776 - 1,776 = 0$. Later, someone breaks the news that you *can* take away more than you have. When you do, though, you need to place a negative sign in front of the difference to show that you have a negative number, a number below 0:

$$4 - 5 = -1$$
$$10 - 13 = -3$$
$$88 - 99 = -11$$

TIP

When subtracting a larger number from a smaller number, remember the words *switch* and *negate*: You *switch* the order of the two numbers and do the subtraction as you normally would, but at the end, you *negate* the result by attaching a negative sign. For example, to find 10 – 13, you switch the order of these two numbers, giving you 13 – 10, which equals 3; then you negate this result to get –3. That's why 10 – 13 = –3.

WARNING

The negative sign does double duty, so don't get confused. When you stick a negative sign between two numbers, it means the first number minus the second number. But when you attach it to the front of a number, it means that this number is a negative number.

A sign of the times: Multiplication

Multiplication is often described as a sort of shorthand for repeated addition. For example,

4×3 means add 4 to itself 3 times: $4 + 4 + 4 = 12$

9×6 means add 9 to itself 6 times: $9 + 9 + 9 + 9 + 9 + 9 = 54$

100×2 means add 100 to itself 2 times: $100 + 100 = 200$

Although multiplication isn't as warm and fuzzy as addition, it's a great timesaver. For example, suppose you coach a Little League baseball team, and you've just won a game against the toughest team in the league. As a reward, you promised to buy three hot dogs for each of the nine players on the team. To find out how many hot dogs you need, you can add 3 together 9 times. Or you can save time by multiplying 3 times 9, which gives you 27. Therefore, you need 27 hot dogs (plus a whole lot of mustard and sauerkraut).

REMEMBER

When you multiply two numbers, the two numbers that you're multiplying are called *factors*, and the result is the *product*.

When you're first introduced to multiplication, you use the times sign ($\times$). As you move onward and upward on your math journey, you need to be aware of the conventions I discuss in the following sections.

REMEMBER

The symbol $\cdot$ is sometimes used to replace the symbol $\times$. For example,

$4 \cdot 2 = 8$ means $4 \times 2 = 8$

$6 \cdot 7 = 42$ means $6 \times 7 = 42$

$53 \cdot 11 = 583$ means $53 \times 11 = 583$

In Units 1 through 4 of this book, I stick to the tried-and-true symbol $\times$ for multiplication. Just be aware that the symbol $\cdot$ exists so that you won't be stumped if your teacher or textbook uses it.

REMEMBER

In math beyond arithmetic, using parentheses without another operator stands for multiplication. The parentheses can enclose the first number, the second number, or both numbers. For example,

$3(5) = 15$ means $3 \times 5 = 15$

$(8)7 = 56$ means $8 \times 7 = 56$

$(9)(10) = 90$ means $9 \times 10 = 90$

This switch makes sense when you stop to consider that the letter *x*, which is often used in algebra, looks a lot like the multiplication sign ×. So in this book, when I start using *x* to discuss algebra topics in Unit 7, I also stop using × and begin using parentheses without another sign to indicate multiplication.

Doing math lickety-split: Division

The last of the Big Four operations is division. Division literally means splitting things up. For example, suppose you're a parent on a picnic with your three children. You've brought along 12 pretzel sticks as snacks, and want to split them fairly so that each child gets the same number (don't want to cause a fight, right?).

Each child gets four pretzel sticks. This problem tells you that

$$12 \div 3 = 4$$

As with multiplication, division also has more than one sign: the division sign (÷) and the fraction slash (/) or fraction bar (—). So, some other ways to write the same information are

$$12 / 3 = 4 \text{ and } \frac{12}{3} = 4$$

Whichever way you write it, the idea is the same: When you divide 12 pretzel sticks equally among three people, each person gets 4 of them.

REMEMBER

When you divide one number by another, the first number is called the *dividend*, the second is called the *divisor*, and the result is the *quotient*. For example, in the division from the earlier example, the dividend is 12, the divisor is 3, and the quotient is 4.

Applying the Big Four Operations to Larger Numbers

So, be honest. Are you're feeling a bit shaky about how to apply the Big Four operations to larger numbers, especially long division? If so, don't worry. Just use this section as a handy reference for remembering how to do stacked addition, subtraction, and multiplication, as well as everybody's favorite, long division.

Calculating stacked addition

Stacked addition allows you to add large numbers in a systematic way.

For example, to add $323 + 425$, stack the numbers up in a column format as follows:

$$\begin{array}{r} 323 \\ +\,425 \\ \hline \end{array}$$

Next, starting in the ones' column and working from right to left, add the numbers and place each result below the line:

$$
\begin{array}{r}
323 \\
+425 \\
\hline
748
\end{array}
$$

Thus, $323 + 425 = 748$.

In most cases when you add, the calculation will involve *carrying* at least one digit from one column to the next.

For example, suppose you want to add $391 + 67 + 784$. Begin by adding $1 + 7 + 4 = 12$ in the ones' column, then write the 2 below the line and carry the 1 to the next column:

$$
\begin{array}{r}
\overset{1}{3}91 \\
67 \\
+784 \\
\hline
2
\end{array}
$$

Next, add $1 + 9 + 6 + 8 = 24$ in the tens' column, then write the 4 below the line and carry the 2:

$$
\begin{array}{r}
\overset{2\,1}{3}91 \\
67 \\
+784 \\
\hline
42
\end{array}
$$

To complete the problem, add $2 + 3 + 7 = 12$:

$$
\begin{array}{r}
\overset{2\,1}{3}91 \\
67 \\
+784 \\
\hline
1242
\end{array}
$$

Therefore, $391 + 67 + 784 = 1,242$.

Performing stacked subtraction

As with addition, *stacked subtraction* allows you to subtract larger numbers systematically.

For example, to subtract $768 - 512$, stack the numbers as follows:

$$
\begin{array}{r}
768 \\
-512 \\
\hline
\end{array}
$$

Now, as with addition, start in the ones' column and work right to left, this time subtracting each pair of numbers in the column, and place each result below the line:

$$
\begin{array}{r}
768 \\
-512 \\
\hline
256
\end{array}
$$

So $768 - 512 = 256$.

In most cases when you subtract, the calculation will involve *borrowing* at least one digit from one column to the next.

For example, suppose you want to subtract $853 - 164$:

$$
\begin{array}{r}
853 \\
-164 \\
\hline
\end{array}
$$

To begin, you want to subtract $3 - 4$ in the ones' column. However, 3 is less than 4, so to subtract, you need to *borrow* 1 from the tens' column, changing the 5 to 4. Then, this 1 that you borrowed is worth 10 in the ones' column, so add it to the 3, changing it to 13:

$$
\begin{array}{r}
8\,\overset{4}{\cancel{5}}\,{}^1 3 \\
-1\ 6\ \ 4 \\
\hline
\end{array}
$$

Now, you can subtract $13 - 4 = 9$, and place this result in the ones' column below the line:

$$
\begin{array}{r}
8\,\overset{4}{\cancel{5}}\,{}^1 3 \\
-1\ 6\ \ 4 \\
\hline
9
\end{array}
$$

Next, you want to subtract $4 - 6$ in the tens' column. Again, you need to borrow 1 from the column to the left, changing 8 to 7, and adding 10 to the 4, making it 14:

$$
\begin{array}{r}
\overset{7}{\cancel{8}}\,\overset{14}{\cancel{5}}\,{}^1 3 \\
-1\ 6\ \ 4 \\
\hline
9
\end{array}
$$

Now, you can subtract $14 - 6 = 8$, placing 8 in the tens' column below the line:

$$
\begin{array}{r}
\overset{7}{\cancel{8}}\,\overset{14}{\cancel{5}}\,{}^1 3 \\
-1\ 6\ \ 4 \\
\hline
8\ 9
\end{array}
$$

To complete the problem, subtract $7 - 1 = 6$ in the hundreds' column:

$$
\begin{array}{r}
\overset{7}{\cancel{8}}\overset{14}{\cancel{5}}{}^{1}3 \\
-1\ 6\ 4 \\
\hline
6\ 8\ 9
\end{array}
$$

Therefore, $853 - 164 = 689$.

Calculating with stacked multiplication

Stacked multiplication provides a systematic way to multiply a pair of larger numbers.

For example, to multiply $8,732 \times 4$, set up the multiplication as shown here, then begin by multiplying $2 \times 4 = 8$ as follows:

$$
\begin{array}{r}
8732 \\
\times\ \ \ 4 \\
\hline
8
\end{array}
$$

Next, multiply $3 \times 4 = 12$. Because this result is more than one digit, place the 2 next to the 8 and carry the 1, as you would when adding:

$$
\begin{array}{r}
\overset{1}{8}732 \\
\times\ \ \ 4 \\
\hline
28
\end{array}
$$

Now, multiply $7 \times 4 = 28$, then add the 1 you carried, $28 + 1 = 29$. This time, write the 9 below the line and carry the 2:

$$
\begin{array}{r}
\overset{2\ 1}{8}732 \\
\times\ \ \ 4 \\
\hline
928
\end{array}
$$

To finish the problem, multiply $8 \times 4 = 32$, then add the 2 you carried, $32 + 2 = 34$, completing the problem as follows:

$$
\begin{array}{r}
\overset{2\ 1}{8}732 \\
\times\ \ \ 4 \\
\hline
34928
\end{array}
$$

Therefore, $8,732 \times 4 = 34,928$.

When multiplying by a number that has more than one digit, multiply each digit in turn by the top number, and then finish by adding the results. For example, to multiply 94×78, begin by multiplying 94×8 as follows:

$$
\begin{array}{r}
\overset{3}{9}4 \\
\times\ 78 \\
\hline
752
\end{array}
$$

Next, multiply 94×7:

$$
\begin{array}{r}
\overset{2}{9}4 \\
\times\ 78 \\
\hline
752 \\
658
\end{array}
$$

Before moving on, notice that I have lined up this result, 658, with the multiplier, 7. To complete the problem, add these two results as follows:

$$
\begin{array}{r}
\overset{2}{9}4 \\
\times\ 78 \\
\hline
752 \\
+\ 658 \\
\hline
7332
\end{array}
$$

Therefore, $94 \times 78 = 7{,}332$.

Understanding long division

Long division allows you to divide larger numbers in a systematic way.

For example, to divide $347 \div 6$, set up the problem as follows:

$$6\overline{)347}$$

Begin by trying to divide 6 into the first digit of 347 by itself. Because 3 is less than 6, this division doesn't work. Next, try dividing 6 into 34. In this case, 6 goes into 34 as many as 5 times, so record this answer over the 4:

$$
\begin{array}{r}
5 \\
6\overline{)347}
\end{array}
$$

Now, multiply $5 \times 6 = 30$, placing this result below 34:

$$
\begin{array}{r}
5 \\
6\overline{)347} \\
\underline{30}
\end{array}
$$

To complete this step, subtract $34 - 30 = 4$ and bring down the 7 next to this result:

$$
\begin{array}{r}
5 \\
6{\overline{)347}} \\
\underline{30} \\
47
\end{array}
$$

At this point, you have a new number to try dividing 6 into: 6 goes into 47 no more than 7 times. Now, multiply $7 \times 6 = 42$:

$$
\begin{array}{r}
57 \\
6{\overline{)347}} \\
\underline{30} \\
47 \\
\underline{42}
\end{array}
$$

To finish up the problem, subtract $47 - 42 = 5$:

$$
\begin{array}{r}
57 \\
6{\overline{)347}} \\
\underline{30} \\
47 \\
\underline{42} \\
5
\end{array}
$$

Thus, $347 \div 6 = 57$ r 5 — that is, 57 with a remainder of 5.

As another example, suppose you want to divide $745,853 \div 1,006$. Set up the problem as follows:

$$
1006{\overline{)745853}}
$$

To begin, note that you can divide 1,006 into 7,458 a maximum of 7 times. Thus, write 7 above the 8, multiply $7 \times 1,006 = 7,042$, then subtract and bring down the 5:

$$
\begin{array}{r}
7 \\
1006{\overline{)745853}} \\
\underline{7042} \\
4165
\end{array}
$$

Now, you can divide 1,006 into 4,165 a maximum of 4 times. So write 4 above the line after the 7, multiply $4 \times 1,006 = 4,024$, then subtract and bring down the 3:

$$
\begin{array}{r}
74 \\
1006{\overline{)745853}} \\
\underline{7042} \\
4165 \\
\underline{4024} \\
1413
\end{array}
$$

To finish the problem, you can divide 1,006 into 1,413 just 1 time. So this time, write 1 above the line after the 4, multiply $1 \times 1,006 = 1,006$, and subtract:

$$
\begin{array}{r}
741 \\
1006\overline{)745853} \\
\underline{7042} \\
4165 \\
\underline{4024} \\
1413 \\
\underline{1006} \\
407
\end{array}
$$

Therefore, $745,853 \div 1,006 = 741 \text{ r } 407$ — that is, 741 with a remainder of 407.

2

The Big Four Operations: Addition, Subtraction, Multiplication, and Division

In This Unit . . .

Chapter **3**

Counting on Success: Numbers and Digits

When you're counting, ten seems to be a natural stopping point — a nice, round number. The fact that our ten fingers match up so nicely with numbers may seem like a happy accident. But of course, it's no accident at all. Fingers were the first calculator that humans possessed. Our number system — the Hindu-Arabic numbers, also called the decimal numbers — is based on the number ten because humans have 10 fingers instead of 8 or 12. In fact, the very word *digit* has two meanings: numerical symbol and finger.

In this chapter, I show you how place value turns digits into numbers. I also show you when 0 is an important placeholder in a number and why leading zeros don't change the value of a number. And I show you how to read and write long numbers. After that, I discuss two important skills: rounding numbers and estimating values.

TELLING THE DIFFERENCE BETWEEN NUMBERS AND DIGITS

Sometimes people confuse numbers and digits. For the record, here's the difference:

- A digit is a single numerical symbol, from 0 to 9.

- A number is a string of one or more digits.

For example, 7 is both a digit and a number. In fact, it's a one-digit number. However, 15 is a string of two digits, so it's a number — a two-digit number. And 426 is a three-digit number. You get the idea.

In a sense, a digit is like a letter of the alphabet. By themselves, the uses of 26 letters, A through Z, are limited. (How much can you do with a single letter such as K or W?) Only when you begin using strings of letters as building blocks to spell words does the power of letters become apparent. Similarly, the ten digits, 0 through 9, have limited usefulness until you begin building strings of digits — that is, numbers.

Knowing Your Place Value

The number system you're most familiar with — Hindu-Arabic numbers — has ten familiar digits:

$$0 \quad 1 \quad 2 \quad 3 \quad 4 \quad 5 \quad 6 \quad 7 \quad 8 \quad 9$$

Yet with only ten digits, you can express numbers as high as you care to go. In this section, I show you how it happens.

Counting to ten and beyond

The ten digits in our number system allow you to count from 0 to 9. All higher numbers are produced using place value. Place value assigns a digit a greater or lesser value, depending on where it appears in a number. Each place in a number is ten times greater than the place to its immediate right.

To understand how a whole number gets its value, suppose you write the number 45,019 all the way to the right in Table 3-1, one digit per cell, and add up the numbers you get.

Table 3-1   45,019 Displayed in a Place-Value Chart

Millions			Thousands			Ones		
Hundred Millions	Ten Millions	Millions	Hundred Thousands	Ten Thousands	Thousands	Hundreds	Tens	Ones
				4	5	0	1	9

You have 4 ten thousands, 5 thousands, 0 hundreds, 1 ten, and 9 ones. The chart shows you that the number breaks down as follows:

$$45,019 = 40,000 + 5,000 + 0 + 10 + 9$$

In this example, notice that the presence of the digit 0 in the hundreds place means that zero hundreds are added to the number.

Telling placeholders from leading zeros

Although the digit 0 adds no value to a number, it acts as a placeholder to keep the other digits in their proper places. For example, the number 5,001,000 breaks down into 5,000,000 + 1,000. Suppose, however, you decide to leave all the 0s out of the chart. Table 3-2 shows what you'd get.

Table 3-2   5,001,000 Displayed Incorrectly without Placeholding Zeros

Millions			Thousands			Ones		
Hundred Millions	Ten Millions	Millions	Hundred Thousands	Ten Thousands	Thousands	Hundreds	Tens	Ones
							5	1

The chart tells you that 5,001,000 = 50 + 1. Clearly, this answer is wrong!

REMEMBER

As a rule, when a 0 appears to the right of at least one digit other than 0, it's a placeholder. Placeholding zeros are important — always include them when you write a number. However, when a 0 appears to the left of every digit other than 0, it's a leading zero. Leading zeros serve no purpose in a number, so dropping them is customary. For example, place the number 003,040,070 on the chart (see Table 3-3).

Table 3-3   3,040,070 Displayed with Two Leading Zeros

Millions			Thousands			Ones		
Hundred Millions	Ten Millions	Millions	Hundred Thousands	Ten Thousands	Thousands	Hundreds	Tens	Ones
0	0	3	0	4	0	0	7	0

The first two 0s in the number are leading zeros because they appear to the left of the 3. You can drop these 0s from the number, leaving you with 3,040,070. The remaining 0s are all to the right of the 3, so they're placeholders — be sure to write them in.

Reading long numbers

When you write a long number, you use commas to separate groups of three numbers. For example, here's about as long a number as you'll ever see:

234,845,021,349,230,467,304

Table 3-4 shows a larger version of the place-value chart.

Table 3-4   A Place-Value Chart Separated by Commas

Quintillions	Quadrillions	Trillions	Billions	Millions	Thousands	Ones
234	845	021	349	230	467	304

This version of the chart helps you read the number. Begin all the way to the left and read, "Two hundred thirty-four quintillion, eight hundred forty-five quadrillion, twenty-one trillion, three hundred forty-nine billion, two hundred thirty million, four hundred sixty-seven thousand, three hundred four."

REMEMBER

When you read and write whole numbers, don't say the word *and*. In math, the word *and* means you have a decimal point. That's why, when you write a check (does anyone still write checks?), you save the word *and* for the number of cents, which is usually expressed as a decimal or sometimes as a fraction. (I discuss decimals in Chapter 13.)

Close Enough for Rock 'n' Roll: Rounding and Estimating

As numbers get longer, calculations become tedious, and you're more likely to make a mistake or just give up. When you're working with long numbers, simplifying your work by rounding numbers and estimating values is sometimes helpful.

When you round a number, you change some of its digits to placeholding zeros. And when you estimate a value, you work with rounded numbers to find an approximate answer to a problem. In this section, you build both skills.

Rounding numbers

Rounding numbers makes long numbers easier to work with. In this section, I show you how to round numbers to the nearest ten, hundred, thousand, and beyond.

Rounding numbers to the nearest ten

The simplest kind of rounding you can do is with two-digit numbers. When you round a two-digit number to the nearest ten, you simply bring it up or down to the nearest number that ends in 0. For example:

$39 \rightarrow 40 \qquad 51 \rightarrow 50 \qquad 73 \rightarrow 70$

Even though numbers ending in 5 are in the middle, always round them up to the next-highest number that ends in 0:

$$15 \to 20 \qquad 35 \to 40 \qquad 85 \to 90$$

Numbers in the upper 90s get rounded up to 100:

$$99 \to 100 \qquad 95 \to 100 \qquad 94 \to 90$$

When you know how to round a two-digit number, you can round just about any number. For example, to round most longer numbers to the nearest ten, just focus on the ones and tens digits:

$$7\underline{34} \to 730 \qquad 1,4\underline{88} \to 1,490 \qquad 12,3\underline{45} \to 12,350$$

Occasionally, a small change to the ones and tens digits affects the other digits. (This situation is a lot like when the odometer in your car rolls a bunch of 9s over to 0s.) For example:

$$8\underline{99} \to 900 \qquad 1,0\underline{97} \to 1,100 \qquad 9,9\underline{95} \to 10,000$$

Rounding numbers to the nearest hundred and beyond

To round numbers to the nearest hundred, thousand, or beyond, focus only on two digits: the digit in the place you're rounding to and the digit to its immediate right. Change all other digits to the right of these two digits to 0s. For example, suppose you want to round 642 to the nearest hundred. Focus on the hundreds digit (6) and the digit to its immediate right (4):

642

I've underlined these two digits. Now just round these two digits as if you were rounding to the nearest ten, and change the digit to the right of them to a 0:

$$\underline{64}2 \to 600$$

Here are a few more examples of rounding numbers to the nearest hundred:

$$7,\underline{89}1 \to 7,900 \qquad 15,\underline{75}3 \to 15,800 \qquad 99,\underline{96}1 \to 100,000$$

When rounding numbers to the nearest thousand, underline the thousands digit and the digit to its immediate right. Round the number by focusing only on the two underlined digits and, when you're done, change all digits to the right of these to 0s:

$$\underline{4,9}84 \to 5,000 \qquad 7\underline{8,5}21 \to 79,000 \qquad 1,0\underline{99,3}04 \to 1,099,000$$

Even when rounding to the nearest million, the same rules apply:

$$\underline{1,2}34,5674 \to 1,000,000 \qquad 7\underline{8,8}83,958 \to 79,000,000$$

1 Round the following numbers to the nearest ten:

(a) 83

(b) 217

(c) 1,885

(d) 6,496

2 Round the following numbers to the nearest hundred:

(a) 347

(b) 2,251

(c) 7,950

(d) 39,974

3 Round the following numbers to the nearest thousand:

(a) 6,543

(b) 9,287

(c) 22,501

(d) 799,643

4 Round the following numbers to the nearest million:

(a) 5,454,545

(b) 10,730,421

(c) 98,765,432

(d) 5,555,555,555

Estimating value to make problems easier

When you know how to round numbers, you can use this skill in estimating values. Estimating saves you time by allowing you to avoid complicated computations and still get an approximate answer to a problem.

REMEMBER

When you get an approximate answer, you don't use an equals sign; instead, you use this wavy symbol, which means *is approximately equal to*: ≈.

Suppose you want to add these number: $722 + 506 + 383 + 1,279 + 91 + 811$. This computation is tedious, and you may make a mistake. But you can make the addition easier by first rounding all the numbers to the nearest hundred and then adding:

$$\approx 700 + 500 + 400 + 1,300 + 100 + 800 = 3,800$$

The approximate answer is 3,800. This answer isn't far off from the exact answer, which is 3,792.

 5 Estimate $63 + 39 + 84 + 71 + 28$ by rounding all values to the nearest ten before adding.

6 Estimate $724 + 531 + 887 + 1,245 + 2,191$ by rounding all values to the nearest hundred before adding.

7 Estimate $3,288 + 640 - 1,192 + 327 - 1,556$ by rounding all values to the nearest hundred before adding and subtracting.

8 Estimate $8,790 + 21,234 - 16,215 - 4,444 - 2,529$ by rounding all values to the nearest thousand before adding and subtracting.

Practice Questions Answers and Explanations

(1)

(a) **80.** $\underline{83} \rightarrow 80$

(b) **220.** $2\underline{17} \rightarrow 220$

(c) **1,890.** $1,8\underline{85} \rightarrow 1,890$

(d) **6,500.** $6,\underline{496} \rightarrow 6,500$

(2)

(a) **300.** $\underline{347} \rightarrow 300$

(b) **2,300.** $2,\underline{251} \rightarrow 2,300$

(c) **8,000.** $7,\underline{950} \rightarrow 8,000$

(d) **40,000.** $39,\underline{974} \rightarrow 40,000$

(3)

(a) **7,000.** $\underline{6},543 \rightarrow 7,000$

(b) **9,000.** $\underline{9},287 \rightarrow 9,000$

(c) **23,000.** $2\underline{2},501 \rightarrow 23,000$

(d) **800,000.** $7\underline{99},643 \rightarrow 800,000$

(4)

(a) **5,000,000.** $\underline{5},454,545 \rightarrow 5,000,000$

(b) **11,000,000.** $1\underline{0},730,421 \rightarrow 11,000,000$

(c) **99,000,000.** $9\underline{8},765,432 \rightarrow 99,000,000$

(d) **5,556,000,000.** $5,55\underline{5},555,555 \rightarrow 5,556,000,000$

(5) **280.** $63 + 39 + 84 + 71 + 28 \approx 60 + 40 + 80 + 70 + 30 = 280$. The answer before rounding? It's 285!

(6) **5,500.** $724 + 531 + 887 + 1,245 + 2,191 \approx 700 + 500 + 900 + 1,200 + 2,200 = 5,500$. The answer before rounding is 5,578.

(7) **1,400.** $3,288 + 640 - 1,192 + 327 - 1,556 \approx 3,300 + 600 - 1,200 + 300 - 1,600 = 1,400$. The answer before rounding: 1,507.

(8) **7,000.** $8,790 + 21,234 - 16,215 - 4,444 - 2,529 \approx 9,000 + 21,000 - 16,000 - 4,000 - 3,000 = 7,000$. The answer before rounding? It's 6,836!

In the section that follows, you'll find a quiz that tests your rounding and estimating skills from this chapter.

Whaddya Know? Chapter 3 Quiz

Answer these 12 questions to test the skills you learned in this chapter. When you're done, flip to the next section for answers and explanations.

1. Which digit is in the ten-thousands place of the number 34,921,706,488?

2. Which digit is in the hundred-thousands place of the number 987,654,321,000?

3. Which digit is in the millions place of the number 7,261,945,803?

4. Round the number 875,921 to the nearest hundred.

5. Round the number 875,921 to the nearest thousand.

6. Round the number 805,921 to the nearest hundred-thousand.

7. Round the number 805,921 to the nearest million.

8. Round the number 55,555 to the nearest ten.

9. Estimate the sum by rounding each number to the nearest ten before adding: $142 + 857 + 555 + 323$

10. Estimate the sum by rounding each number to the nearest hundred before adding: $142 + 857 + 555 + 323$

11. Estimate the result by rounding each number to the nearest ten before adding and subtracting: $952 - 857 + 509 - 313$

12. Estimate the result by rounding each number to the nearest million before adding and subtracting: $87,435,212 + 123,556,432 - 16,758,000 - 1,412,123$

Answers to Chapter 3 Quiz

(1) **0.** Ten-thousands place: 34,921,7**0**6,488?

(2) **3.** Hundred-thousands place: 987,654,**3**21,000?

(3) **1.** Millions place: 7,26**1**,945,803?

(4) **875,900.** 875,$\underline{9}$21 → 875,900

(5) **876,000.** 875,$\underline{9}$21 → 876,000

(6) **800,000.** $\underline{8}$05,921 → 800,000

(7) **1,000,000.** Add a leading 0 to the beginning of the number, in the millions place. $\underline{0}$,805,921 → 1,000,000

(8) **555,560.** 555,5$\underline{5}$5 → 555,560

(9) **1,880.** 1$\underline{4}$2 + 8$\underline{5}$7 + 5$\underline{5}$5 + 3$\underline{2}$3 → 140 + 860 + 560 + 320 = 1,880. What is the sum if you don't round? It's 1,877!

(10) **1,900.** $\underline{1}$42 + $\underline{8}$57 + $\underline{5}$55 + $\underline{3}$23 → 100 + 900 + 600 + 300 = 1,900

(11) **290.** 9$\underline{5}$2 − 8$\underline{5}$7 + 5$\underline{0}$9 − 3$\underline{1}$3 → 950 − 860 + 510 − 310 = 290. The answer before rounding? It's 291!

(12) **193,000,000.** 87,$\underline{4}$35,212 + 123,$\underline{5}$56,432 − 16,$\underline{7}$58,000 − 1,$\underline{4}$12,123

→ 87,000,000 + 124,000,000 − 17,000,000 − 1,000,000 = 193,000,000. The answer before rounding is 192,821,521.

Chapter **4**

Staying Positive with Negative Numbers

I n this chapter, you work with negative numbers — that is, numbers that are less than zero. To begin, you see how negative numbers arise when you subtract a smaller number minus a greater one. Next, you discover how to negate a number by flipping its sign. You also work with absolute value, which is the positive value of a number.

When you're comfortable working with negative numbers, you begin to use them with the Big Four operations — addition, subtraction, multiplication, and division.

Understanding Where Negative Numbers Come From

When you first discovered subtraction, you were probably told that you can't take a small number minus a greater number. For example, if you start with four marbles, you can't subtract six because you can't take away more marbles than you have. This rule is true for marbles, but in other situations, you *can* subtract a big number from a small one.

In real-world applications, negative numbers can represent debt. For example, if you have only five chairs to sell but a customer pays for eight of them, you owe them three more chairs. Even though you may have trouble picturing –3 chairs, you still need to account for this debt, and negative numbers are the right tool for the job.

As another example, if you have $4 and you buy something that costs $6, you end up with less than $0 dollars — that is, –$2, which means a debt of $2.

A number with a negative sign in front of it, like –2, is called a *negative number.* You call the number –2 either *negative two* or *minus two.* Negative numbers appear on the number line to the left of 0, as shown in Figure 4-1.

FIGURE 4-1:
Negative numbers on the number line.

When you don't have a number line to work with, here's a simple rule for subtracting a large number from a small number: Switch the two numbers around and take the small number from large number; then attach a negative sign to the result.

REMEMBER

EXAMPLE

Q. Use the number line to subtract 5 – 8.

A. **–3.** On the number line, 5 – 8 means

Q. What is $11 - 19$?

A. **–8.** Because 11 is less than 19, subtract $19 - 11$, which equals 8, and attach a negative sign to the result. Therefore, $11 - 19 = -8$.

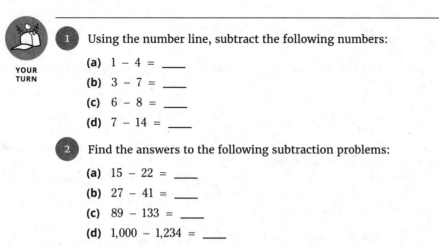

YOUR TURN

1 Using the number line, subtract the following numbers:

(a) $1 - 4 =$ ____

(b) $3 - 7 =$ ____

(c) $6 - 8 =$ ____

(d) $7 - 14 =$ ____

2 Find the answers to the following subtraction problems:

(a) $15 - 22 =$ ____

(b) $27 - 41 =$ ____

(c) $89 - 133 =$ ____

(d) $1,000 - 1,234 =$ ____

Sign-Switching: Understanding Negation and Absolute Value

REMEMBER

When you attach a negative sign to any number, you *negate* that number. Negating a number means changing its sign to the opposite sign, so

>> Attaching a negative sign to a positive number makes it negative.

>> Attaching a negative sign to a negative number makes it positive. The two *adjacent* (side-by-side) negative signs cancel each other out.

>> Attaching a negative sign to 0 doesn't change its value, so $-0 = 0$.

In contrast to negation, placing two bars around a number gives you the absolute value of that number. *Absolute value* is the number's distance from 0 on the number line — that is, it's the positive value of a number, regardless of whether you started out with a negative or positive number:

>> The absolute value of a positive number is the same number.

>> The absolute value of a negative number makes it a positive number.

>> Placing absolute value bars around 0 doesn't change its value, so $|0| = 0$.

>> Placing a negative sign outside absolute value bars gives you a negative result — for example, $-|6| = -6$, and $-|-6| = -6$.

EXAMPLE

Q. Negate the number 7.

A. **–7.** Negate 7 by attaching a negative sign to it: –7.

Q. Find the negation of –3.

A. **3.** The negation of –3 is $-(-3)$. The two adjacent negative signs cancel out, which gives you 3.

Q. What's the negation of 7 – 12?

A. **5.** First, do the subtraction, which tells you $7 - 12 = -5$. To find the negation of –5, attach a negative sign to the answer: $-(-5)$. The two *adjacent* negative signs cancel out, which gives you 5.

Q. What number does $|9|$ equal?

A. **9.** The number 9 is already positive, so the absolute value of 9 is also 9.

Q. What number does $|-17|$ equal?

A. **17.** Because –17 is negative, the absolute value of –17 is 17.

Q. Solve this absolute value problem: $-|9 - 13| = ?$

A. **–4.** Do the subtraction first: $9 - 13 = -4$, which is negative, so the absolute value of –4 is 4. But the negative sign on the left (outside the absolute value bars in the original expression) negates this result, so the answer is –4.

YOUR TURN

3 Negate each of the following numbers and expressions by attaching a negative sign and then canceling out negative signs when possible:

(a) 6

(b) –29

(c) 0

(d) 10 + 4

(e) 15 – 7

(f) 9 – 10

4 Solve the following absolute value problems:

(a) $|7| = ?$

(b) $|-11| = ?$

(c) $|3 + 15| = ?$

(d) $-|10 - 1| = ?$

(e) $|1 - 10| = ?$

(f) $|0| = ?$

Addition and Subtraction with Negative Numbers

The great secret to adding and subtracting negative numbers is to turn every problem into a series of ups and downs on the number line. When you know how to do this, you find that all these problems are quite simple.

So in this section, I explain how to add and subtract negative numbers on the number line. Don't worry about memorizing every little bit of this procedure. Instead, just follow along so you get a sense of how negative numbers fit onto the number line. (If you need a quick refresher on how the number line works, see Chapter 1.)

Starting with a negative number

When you're adding and subtracting on the number line, starting with a negative number isn't much different from starting with a positive number. For example, suppose you want to calculate $-3 + 4$. Using the up and down rule, you start at -3 and go up 4:

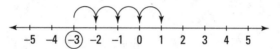

So $-3 + 4 = 1$.

Similarly, suppose you want to calculate $-2 - 5$. Again, the up and down rule helps you out. You're subtracting, so move to the left: start at -2, down 5:

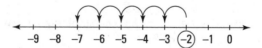

So $-2 - 5 = -7$.

Adding a negative number

Suppose you want to calculate $-2 + (-4)$. You already know to start at -2, but where do you go from there? Here's the up and down rule for adding a negative number:

REMEMBER

Adding a negative number is the same as subtracting a positive number — go *down* on the number line.

By this rule, $-2 + (-4)$ is the same as $-2 - 4$, so start at -2, down 4:

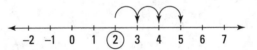

So $-2 + (-4) = -6$.

TIP

If you rewrite a subtraction problem as an addition problem — for instance, rewriting $3 - 7$ as $3 + (-7)$ — you can use the commutative and associative properties of addition, which I discuss earlier in this chapter. Just remember to keep the negative sign attached to the number when you rearrange the problem: $(-7) + 3$.

Subtracting a negative number

The last rule you need to know is how to subtract a negative number. For example, suppose you want to calculate $2 - (-3)$. Here's the up and down rule:

REMEMBER

Subtracting a negative number is the same as adding a positive number — go *up* on the number line.

This rule tells you that $2 - (-3)$ is the same as $2 + 3$, so start at 2, up 3:

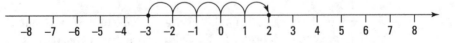

So $2 - (-3) = 5$.

TIP

When subtracting negative numbers, you can think of the two negative signs canceling each other out to create a positive.

EXAMPLE

Q. Use the number line to add $-3 + 5$.

A. **2.** On the number line, $-3 + 5$ means *start at* -3, *up* 5, which brings you to 2:

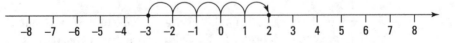

Q. Use the number line to add $6 + (-2)$.

A. **4.** On the number line, $6 + (-2)$ means *start at 6, down 2*, which brings you to 4:

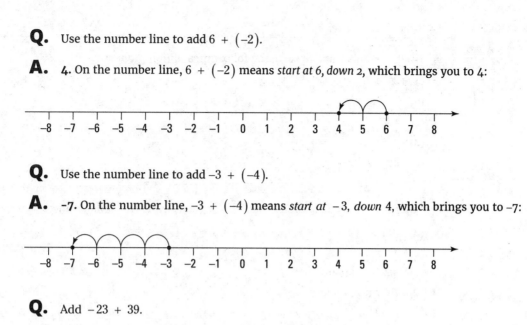

Q. Use the number line to add $-3 + (-4)$.

A. **–7.** On the number line, $-3 + (-4)$ means *start at* -3, *down* 4, which brings you to –7:

Q. Add $-23 + 39$.

A. **16.** Switch around the two numbers with the signs attached:

$-23 + 39 = +39 - 23$

Now you can drop the plus sign and use the negative sign for subtraction:

$39 - 23 = 16$

Q. Use the number line to subtract $-1 - 4$.

A. **–5.** On the number line, $-1 - 4$ means *start at* –1, *down* 4, which brings you to –5:

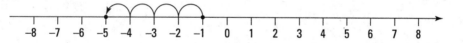

Q. Use the number line to subtract $-5 - (-3)$.

A. **–2.** The expression $-5 - (-3)$ means *start at* –5, *up* 3 (because the two negative signs cancel each other out), which brings you to –2:

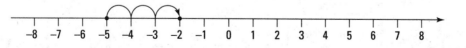

YOUR TURN

5 Use the number line to solve the following addition problems:

(a) $-5 + 6$

(b) $-1 + (-7)$

(c) $4 + (-6)$

(d) $-3 + 9$

(e) $2 + (-1)$

(f) $-4 + (-4)$

6 Solve the following addition problems without using the number line:

(a) $-17 + 35$

(b) $29 + (-38)$

(c) $-61 + (-18)$

(d) $70 + (-63)$

(e) $-112 + 84$

(f) $-215 + (-322)$

7 Use the number line to solve the following subtraction problems:

(a) $-3 - 4$

(b) $5 - (-3)$

(c) $-1 - (-8)$

(d) $-2 - 4$

(e) $-4 - 2$

(f) $-6 - (-10)$

8 Solve the following subtraction problems without using the number line:

(a) $17 - (-26)$

(b) $-21 - 45$

(c) $-42 - (-88)$

(d) $-67 - 91$

(e) $75 - (-49)$

(f) $-150 - (-79)$

Knowing Signs of the Times (and Division) for Negative Numbers

Multiplication and division with negative numbers is virtually the same as with positive numbers. The presence of one or more negative signs (–) doesn't change the numerical part of the answer. The only question is whether the sign is positive or negative:

Just remember that when you multiply or divide two numbers,

>> If the numbers have the *same sign,* the result is always positive.

>> If the numbers have *opposite signs,* the result is always negative.

For example,

$$2 \times 3 = 6 \qquad 2 \times (-3) = -6$$
$$-2 \times (-3) = 6 \qquad -2 \times 3 = -6$$

As you can see, the numerical portion of the answer is always 6. The only question is whether the complete answer is 6 or –6. That's where the rule of same or opposite signs comes in.

Another way of thinking of this rule is that the two negatives cancel each other out to make a positive.

Similarly, look at these four division equations:

$$10 \div 2 = 5 \qquad 10 \div (-2) = -5$$
$$-10 \div (-2) = 5 \qquad -10 \div 2 = -5$$

In this case, the numerical portion of the answer is always 5. When the signs are the same, the result is positive, and when the signs are different, the result is negative.

Q. Solve the following four multiplication problems:

$$5 \times 6 = \underline{\hspace{1cm}}$$
$$-5 \times 6 = \underline{\hspace{1cm}}$$
$$5 \times (-6) = \underline{\hspace{1cm}}$$
$$-5 \times (-6) = \underline{\hspace{1cm}}$$

A. As you can see from this example, the numerical part of the answer (30) doesn't change. Only the sign of the answer changes, depending on the signs of the two numbers in the problem.

$$5 \times 6 = \mathbf{30}$$
$$-5 \times 6 = \mathbf{-30}$$
$$5 \times (-6) = \mathbf{-30}$$
$$-5 \times (-6) = \mathbf{30}$$

Q. Solve the following four division problems:

$$18 \div 3 = \underline{\hspace{1cm}}$$
$$-18 \div 3 = \underline{\hspace{1cm}}$$
$$18 \div (-3) = \underline{\hspace{1cm}}$$
$$-18 \div (-3) = \underline{\hspace{1cm}}$$

A. The numerical part of the answer (6) doesn't change. Only the sign of the answer changes, depending on the signs of the two numbers in the problem.

$$18 \div 3 = \mathbf{6}$$
$$-18 \div 3 = \mathbf{-6}$$
$$18 \div (-3) = \mathbf{-6}$$
$$-18 \div (-3) = \mathbf{6}$$

Q. What is -84×21?

A. **–1,764.** First, drop the signs and multiply:

$$84 \times 21 = 1{,}764$$

The numbers -84 and 21 have different signs, so the answer is negative: $-1{,}764$.

Q. What is $-580 \div (-20)$?

A. **29.** Drop the signs and divide:

$$580 \div 20 = 29$$

The numbers -580 and -20 have the same signs, so the answer is positive: 29.

YOUR TURN

9 Solve the following multiplication problems:

   (a) $7 \times 11 = $ ____

   (b) $-7 \times 11 = $ ____

   (c) $7 \times (-11) = $ ____

   (d) $-7 \times (-11) = $ ____

10 Solve the following division problems:

   (a) $32 \div (-8) = $ ____

   (b) $-32 \div (-8) = $ ____

   (c) $-32 \div 8 = $ ____

   (d) $32 \div 8 = $ ____

11 What is -65×23?

12 Find $-143 \times (-77)$.

13 Calculate $216 \div (-9)$.

14 What is $-3{,}375 \div (-25)$?

Practice Questions Answers and Explanations

1

    (a) −**3.** Start at 1, down 4.

    (b) −**4.** Start at 3, down 7.

    (c) −**2.** Start at 6, down 8.

    (d) −**7.** Start at 7, down 14.

2

    (a) −**7.** Fifteen is less than 22, so subtract $22 - 15 = 7$ and attach a negative sign to the result: −7.

    (b) −**14.** Twenty-seven is less than 41, so subtract $41 - 27 = 14$ and attach a negative sign to the result: −14.

    (c) −**44.** Eighty-nine is less than 133, so subtract $133 - 89 = 44$ and attach a negative sign to the result: −44.

    (d) −**234.** One thousand is less than 1,234, so subtract $1{,}234 - 1{,}000 = 234$ and attach a negative sign to the result: −234.

3

    (a) −**6.** To negate 6, attach a negative sign: −6.

    (b) **29.** To negate −29, attach a negative sign: $-(-29)$. Now cancel adjacent negative signs: $-(-29) = 29$.

    (c) **0.** Zero is its own negation.

    (d) −**14.** Add first: $10 + 4 = 14$, and the negation of 14 is -14.

    (e) −**8.** Subtract first: $15 - 7 = 8$, and the negation of 8 is -8.

    (f) **1.** Begin by subtracting: $9 - 10 = -1$, and the negation of −1 is 1.

4

    (a) **7.** The number 7 is already positive, so the absolute value of 7 is also 7.

    (b) **11.** The number −11 is negative, so the absolute value of −11 is 11.

    (c) **18.** First, do the addition inside the absolute value bars: $3 + 15 = 18$, which is positive. The absolute value of 18 is 18.

    (d) −**9.** First, do the subtraction: $10 - 1 = 9$, which is positive. The absolute value of 9 is 9. You have a negative sign outside the absolute value bars, so negate your answer to get −9.

    (e) **9.** Begin by subtracting: $1 - 10 = -9$, which is negative. The absolute value of −9 is 9.

    (f) **0.** The absolute value of 0 is 0.

(5)

(a) 1. Start at –5, up 6.

(b) –8. Start at –1, down 7.

(c) –2. Start at 4, down 6.

(d) 6. Start at –3, up 9.

(e) 1. Start at 2, down 1.

(f) –8. Start at –4, down 4.

(6)

(a) 18. Switch around the numbers (with their signs) to turn the problem into subtraction:

$$-17 + 35 = 35 - 17 = 18$$

(b) –9. Drop the plus sign to turn the problem into subtraction:

$$29 + (-38) = 29 - 38 = -9$$

(c) –79. Drop the signs, add the numbers, and negate the result:

$$61 + 18 = 79, \text{so} - 61 + (-18) = -79$$

(d) 7. Turn the problem into subtraction:

$$70 + (-63) = 70 - 63 = 7$$

(e) –28. Turn the problem into subtraction:

$$-112 + 84 = 84 - 112 = -28$$

(f) –537. Drop the signs, add the numbers, and negate the result:

$$215 + 322 = 537, \text{so} - 215 + (-322) = -537$$

(7)

(a) –7. Start at –3, down 4.

(b) 8. Start at 5, up 3.

(c) 7. Start at –1, up 8.

(d) –6. Start at –2, down 4.

(e) –6. Start at –4, down 2.

(f) 4. Start at –6, up 10.

8

(a) 43. Cancel the adjacent negative signs to turn the problem into addition:

$$17-(-26)=17+26=43$$

(b) –66. Drop the signs, add the numbers, and negate the result:

$$21+45=66, \text{ so}-21-45=-66$$

(c) 46. Cancel the adjacent negative signs to turn the problem into addition:

$$-42-(-88)=-42+88$$

Now switch around the numbers (with their signs) to turn the problem back into subtraction:

$$88-42=46$$

(d) –158. Drop the signs, add the numbers, and negate the result:

$$67+91=158, \text{ so }-67-91=-158$$

(e) 124. Cancel the adjacent negative signs to turn the problem into addition:

$$75-(-49)=75+49=124$$

(f) –71. Cancel the adjacent negative signs to turn the problem into addition:

$$-150-(-79)=-150+79$$

Now switch around the numbers (with their signs) to turn the problem back into subtraction:

$$79-150=-71$$

9

(a) $7 \times 11 = \mathbf{77}$

(b) $-7 \times 11 = \mathbf{-77}$

(c) $7 \times (-11) = \mathbf{-77}$

(d) $-7 \times (-11) = \mathbf{77}$

10

(a) $32 \div (-8) = \mathbf{-4}$

(b) $-32 \div (-8) = \mathbf{4}$

(c) $-32 \div 8 = \mathbf{-4}$

(d) $32 \div 8 = \mathbf{4}$

(11) **−1,495.** First, drop the signs and multiply:

$$65 \times 23 = 1,495$$

The numbers −65 and 23 have different signs, so the answer is negative: −1,495.

(12) **11,011.** Drop the signs and multiply:

$$143 \times 77 = 11,011$$

The numbers −143 and −77 have the same sign, so the answer is positive: 11,011.

(13) **−24.** Drop the signs and divide (use long division, as I show you in Chapter 2):

$$216 \div 9 = 24$$

The numbers 216 and −9 have different signs, so the answer is negative: −24.

(14) **135.** First, drop the signs and divide:

$$3,375 \div 25 = 135$$

The numbers −3,375 and −25 have the same sign, so the answer is positive: 135.

The chapter quiz in the next section tests you on your skills working with negative numbers.

Whaddya Know? Chapter 4 Quiz

In the 26 quiz questions that follow, evaluate by performing the operations as indicated.

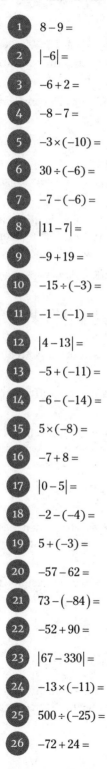

1. $8 - 9 =$

2. $|-6| =$

3. $-6 + 2 =$

4. $-8 - 7 =$

5. $-3 \times (-10) =$

6. $30 \div (-6) =$

7. $-7 - (-6) =$

8. $|11 - 7| =$

9. $-9 + 19 =$

10. $-15 \div (-3) =$

11. $-1 - (-1) =$

12. $|4 - 13| =$

13. $-5 + (-11) =$

14. $-6 - (-14) =$

15. $5 \times (-8) =$

16. $-7 + 8 =$

17. $|0 - 5| =$

18. $-2 - (-4) =$

19. $5 + (-3) =$

20. $-57 - 62 =$

21. $73 - (-84) =$

22. $-52 + 90 =$

23. $|67 - 330| =$

24. $-13 \times (-11) =$

25. $500 \div (-25) =$

26. $-72 + 24 =$

Answers to Chapter 4 Quiz

(1) **-1.** Find the difference; the answer is negative because 9 is farther from 0 than 8 on the number line: $8 - 9 = -(9 - 8) = -1$.

(2) **6.** Drop the negative sign: $|-6| = 6$.

(3) **-4.** Find the difference; the answer is negative because 6 is farther from 0 than 2 on the number line: $-6 + 2 = -(6 - 2) = -4$.

(4) **-15.** The signs are the same, so drop the negative signs and find the sum. The answer is negative: $-8 - 7 = -(8 + 7) = -15$.

(5) **30.** Drop the negative signs and multiply. The product is positive, because the signs are the same: $-3 \times (-10) = 3 \times 10 = 30$.

(6) **-5.** Drop the negative sign and divide. The answer is negative, because the signs are different: $30 \div (-6) = -(30 \div 6) = -5$.

(7) **-1.** Change the subtraction to addition and the negative 6 to +6. Then find the difference. The answer is negative, because the newly adjusted signs are different and 7 is farther from 0 than 6 on the number line: $-7 - (-6) = -7 + (+6) = -(7 - 6) = -1$.

(8) **4.** Find the difference. The result is positive, so applying the absolute value keeps it a positive answer: $|11 - 7| = |4| = 4$.

(9) **10.** Find the difference, because the signs are different. The answer is positive, because 19 is farther from 0 on the number line: $-9 + 19 = +(19 - 9) = 10$.

(10) **5.** Drop the negative signs and divide. The answer is positive, because the signs are the same: $-15 \div (-3) = +(15 \div 3) = +5$.

(11) **0.** Change the subtraction to addition and the second –1 to +1. Then find the difference: $-1 - (-1) = -1 + (+1) = (1 - 1) = 0$.

(12) **9.** First, find the difference of the numbers in the absolute value symbols. The sign of the difference is negative because 13 is farther from 0 than 4 on the number line. Evaluating the absolute value gives you a positive result: $|4 - 13| = |-(13 - 4)| = |-9| = 9$.

(13) **-16.** Drop the negative signs and find the sum of the two numbers. The answer is negative, because the signs are the same: $-5 + (-11) = -(5 + 11) = -16$.

(14) **8.** Change the problem to addition and the negative 14 to a positive number. Then, because the signs are different, find the difference. The answer is positive because 14 is farther from 0 than 6 on the number line: $-6 - (-14) = -6 + (+14) = +(14 - 6) = 8$.

(15) **-40.** Drop the negative sign and multiply. The answer is negative, because the signs are different: $5 \times (-8) = -(5 \times 8) = -40$.

(16) **1.** Find the difference between the two numbers, because the signs are different. The answer is positive, because 8 is farther from 0 than 7 on the number line: $-7 + 8 = +(8 - 7) = 1$.

(17) **5.** First, find the difference. Then apply the absolute value operation: $|0 - 5| = |-5| = 5$.

(18) **2.** Change the subtraction to addition and the negative 4 to +4. Then find the difference, because the signs are different. The answer is positive, because 4 is farther from 0 than 2 on the number line: $-2 - (-4) = -2 + (+4) = +(4 - 2) = 2$.

(19) **2.** Find the difference, because the signs are different. The answer is positive, because 5 is farther from 0 than 3 on the number line: $5 + (-3) = +(5 - 3) = 2$.

(20) **–119.** Change the subtraction to addition and make the 62 a negative number. Ignore the signs when you add the two numbers with the same sign. The sum is negative, because the signs are the same: $-57 - 62 = -57 + (-62) = -(57 + 62) = -119$.

(21) **157.** Change the problem to addition and the negative 84 to a positive number. Then, because the signs are different, find the sum: $73 - (-84) = 73 + 84 = 157$.

(22) **38.** Find the difference, because the signs are different. The answer is positive, because 90 is farther from 0 on the number line: $-52 + 90 = +(90 - 52) = 38$.

(23) **263.** First, find the difference, and then apply the absolute value operation. The difference will be negative, because 330 is negative and it's farther from 0 on the number line. The absolute value operation makes the result positive: $|67 - 330| = |-(330 - 67)| = |-263| = 263$.

(24) **143.** Drop the negative signs and multiply. The answer is positive, because the signs are the same: $-13 \times (-11) = +(13 \times 11) = 143$.

(25) **–20.** Drop the negative sign and divide. The answer is negative, because the signs are different: $500 \div (-25) = -(500 \div 25) = -20$.

(26) **–48.** Find the difference, because the signs are different. The answer is negative, because 72 is farther from 0 on the number line: $-72 + 24 = -(72 - 24) = -48$.

Chapter **5**

Putting the Big Four Operations to Work

When you understand the Big Four operations — adding, subtracting, multiplying, and dividing — you can begin to look at math on a whole new level. In this chapter, you extend your understanding of the Big Four operations and move beyond them. I begin by focusing on four important properties of the Big Four operations: inverse operations, commutative operations, associative operations, and distribution. Then I show you how to perform the Big Four on negative numbers.

I continue by introducing you to some important symbols for inequality. Finally, you're ready to move beyond the Big Four by discovering three more advanced operations: exponents (also called *powers*), square roots (also called *radicals*), and absolute values.

Switching Things Up with Inverse Operations and the Commutative Property

The Big Four operations are actually two pairs of *inverse operations*, which means the operations can undo each other.

>> **Addition and subtraction:** Subtraction undoes addition. For example, if you start with 3 and add 4, you get 7. Then, when you subtract 4, you undo the original addition and arrive back at 3:

$$3 + 4 = 7 \quad \rightarrow \quad 7 - 4 = 3$$

This idea of inverse operations makes a lot of sense when you look at the number line. On a number line, 3 + 4 means *start at 3, up 4*. And 7 − 4 means *start at 7, down 4*. So when you add 4 and then subtract 4, you end up back where you started.

>> **Multiplication and division:** Division undoes multiplication. For example, if you start with 6 and multiply by 2, you get 12. Then, when you divide by 2, you undo the original multiplication and arrive back at 6:

$$6 \times 2 = 12 \quad \rightarrow \quad 12 \div 2 = 6$$

REMEMBER

The *commutative property of addition* tells you that you can change the order of the numbers in an addition problem without changing the result, and the *commutative property of multiplication* says you can change the order of the numbers in a multiplication problem without changing the result. For example,

$$2 + 5 = 7 \quad \rightarrow \quad 5 + 2 = 7$$
$$3 \times 4 = 12 \quad \rightarrow \quad 4 \times 3 = 12$$

TIP

In contrast, subtraction and division are *non-commutative* operations. When you switch the order of the numbers, the result changes.

Here's an example of how subtraction is non-commutative:

$$6 - 4 = 2 \quad \text{but} \quad 4 - 6 = -2$$

Subtraction is non-commutative, so if you have \$6 and spend \$4, the result is *not* the same as if you have \$4 and spend \$6. In the first case, you still have \$2 left over. In the second case, you *owe* \$2. In other words, switching the numbers turns the result into a negative number. (I discuss negative numbers in Chapter 4.)

And here's an example of how division is non-commutative:

$$5 \div 2 = 2 \text{ r } 1 \quad \text{but} \quad 2 \div 5 = 0 \text{ r } 2$$

For example, when you have five dog biscuits to divide between two dogs, each dog gets two biscuits and you have one biscuit left over. But when you switch the numbers and try to divide two biscuits among five dogs, you don't have enough biscuits to go around, so each dog gets none and you have two left over.

Through the commutative property and inverse operations, every equation has four alternative forms that contain the same information expressed in slightly different ways. For example, $2 + 3 = 5$ and $3 + 2 = 5$ are alternative forms of the same equation but tweaked using the commutative property. And $5 - 3 = 2$ is the inverse of $2 + 3 = 5$. Finally, $5 - 2 = 3$ is the inverse of $3 + 2 = 5$.

TIP

You can use alternative forms of equations to solve fill-in-the-blank problems. As long as you know two numbers in an equation, you can always find the remaining number. Just figure out a way to get the blank to the other side of the equals sign.

>> When the *first* number is missing in any problem, use the inverse to turn the problem around:

$$\_\_\_ + 6 = 10 \quad \rightarrow \quad 10 - 6 = \_\_\_$$

>> When the *second* number is missing in an addition or multiplication problem, use the commutative property and then the inverse:

$$9 + \_\_\_ = 17 \quad \rightarrow \quad \_\_\_ + 9 = 17 \quad \rightarrow \quad 17 - 9 = \_\_\_$$

>> When the *second* number is missing in a subtraction or division problem, just switch around the two values that are next to the equals sign (that is, the blank and the equals sign):

$$15 - \_\_\_ = 8 \quad \rightarrow \quad 15 - 8 = \_\_\_$$

EXAMPLE

Q. What's the inverse equation to $16 - 9 = 7$?

A. $7 + 9 = 16$. In the equation $16 - 9 = 7$, you start at 16 and subtract 9, which brings you to 7. The inverse equation undoes this process, so you start at 7 and add 9, which brings you back to 16:

$$16 - 9 = 7 \quad \rightarrow \quad 7 + 9 = 16$$

Q. What's the inverse equation to $6 \times 7 = 42$?

A. $42 \div 7 = 6$. In the equation $6 \times 7 = 42$, you start at 6 and multiply by 7, which brings you to 42. The inverse equation undoes this process, so you start at 42 and divide by 7, which brings you back to 6:

$$6 \times 7 = 42 \quad \rightarrow \quad 42 \div 7 = 6$$

Q. Use inverse operations and the commutative property to find three alternative forms of the equation $7 - 2 = 5$.

A. $5 + 2 = 7$, $2 + 5 = 7$, and $7 - 5 = 2$. First, use inverse operations to change subtraction to addition:

$$7 - 2 = 5 \quad \rightarrow \quad 5 + 2 = 7$$

Now use the commutative property to change the order of this addition:

$$5 + 2 = 7 \quad \rightarrow \quad 2 + 5 = 7$$

Finally, use inverse operations to change addition to subtraction:

$$2 + 5 = 7 \quad \rightarrow \quad 7 - 5 = 2$$

Q. Fill in the blank: ____ ÷ 3 = 13.

A. **39.** Use inverse operations to turn the problem from division to multiplication:

$$\_\_\_ \div 3 = 13 \quad \rightarrow \quad 13 \times 3 = \_\_\_$$

Now you can solve the problem by multiplying $13 \times 3 = 39$.

Q. Solve this problem by filling in the blank: 16 + ____ = 47.

A. **31.** First, use the commutative property to reverse the addition:

$$16 + \_\_\_ = 47 \quad \rightarrow \quad \_\_\_ + 16 = 47$$

Now use inverse operations to change the problem from addition to subtraction:

$$\_\_\_ + 16 = 47 \quad \rightarrow \quad 47 - 16 = \_\_\_$$

At this point, you can solve the problem by subtracting $47 - 16 = 31$.

Q. Fill in the blank: 64 − ____ = 15.

A. **49.** Switch around the last two numbers in the problem:

$$64 - \_\_\_ = 15 \quad \rightarrow \quad 64 - 15 = \_\_\_$$

Now you can solve the problem by subtracting $64 - 15 = 49$.

YOUR TURN

1 Using inverse operations, write down an alternative form of each equation:

(a) 8 + 9 = 17

(b) 23 − 13 = 10

(c) 15 × 5 = 75

(d) 132 ÷ 11 = 12

2 Use the commutative property to write down an alternative form of each equation:

(a) 19 + 35 = 54

(b) 175 + 88 = 263

(c) 22 × 8 = 176

(d) 101 × 99 = 9,999

3 Use inverse operations and the commutative property to find all three alternative forms for each equation:

(a) $7 + 3 = 10$

(b) $12 - 4 = 8$

(c) $6 \times 5 = 30$

(d) $18 \div 2 = 9$

 4 Fill in the blanks in each question:

(a) ____ $- 74 = 36$

(b) ____ $\times 7 = 105$

(c) $45 + $ ____ $= 132$

(d) $273 - $ ____ $= 70$

(e) $8 \times $ ____ $= 648$

(f) $180 \div $ ____ $= 9$

Getting with the In-Group: Parentheses and the Associative Property

REMEMBER

Parentheses allow you to group operations together, telling you to do any operations inside a set of parentheses *before* you do operations outside of it. Parentheses can make a big difference in the result you get when solving a problem, especially in a problem with mixed operations. In two important cases, however, moving parentheses doesn't change the answer to a problem:

>> The *associative property of addition* says that when every operation is addition, you can group numbers however you like and choose which pair of numbers to add first; you can move parentheses without changing the answer.

>> The *associative property of multiplication* says you can choose which pair of numbers to multiply first, so when every operation is multiplication, you can move parentheses without changing the answer.

In contrast, subtraction and division are non-associative operations. This means that grouping them in different ways changes the result.

WARNING

Don't confuse the commutative property with the associative property. The commutative property tells you that it's okay *to switch* two numbers that you're adding or multiplying. The associative property tells you that it's okay to *regroup* three (or more) numbers using parentheses.

TIP

Taken together, the associative property and the commutative property (which I discuss in the preceding section) allow you to completely rearrange all the numbers in any problem that's either all addition or all multiplication.

Q. What's $(21 - 6) \div 3$? What's $21 - (6 \div 3)$?

EXAMPLE **A.** **5 and 19.** To calculate $(21 - 6) \div 3$, first do the operation inside the parentheses — that is, $21 - 6 = 15$:

$$(21 - 6) \div 3 = 15 \div 3$$

Now finish the problem by dividing $15 \div 3 = 5$.

To solve $21 - (6 \div 3)$, first do the operation inside the parentheses — that is, $6 \div 3 = 2$:

$$21 - (6 \div 3) = 21 - 2$$

Finish up by subtracting $21 - 2 = 19$. Notice that the placement of the parentheses changes the answer.

Q. Solve $1 + (9 + 2)$ and $(1 + 9) + 2$.

A. **12 and 12.** To solve $1 + (9 + 2)$, first do the operation inside the parentheses — that is, $9 + 2 = 11$:

$$1 + (9 + 2) = 1 + 11$$

Finish up by adding $1 + 11 = 12$.

To solve $(1 + 9) + 2$, first do the operation inside the parentheses — that is, $1 + 9 = 10$:

$$(1 + 9) + 2 = 10 + 2$$

Finish up by adding $10 + 2 = 12$. Notice that the only difference between the two problems is the placement of the parentheses, but because both operations are addition, moving the parentheses doesn't change the answer.

Q. Solve $2 \times (4 \times 3)$ and $(2 \times 4) \times 3$.

A. **24 and 24.** To solve $2 \times (4 \times 3)$, first do the operation inside the parentheses — that is, $4 \times 3 = 12$:

$$2 \times (4 \times 3) = 2 \times 12$$

Finish by multiplying $2 \times 12 = 24$.

To solve $(2 \times 4) \times 3$, first do the operation inside the parentheses — that is, $2 \times 4 = 8$:

$$(2 \times 4) \times 3 = 8 \times 3$$

Finish by multiplying $8 \times 3 = 24$. No matter how you group the multiplication, the answer is the same.

Q. Solve $41 \times 5 \times 2$.

A. **410.** The last two numbers are small, so place parentheses around these numbers:

$$41 \times 5 \times 2 = 41 \times (5 \times 2)$$

First, do the multiplication inside the parentheses:

$$41 \times (5 \times 2) = 41 \times 10$$

Now you can easily multiply $41 \times 10 = 410$.

YOUR TURN

5 Find the value of $(8 \times 6) + 10$.

6 Find the value of $123 \div (145 - 144)$.

7 Solve the following two problems:

(a) $(40 \div 2) + 6 = ?$

(b) $40 \div (2 + 6) = ?$

Do the parentheses make a difference in the answers?

8 Solve the following two problems:

(a) $(16 + 24) + 19$

(b) $16 + (24 + 19)$

Do the parentheses make a difference in the answers?

9 Solve the following two problems:

(a) $(18 \times 25) \times 4$

(b) $18 \times (25 \times 4)$

Do the parentheses make a difference in the answers?

10 Find the value of $93,769 \times 2 \times 5$. (*Hint:* Use the associative property for multiplication to make the problem easier.)

Distribution to lighten the load

If you've ever tried to carry a heavy bag of groceries, you may have found that distributing the contents into two smaller bags is helpful. This same concept also works for multiplication.

In math, *distribution* (also called the *distributive property of multiplication over addition*) allows you to split a large multiplication problem into two smaller ones and add the results to get the answer.

For example, suppose you want to multiply these two numbers:

$$17 \times 101$$

You can go ahead and just multiply them, but distribution provides a different way to think about the problem that you may find easier. Because 101 = 100 + 1, you can split this problem into two easier problems, as follows:

$$= 17 \times (100 + 1)$$
$$= (17 \times 100) + (17 \times 1)$$

You take the number outside the parentheses, multiply it by each number inside the parentheses one at a time, and then add the products. At this point, you may be able to calculate the two multiplications in your head and then add them up easily:

$$= 1,700 + 17 = 1,717$$

Distribution becomes even more useful when you get to algebra in Unit 7.

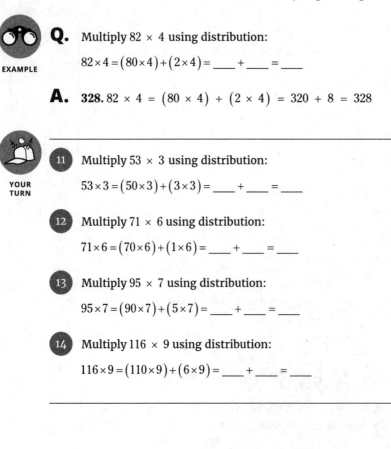

Q. Multiply 82 × 4 using distribution:

EXAMPLE

$$82 \times 4 = (80 \times 4) + (2 \times 4) = \underline{\quad} + \underline{\quad} = \underline{\quad}$$

A. **328.** 82 × 4 = $(80 \times 4) + (2 \times 4)$ = 320 + 8 = 328

---

11 Multiply 53 × 3 using distribution:

YOUR
TURN

$$53 \times 3 = (50 \times 3) + (3 \times 3) = \underline{\quad} + \underline{\quad} = \underline{\quad}$$

12 Multiply 71 × 6 using distribution:

$$71 \times 6 = (70 \times 6) + (1 \times 6) = \underline{\quad} + \underline{\quad} = \underline{\quad}$$

13 Multiply 95 × 7 using distribution:

$$95 \times 7 = (90 \times 7) + (5 \times 7) = \underline{\quad} + \underline{\quad} = \underline{\quad}$$

14 Multiply 116 × 9 using distribution:

$$116 \times 9 = (110 \times 9) + (6 \times 9) = \underline{\quad} + \underline{\quad} = \underline{\quad}$$

Understanding Inequalities

Sometimes you want to talk about when two quantities are different. These statements are called *inequalities*. In this section, I discuss six types of inequalities: ≠ (doesn't equal), < (less than), > (greater than), ≤ (less than or equal to), ≥ (greater than or equal to), and ≈ (approximately equals).

Doesn't equal (≠)

The simplest inequality is ≠, which you use when two quantities are not equal. For example,

$$2 + 2 \neq 5$$
$$3 \times 4 \neq 34$$
$$999,9000 \neq 1,000,000$$

You can read ≠ as "doesn't equal" or "is not equal to." Therefore, read $2 + 2 \neq 5$ as "two plus two doesn't equal five."

Less than (<) and greater than (>)

The symbol < means *less than*. For example, the following statements are true:

$$4 < 5$$
$$100 < 1,000$$
$$2 + 2 < 5$$

Similarly, the symbol > means *greater than*. For example,

$$5 > 4$$
$$100 > 99$$
$$2 + 2 > 3$$

TIP

The two symbols < and > are similar and easily confused. Here are two simple ways to remember which is which:

>> Notice that the < looks sort of like an *L*. This *L* should remind you that it means *less than*.

>> Remember that, in any true statement, the *large* open mouth of the symbol is on the side of the *greater* amount, and the *small* point is on the side of the *lesser* amount.

Less than or equal to (≤) and greater than or equal to (≥)

The symbol ≤ means *less than or equal to*. For example, the following statements are true:

$$100 \leq 1,000$$
$$2 + 2 \leq 5$$
$$2 + 2 \leq 4$$

Similarly, the symbol ≥ means *greater than or equal to*. For example,

$$100 \geq 99$$
$$2 + 2 \geq 3$$
$$2 + 2 \geq 4$$

TIP

The symbols ≤ and ≥ are called *inclusive inequalities* because they *include* (allow) the possibility that both sides are equal. In contrast, the symbols < and > are called *exclusive inequalities* because they *exclude* (don't allow) this possibility.

Approximately equals (≈)

In Chapter 3, I show you how rounding numbers makes large numbers easier to work with. In that chapter, I also introduce ≈, which means *approximately equals.*

For example,

$$49 \approx 50$$
$$1,024 \approx 1,000$$
$$999,999 \approx 1,000,000$$

You can also use ≈ when you estimate the answer to a problem:

$$1,000,487 + 2,001,932 + 5,000,032$$
$$\approx 1,000,000 + 2,000,000 + 5,000,000$$
$$= 8,000,000$$

EXAMPLE

Q. Place the correct symbol (=, >, or <) in the blank: 2 + 2 ____ 5.

A. <. Because 2 + 2 = 4 and 4 is less than 5, use the symbol that means *is less than.*

Q. Place the correct symbol (=, >, or <) in the blank: 42 − 19 ____ 5 × 4.

A. >. Because 42 − 19 = 23 and 5 × 4 = 20, and 23 is greater than 20, use the symbol that means *is greater than.*

Q. Sam worked 7 hours for his parents at $8 an hour, and his parents paid him with a $50 bill. Use the symbol ≠ to point out why Sam was upset.

A. **$50 ≠ $56.** Sam worked 7 hours for $8 an hour, so here's how much he earned:

$7 \times \$8 = \$56$

He was upset because his parents didn't pay him the correct amount: $50 ≠ $56.

Q. Find an approximate solution to 2,000,398 + 6,001,756.

A. **8,000,000.** The two numbers are both in the millions, so you can use ≈ to round them to the nearest million:

$2,000,398 + 6,001,756 \approx 2,000,000 + 6,000,000$

Now it's easy to add 2,000,000 + 6,000,000 = 8,000,000.

YOUR TURN

15 Place the correct symbol $(=, >, or <)$ in the blanks:

(a) 4 + 6 _____ 13

(b) 9 × 7 _____ 62

(c) 33 − 16 _____ 60 ÷ 3

(d) 100 ÷ 5 _____ 83 − 63

16 Change the ≠ signs to either > or <:

(a) 17 + 14 ≠ 33

(b) 144 − 90 ≠ 66

(c) 11 × 14 ≠ 98

(d) 150 ÷ 6 ≠ 20

17 Tim's boss paid him for 40 hours of work last week. Tim accounted for his time by saying that he spent 19 hours with clients, 11 hours driving, and 7 hours doing paperwork. Use ≠ to show why Tim's boss was unhappy with Tim's work.

18 Find an approximate solution to 10,002 − 6,007.

Moving Beyond the Big Four: Exponents and Square Roots

In this section, I introduce you to three new operations that you need as you move on with math: exponents, square roots, and absolute value. As with the Big Four operations, these three operations tweak numbers in various ways.

To tell the truth, these three operations have fewer everyday applications than the Big Four. But you'll be seeing a lot more of them as you progress in your study of math. Fortunately, they aren't difficult, so this is a good time to become familiar with them.

Understanding exponents

Exponents (also called *powers*) are shorthand for repeated multiplication. For example, 2^3 means to multiply 2 by itself three times. To do that, use the following notation:

$$2^3 = 2 \times 2 \times 2 = 8$$

In this example, 2 is the *base number* and 3 is the *exponent.* You can read 2^3 as "2 to the third power" or "2 to the power of 3" (or even "2 cubed," which has to do with the formula for finding the volume of a cube — see Chapter 19 for details).

Here's another example:

10^5 means to multiply 10 by itself five times.

That works out like this:

$$10^5 = 10 \times 10 \times 10 \times 10 \times 10 = 100,000$$

This time, 10 is the base number and 5 is the exponent. Read 10^5 as "10 to the fifth power" or "10 to the power of 5."

TIP

When the base number is 10, figuring out any exponent is easy. Just write down a 1 and that many 0s after it:

1 with two 0s	1 with seven 0s	1 with twenty 0s
$10^2 = 100$	$10^7 = 10,000,000$	$10^{20} = 100,000,000,000,000,000,000$

Exponents with a base number of 10 are important in scientific notation, which I cover in Chapter 17.

The most common exponent is the number 2. When you take any whole number to the power of 2, the result is a square number. (For more information on square numbers, see Chapter 1.) For this reason, taking a number to the power of 2 is called *squaring* that number. You can read 3^2 as "three squared," 4^2 as "four squared," and so forth. Here are some squared numbers:

$$3^2 = 3 \times 3 = 9$$
$$4^2 = 4 \times 4 = 16$$
$$5^2 = 5 \times 5 = 25$$

REMEMBER

Any number (except 0) raised to the 0 power equals 1. So 1^0, 37^0, and $999,999^0$ are equivalent, or equal, because they all equal 1.

Discovering your roots

Earlier in this chapter, in the section, "Switching Things Up with Inverse Operations and the Commutative Property," I show you how addition and subtraction are inverse operations. I also show you how multiplication and division are inverse operations. In a similar way, roots are the inverse operation of exponents.

The most common root is the square root. A *square root* undoes an exponent of 2. For example,

$$3^2 = 3 \times 3 = 9 \quad \rightarrow \quad \sqrt{9} = 3$$
$$4^2 = 4 \times 4 = 16 \quad \rightarrow \quad \sqrt{16} = 4$$
$$5^2 = 5 \times 5 = 25 \quad \rightarrow \quad \sqrt{25} = 5$$

You can read the symbol either as "the square root of" or as "radical." So read $\sqrt{9}$ as either "the square root of 9" or "radical 9."

As you can see, when you take the square root of any square number, the result is the number that you multiplied by itself to get that square number in the first place. For example, to find $\sqrt{100}$, you ask the question, "What number when multiplied by itself equals 100?" The answer here is 10 because

$$10^2 = 10 \times 10 = 100 \quad \rightarrow \quad \sqrt{100} = 10$$

You probably won't use square roots much until you get to algebra, but at that point, they become handy.

EXAMPLE

Q. What is 3^4?

A. **81.** The expression 3^4 tells you to multiply 3 by itself 4 times:

$$3 \times 3 \times 3 \times 3 = 81$$

Q. What is 10^6?

A. **1,000,000.** Using the power of ten rule, 10^6 is 1 followed by six 0s, so $10^6 = 1,000,000$.

Q. What is $\sqrt{36}$?

A. **6.** To find $\sqrt{36}$, you want to find a number that, when multiplied by itself, equals 36. You know that $6 \times 6 = 36$, so $\sqrt{36} = 6$.

Q. What is $\sqrt{256}$?

A. **16.** To find $\sqrt{256}$, you want to find a number that, when multiplied by itself, equals 256. Try guessing to narrow down the possibilities. Start by guessing 10:

$$10 \times 10 = 100$$

$256 > 100$, so $\sqrt{256}$ is greater than 10. Guess 20:

$20 \times 20 = 400$

$256 < 400$, so $\sqrt{256}$ is between 10 and 20. Guess 15:

$15 \times 15 = 225$

$256 > 225$, so $\sqrt{256}$ is between 15 and 20. Guess 16:

$16 \times 16 = 256$

This is correct, so $\sqrt{256} = 16$.

19 Solve these exponents.

    (a) 6^2

    (b) 3^5

    (c) 2^7

    (d) 2^8 (*Hint:* You can make your work easier by using the answer to *c*.)

20 Solve these exponents.

    (a) 10^4

    (b) 10^{10}

    (c) 10^{15}

    (d) 10^1

Practice Questions Answers and Explanations

1

    (a) $17 - 9 = 8$

    (b) $10 + 13 = 23$

    (c) $75 \div 5 = 15$

    (d) $12 \times 11 = 132$

2

    (a) $35 + 19 = 54$

    (b) $88 + 175 = 263$

    (c) $8 \times 22 = 176$

    (d) $99 \times 101 = 9{,}999$

3

    (a) $10 - 3 = 7$, $3 + 7 = 10$, and $10 - 7 = 3$

    (b) $8 + 4 = 12$, $4 + 8 = 12$, and $12 - 8 = 4$

    (c) $30 \div 5 = 6$, $5 \times 6 = 30$, and $30 \div 6 = 5$

    (d) $9 \times 2 = 18$, $2 \times 9 = 18$, $18 \div 9 = 2$

4

    (a) **110.** Rewrite ____ $- 74 = 36$ as its inverse: $36 + 74 =$ ____. Therefore, $36 + 74 = 110$.

    (b) **15.** Rewrite ____ $\times 7 = 105$ as its inverse: $105 \div 7 =$ ____. So, $105 \div 7 = 15$.

    (c) **87.** Rewrite $45 +$ ____ $= 132$ using the commutative property: ____ $+ 45 = 132$. Now rewrite this equation as its inverse: $132 - 45 =$ ____. Therefore, $132 - 45 = 87$.

    (d) **203.** Rewrite $273 -$ ____ $= 70$ by switching around the two numbers next to the equal sign: $273 - 70 =$ ____. So, $273 - 70 = 203$.

    (e) **81.** Rewrite $8 \times$ ____ $= 648$ using the commutative property: ____ $\times 8 = 648$. Now rewrite this equation as its inverse: $648 \div 8 =$ ____. So, $648 \div 8 = 81$.

    (f) **20.** Rewrite $180 \div$ ____ $= 9$ by switching around the two numbers next to the equals sign: $180 \div 9 =$ ____. So, $180 \div 9 = 20$.

5 **58.** First, do the multiplication inside the parentheses:

$$(8 \times 6) + 10 = 48 + 10$$

Now add: $48 + 10 = 58$.

6 **123.** First, do the subtraction inside the parentheses:

$$123 \div (145 - 144) = 123 \div 1$$

Now simply divide $123 \div 1 = 123$.

(7)

(a) **26.** $(40 \div 2) + 6 = 20 + 6 = 26$

(b) **5.** $40 \div (2 + 6) = 40 \div 8 = 5$

Yes, the placement of parentheses changes the result.

(8)

(a) **59.** $(16 + 24) + 19 = 40 + 19 = 59$

(b) **59.** $16 + (24 + 19) = 16 + 43 = 59$

No, because of the associative property of addition, the placement of parentheses doesn't change the result.

(9)

(a) **1,800.** $(18 \times 25) \times 4 = 450 \times 4 = 1,800$

(b) **1,800.** $18 \times (25 \times 4) = 18 \times 100 = 1,800$

No, because of the associative property of multiplication, the placement of parentheses doesn't change the result.

(10) **937,690.** The problem is easiest to solve by placing parentheses around 2×5:

$$93,769 \times (2 \times 5) = 93,769 \times 10 = 937,690$$

(11) **159.** $53 \times 3 = (50 \times 3) + (3 \times 3) = 150 + 9 = 159$

(12) **426.** $71 \times 6 = (70 \times 6) + (1 \times 6) = 420 + 6 = 426$

(13) **665.** $95 \times 7 = (90 \times 7) + (5 \times 7) = 630 + 35 = 665$

(14) **1,044.** $116 \times 9 = (110 \times 9) + (6 \times 9) = 990 + 54 = 1,044$

(15)

(a) $4 + 6 = 10$, and $10 < 13$

(b) $9 \times 7 = 63$, and $63 > 62$

(c) $33 - 16 = 17$ and $60 \div 3 = 20$, so $17 < 20$.

(d) $100 \div 5 = 20$ and $83 - 63 = 20$, so $20 = 20$.

(16)

(a) $17 + 14 = 31$, and $31 < 33$

(b) $144 - 90 = 54$, and $54 < 66$

(c) $11 \times 14 = 154$, and $154 > 98$

(d) $150 \div 6 = 25$, and $25 > 20$

(17) **$37 \neq 40$**

(18) **4,000**

(19)

   (a) **36.** $6^2 = 6 \times 6 = 36$

   (b) **243.** $3^5 = 3 \times 3 \times 3 \times 3 \times 3 = 243$

   (c) **128.** $2^7 = 2 \times 2 \times 2 \times 2 \times 2 \times 2 \times 2 = 128$

   (d) **256.** $2^8 = 2 \times 2 \times 2 \times 2 \times 2 \times 2 \times 2 \times 2 = 256$. You already know from part c that $2^7 = 128$, so multiply this number by 2 to get your answer: $128 \times 2 = 256$.

(20)

   (a) **10,000.** Write 1 followed by four 0s.

   (b) **10,000,000,000.** Write 1 followed by ten 0s.

   (c) **1,000,000,000,000,000.** Write 1 followed by fifteen 0s.

   (d) **10.** Any number raised to the power of 1 is that number.

In the next section, the chapter quiz gives you a chance to practice the skills you learned in this chapter.

Whaddya Know? Chapter 5 Quiz

Try out these 17 questions, which cover the gamut of skills covered in this chapter. When you're done, the following section contains answers and complete explanations.

1. Change the $\neq$ sign to > or < to create a true statement: $57 - 40 \neq 27$.

2. Evaluate 4^3.

3. Fill in the blank using the inverse operation: $9 \times \underline{\hspace{1cm}} = 45$.

4. Evaluate $(94 \times 5) \times 2$.

5. Multiply using the distributive property: 47×8.

6. What is the inverse equation to $10 - 4 = 6$?

7. Place the correct symbol, =, <, or >, in the blank to create a true statement: $15 - 4 \underline{\hspace{1cm}} 6$.

8. Evaluate $\sqrt{49}$.

9. Fill in the blank using the inverse operation: $\underline{\hspace{1cm}} = 45 \div 9$.

10. What is the inverse equation to $12 \div 4 = 3$?

11. Evaluate $48 \div (8 - 2)$.

12. Place the correct symbol, =, <, or >, in the blank to create a true statement: $12 \div 6 \underline{\hspace{1cm}} 4$.

13. Evaluate $99 \times \left(\dfrac{1}{99} \times 83 \right)$.

14. Change the $\neq$ sign to > or < to create a true statement: $12 \times 5 \neq 50$.

15. Evaluate 10^{13}.

16. Evaluate $(7 + 3) \times 14$.

17. Place the correct symbol, =, <, or >, in the blank to create a true statement: $2^5 \underline{\hspace{1cm}} 8 \times 4$.

Answers to Chapter 5 Quiz

(1) $57 - 40 < 27$. Performing the subtraction, you have $57 - 40 = 17$, and 17 is less than 27.

(2) **64.** The exponent says to multiply the 4 three times: $4^3 = 4 \times 4 \times 4 = 64$.

(3) **5.** Use division, the inverse of multiplication, and write ____ $= 45 \div 9$. This equals 5.

(4) **940.** Use the associative property to regroup: $(94 \times 5) \times 2 = 94 \times (5 \times 2) = 94 \times 10 = 940$.

(5) **376.** Rewrite the number 47 as a sum: $47 \times 8 = (40 + 7) \times 8$. Distribute and find the sum of the products: $(40 + 7) \times 8 = 320 + 56 = 376$.

(6) $6 + 4 = 10$. Determine the inverse equation to $10 - 4 = 6$ by starting with the 6 and adding 4.

(7) $15 - 4 > 6$. Because $15 - 4 = 11$, you want $11 > 6$.

(8) **7.** Determine that $7 \times 7 = 49$ so $\sqrt{49} = 7$.

(9) **5.** Change the division to multiplication and write $9 \times \underline{\ 5\ } = 45$.

(10) $12 = 3 \times 4$. Start with the 3 and change the operation to multiplication: $12 = 3 \times 4$.

(11) **8.** First, perform the subtraction in the parentheses. Then divide: $48 \div (8 - 2) = 48 \div 6 = 8$.

(12) **<.** $12 \div 6 < 4$. Because $12 \div 6 = 2$, you need the less-than sign.

(13) **83.** Use the associative property to regroup the factors: $99 \times \left(\frac{1}{99} \times 83 \right) = \left(99 \times \frac{1}{99} \right) \times 83 = 1 \times 83$.

(14) **>.** $12 \times 5 > 50$. Because $12 \times 5 = 60$, you need the greater-than sign.

(15) **10,000,000,000,000.** There are 13 zeros in this power.

(16) **140.** Perform the addition before multiplying: $(7 + 3) \times 14 = (10) \times 14 = 140$.

(17) $2^5 = 8 \times 4$. They are the same: $2^5 = 32$ and $8 \times 4 = 32$.

3

Getting a Handle on Whole Numbers

In This Unit . . .

Chapter **6**

Please Excuse My Dear Aunt Sally: Evaluating Arithmetic Expressions with PEMDAS

In this chapter, I introduce you to what I call the Three E's of math: equations, expressions, and evaluation. You'll likely find the Three E's of math familiar because, whether you realize it or not, you've been using them for a long time. Whenever you add up the cost of several items at the store, calculate your monthly car payment, or figure out the area of your room, you're evaluating expressions and setting up equations. In this section, I shed light on this stuff and give you a new way to look at it.

You probably already know that an *equation* is a mathematical statement that has an equals sign (=) — for example, $1+1=2$. An *expression* is a string of mathematical symbols that can be placed on one side of an equation — for example, $1 + 1$. And *evaluation* is finding out the *value* of an expression as a number — for example, finding out that the expression $1+1$ is equal to the number 2.

Throughout the rest of the chapter, I show you how to turn expressions into numbers using a set of rules called the *order of operations* (or *order of precedence*). Many students use the mnemonic PEMDAS to help remember these rules. Here, I break them down so you can see for yourself what to do next in any situation.

The Three E's of Math: Equations, Expressions, and Evaluation

In a lot of cases, difficult-looking math words can make math seem a lot harder than it really is. In this chapter, I introduce three words that are related and sometimes cause confusion. Then I show you an easy way to think about them.

Seeking equality for all: Equations

An *equation* is a mathematical statement that tells you that two things have the same value — in other words, it's a statement with an equals sign. The equation is one of the most important concepts in mathematics because it allows you to boil down a bunch of complicated information into a single number.

Mathematical equations come in a lot of varieties: arithmetic equations, algebraic equations, differential equations, partial differential equations, Diophantine equations, and many more. In this book, I look at only two types: arithmetic equations and algebraic equations.

In this chapter, I discuss only *arithmetic equations,* which are equations involving numbers, the Big Four operations, and other basic operations such as absolute values, exponents, and roots. Here are a few examples of simple arithmetic equations:

$$2 + 2 = 4$$
$$3 \times 4 = 12$$
$$20 \div 2 = 10$$

And here are a few examples of more-complicated arithmetic equations:

$$1{,}000 - 1 - 1 - 1 = 997$$
$$(3 + 5) \div (9 - 7) = 4$$
$$4^2 - \sqrt{256} = (791 - 842) \times 0$$

Hey, it's just an expression

An *expression* is any string of mathematical symbols that can be placed on one side of an equation. Mathematical expressions, just like equations, come in a lot of varieties. In this chapter, I focus only on *arithmetic expressions,* which are expressions that contain numbers, the Big Four operations, and a few other basic operations. In Unit 7, I introduce you to algebraic expressions. Here are a few examples of simple expressions:

$$2+2$$
$$-17+(-1)$$
$$14 \div 7$$

And here are a few examples of more-complicated expressions:

$$(88-23) \div 13$$
$$100+2-3 \times 17$$
$$\sqrt{441} + \left| -2^3 \right|$$

Evaluating the situation

At the root of the word *evaluation* is the word *value*. In other words, when you evaluate something, you find its value. Evaluating an arithmetic expression is also referred to as *simplifying, solving,* or *finding the value of an expression.* The words may change, but the idea is the same — boiling down a string of numbers and math symbols to a single number.

When you evaluate an arithmetic expression, you simplify it to a single numerical value — in other words, you find the number that it's equal to. For example, evaluate the following arithmetic expression:

$$7 \times 5$$

How? Simplify it to a single number:

35

Putting the Three E's together

I'm sure you're dying to know how the Three E's — equations, expressions, and evaluation — are all connected. *Evaluation* allows you to take an *expression* containing more than one number and reduce it to a single number. Then you can make an *equation*, using an equals sign, to connect the expression and the number. For example, here's an *expression* containing four numbers:

$$1+2+3+4$$

When you *evaluate* it, you reduce it to a single number:

10

And now you can make an *equation* by connecting the expression and the number with an equals sign:

$$1+2+3+4=10$$

Introducing Order of Operations (PEMDAS)

When you were a kid, did you ever try putting on your shoes first and then your socks? If you did, you probably discovered this simple rule:

1. **Put on socks.**

2. **Put on shoes.**

Thus, you have an order of operations: The socks have to go on your feet before your shoes. So in the act of putting on your shoes and socks, your socks have precedence over your shoes. A simple rule to follow, right?

In this section, I outline a similar set of rules for evaluating expressions, called the *order of operations* (sometimes called *order of precedence*). Don't let the long name throw you. Order of operations is just a set of rules to make sure you get your socks and shoes on in the right order, mathematically speaking, so you always get the right answer.

Note: Through most of this book, I introduce overarching themes at the beginning of each section and then explain them later in the chapter instead of building them and finally revealing the result. But order of operations is a bit too confusing to present that way. Instead, I start with a list of four rules and go into more detail about them later in the chapter. Don't let the complexity of these rules scare you off before you work through them!

Evaluate arithmetic expressions from left to right according to the following order of operations:

REMEMBER 1. **Parentheses**

2. **Exponents (also called powers)**

3. **Multiplication and division**

4. **Addition and subtraction**

Many students use the mnemonic PEMDAS to remember this list. And to help you remember the mnemonic, you can use the sentence, "Please Excuse My Dear Aunt Sally," which contains the six letters of PEMDAS in the correct order.

But don't worry too much about memorizing this list right now. I break it to you slowly in the remaining sections of this chapter, starting from the bottom and working toward the top.

Generally speaking, the Big Four expressions come in the three types in the following table.

Expression	Example	Rule
Contains only addition and subtraction	$12 + 7 - 6 - 3 + 8$	Evaluate left to right.
Contains only multiplication and division	$18 \div 3 \times 7 \div 14$	Evaluate left to right.
Mixed-operator expression: contains a combination of addition/subtraction and multiplication/division	$9 + 6 \div 3$	1. Evaluate multiplication and division left to right. 2. Evaluate addition and subtraction left to right.

In this section, I show you how to identify and evaluate all three types of expressions.

Expressions with only addition and subtraction

Some expressions contain only addition and subtraction. When this is the case, the rule for evaluating the expression is simple.

When an expression contains only addition and subtraction, evaluate it step by step from left to right.

Q. Find $7 + (-3) - 6 - (-10)$.

A. **8.** Start at the left with the first two numbers, $7 + (-3) = 4$:

$$7 + (-3) - 6 - (-10) = \underline{4} - 6 - (-10)$$

Continue with the next two numbers, $4 - 6 = -2$:

$$\underline{4 - 6} - (-10) = \underline{-2} - (-10)$$

Finish up with the last two numbers, remembering that subtracting a negative number is the same thing as adding a positive number:

$$-2 - (-10) = -2 + 10 = 8$$

1. What's $9 - 3 + 8 - 7$?

2. Evaluate $11 - 5 - 2 + 6 - 12$.

3. Find $17 - 11 - (-4) + (-10) - 8$.

4. $-7 + (-3) - (-11) + 8 - 10 + (-20) = ?$

Expressions with only multiplication and division

Some expressions contain only multiplication and division. Unless you've been using your free time to practice your Persian, Arabic, or Hebrew reading skills, you're probably used to reading from left to right. Luckily, that's the direction of choice for multiplication and division problems, too. When an expression has *only* multiplication and division, in any combination, you should have no trouble evaluating it: Just start with the first two numbers and continue from left to right.

EXAMPLE

Q. Evaluate this expression:

$$9 \times 2 \div 6 \div 3 \times 2$$

A. **2.** The expression contains only multiplication and division, so you can move from left to right, starting with 9×2:

$$= 18 \div 6 \div 3 \times 2$$
$$= 3 \div 3 \times 2$$
$$= 1 \times 2$$
$$= 2$$

Notice that the expression shrinks one number at a time until all that's left is 2. So $9 \times 2 \div 6 \div 3 \times 2 = 2$.

Q. Evaluate $-10 \times 2 \times (-3) \div (-4)$.

A. **–15.** The same procedure applies when you have negative numbers (just make sure you use the correct rule for multiplying or dividing by negative numbers, as I explain in Chapter 4). Start at the left with the first two numbers $-10 \times 2 = -20$:

$$\underline{-10 \times 2} \times (-3) \div (-4) = \underline{-20} \times (-3) \div (-4)$$

Continue with the next two numbers, $-20 \times (-3) = 60$:

$$\underline{-20 \times (-3)} \div (-4) = \underline{60} \div (-4)$$

Finish up with the last two numbers:

$$60 \div (-4) = -15$$

YOUR TURN

⑤ Find $18 \div 6 \times 10 \div 6$.

⑥ Evaluate $20 \div 4 \times 8 \div 5 \div (-2)$.

⑦ $12 \div (-3) \times (-9) \div 6 \times (-7) = ?$

⑧ Solve $-90 \div 9 \times (-8) \div (-10) \div 4 \times (-15)$.

Mixed-operator expressions

Often an expression contains

» At least one addition or subtraction operator

» At least one multiplication or division operator

I call these *mixed-operator expressions*. To evaluate them, you need some stronger medicine.

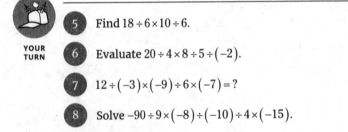

Evaluate mixed-operator expressions as follows:

REMEMBER **1.** **Evaluate the multiplication and division from left to right.**

2. **Evaluate the addition and subtraction from left to right.**

Q. What's $-15 \times 3 \div (-5) - (-3) \times (-4)$?

A. **-3.** Start by underlining all the multiplication and division in the problem; then evaluate all multiplication and division from left to right:

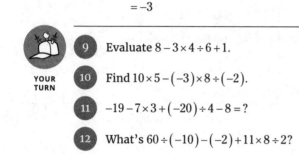

$$\underline{-15 \times 3 \div (-5)} - \underline{(-3) \times (-4)}$$
$$= -45 \div (-5) - (-3) \times (-4)$$
$$= 9 - (-3) \times (-4)$$
$$= 9 - 12$$

Finish up by evaluating the addition and subtraction from left to right:

$$= -3$$

YOUR TURN

9. Evaluate $8 - 3 \times 4 \div 6 + 1$.

10. Find $10 \times 5 - (-3) \times 8 \div (-2)$.

11. $-19 - 7 \times 3 + (-20) \div 4 - 8 = ?$

12. What's $60 \div (-10) - (-2) + 11 \times 8 \div 2$?

Handling Powers Responsibly

You may have heard that power corrupts, but rest assured that when mathematicians deal with powers, the order of operations usually keeps them in line. When an expression contains one or more powers, evaluate all powers from left to right before moving on to the Big Four operators. Here's the breakdown:

1. **Evaluate all powers from left to right.**

   In Chapter 5, I show you that raising a number to a power simply means multiplying the number by itself that many times. For example, $2^3 = 2 \times 2 \times 2 = 8$. Remember that anything raised to the 0 power equals 1.
2. **Evaluate all multiplication and division from left to right.**
3. **Evaluate addition and subtraction from left to right.**

If you compare this numbered list with the one in the preceding section, you'll notice the only difference is that I've now inserted a new rule at the top.

Q. Evaluate $7 - 4^2 \div 2^4 + 9 \times 2^3$.

A. **78.** Evaluate all powers from left to right, starting with $4^2 = 4 \times 4 = 16$:

$$= 7 - 16 \div 2^4 + 9 \times 2^3$$

Move on to evaluate the remaining two powers:

$$= 7 - 16 \div 16 + 9 \times 8$$

Next, evaluate all multiplication and division from left to right:

$$= 7 - 1 + 9 \times 8$$
$$= 7 - 1 + 72$$

Finish up by evaluating the addition and subtraction from left to right:

$$= 6 + 72$$
$$= 78$$

EXAMPLE

YOUR TURN

13. Evaluate $3^2 - 2^3 \div 2^2$.

14. Find $5^2 - 4^2 - (-7) \times 2^2$.

15. $70^1 - 3^4 \div (-9) \times (-7) + 123^0 = ?$

16. What's $11^2 - 2^7 + 3^5 \div 3^3$?

Prioritizing parentheses

Did you ever go to the post office and send a package high-priority so that it'd arrive as soon as possible? Parentheses work just like that. Parentheses — ( ) — allow you to indicate that a piece of an expression is high-priority — that is, it has to be evaluated before the rest of the expression.

When an expression includes parentheses with only Big Four operators, just do the following:

REMEMBER
1. **Evaluate the contents of parentheses, from left to right, removing parentheses as you go.**

2. **Evaluate the rest of the expression.**

Q. Evaluate $(6-2)+(10-15 \div 3)$.

EXAMPLE **A.** **9.** Start by evaluating the contents of the first set of parentheses:

$$(6-2)+(10-15 \div 3) = 4+(10-15 \div 3)$$

Move on to the next set of parentheses. This contains a mixed-operator expression, so evaluate it in two steps:

$$= 4+(10-5)$$
$$= 4+5$$

To finish up, evaluate the addition:

$$= 9$$

17 Evaluate $4 \times (3+4)-(16 \div 2)$.

YOUR TURN

18 What's $\left[5+(-8) \div 2\right]+(3 \times 6)$?

19 Find $(4+12 \div 6 \times 7)-(3+8)$.

20 $\left[2 \times (-5)\right]-(10-7) \times \left[13+(-8)\right] = ?$

Pulling apart parentheses and powers

When an expression has parentheses and powers, evaluate it in the following order:

1. **Contents of parentheses**

2. **Powers from left to right**

After you've evaluated all parentheses and powers, the expression will be left with only the Big Four operations. Finish up by evaluating these four operations as I discuss earlier in the section, "Mixed-operator expressions."

Q. Evaluate $(8+6^2) \div (2^3-4)$.

EXAMPLE **A.** **11.** Begin by evaluating the contents of the first set of parentheses. Inside this set, evaluate the power first and do the addition next:

$$(8+6^2) \div (2^3-4)$$
$$= (8+36) \div (2^3-4)$$
$$= 44 \div (2^3-4)$$

Move to the next set of parentheses, evaluating the power first and then the subtraction:

$$= 44 \div (8 - 4) = 44 \div 4$$

Finish up by evaluating the division: $44 \div 4 = 11$.

Q. Find the value of $-1 + \left(-20 + 3^3\right)^{8-6}$.

A. **48.** When the entire contents of a set of parentheses is raised to a power, evaluate what's inside the parentheses before evaluating the power. *Inside* this set, evaluate the power first and the addition next:

$$-1 + \left(-20 + 3^3\right)^{8-6} = -1 + \left(-20 + 27\right)^{8-6} = -1 + 7^{8-6}$$

Next, evaluate the power. Begin by evaluating the expression in the exponent, $8 - 6$, and then evaluate the power itself:

$$7^{8-6} = 7^2 = 7 \times 7 = 49:$$
$$= -1 + 49$$

Finish up by evaluating the addition: $-1 + 49 = 48$.

21 Find $\left(6^2 - 12\right) \div \left(16 \div 2^3\right)$.

22 Evaluate $-10 - \left[2 + 3^2 \times (-4)\right]$.

23 $7^2 - \left[3 + 3^2 \div (-9)\right]^5 = ?$

24 What is $\left(10 - 1^{14} \times 8\right)^{4 \div 4 + 5}$?

Figuring out nested parentheses

Have you ever seen nested wooden dolls? (They originated in Russia, where their ultra-cool name is *matryoshka*.) This curiosity appears to be a single carved piece of wood in the shape of a doll. But when you open it up, you find a smaller doll nested inside it. And when you open up the smaller doll, you find an even smaller one hidden inside that one — and so on.

Like these Russian dolls, some arithmetic expressions contain sets of *nested* parentheses — one set of parentheses inside another set. To evaluate a set of nested parentheses, start by evaluating the inner set of parentheses and work your way outward.

TIP

Parentheses — ( ) — come in a number of styles, including brackets — [ ] — and braces — { }. These different styles help you keep track of where a statement in parentheses begins and ends. No matter what they look like, to the mathematician these different styles are all parentheses, so they all get treated the same.

Q. Find the value of $\left\{3 \times \left[10 \div (6-4)\right]\right\} + 2$.

EXAMPLE **A.** **17.** Begin by evaluating what's inside the innermost set of parentheses: $6 - 4 = 2$:

$$\left\{3 \times \left[10 \div (6-4)\right]\right\} + 2 = \left\{3 \times \left[10 \div 2\right]\right\} + 2$$

The result is an expression with one set of parentheses inside another set, so evaluate what's inside the inner set: $10 \div 2 = 5$:

$$= \left\{3 \times 5\right\} + 2$$

Now, evaluate what's inside the final set of parentheses:

$$= 15 + 2$$

Finish up by evaluating the addition: $15 + 2 = 17$.

YOUR TURN

25 Evaluate $7 + \left\{\left[(10-6) \times 5\right] + 13\right\}$.

26 Find the value of $\left[(2+3) - (30 \div 6)\right] + (-1 + 7 \times 6)$.

27 $-4 + \left\{\left[-9 \times (5-8)\right] \div 3\right\} = ?$

28 Evaluate $\left\{(4-6) \times \left[18 \div (12 - 3 \times 2)\right]\right\} - (-5)$.

Bringing It All Together: The Order of Operations

REMEMBER Throughout this chapter, you work with a variety of rules for deciding how to evaluate arithmetic expressions. These rules all give you a way to decide the order in which an expression gets evaluated. All together, this set of rules is called the *order of operations* (or sometimes, the *order of precedence*). Here's the complete order of operations for arithmetic:

1. **Contents of parentheses from the inside out**

2. **Powers from left to right**

3. **Multiplication and division from left to right**

4. **Addition and subtraction from left to right**

The only difference between this list and the one in the section "Pulling apart parentheses and powers" is that I've now added a few words to the end of Step 1 to cover nested parentheses (which I discuss in the preceding section).

CHAPTER 6 **Please Excuse My Dear Aunt Sally: Evaluating Arithmetic Expressions with PEMDAS** 87

Q. Evaluate $\left[\left(8\times4+2^3\right)\div10\right]^{7-5}$.

EXAMPLE

A. **16.** Start by focusing on the inner set of parentheses, evaluating the power, then the multiplication, and then the addition:

$$\left[\left(8\times4+2^3\right)\div10\right]^{7-5}$$
$$=\left[\left(8\times4+8\right)\div10\right]^{7-5}$$
$$=\left[\left(32+8\right)\div10\right]^{7-5}$$
$$=\left[40\div10\right]^{7-5}$$

Next, evaluate what's inside the parentheses and the expression that makes up the exponent:

$$=4^{7-5}=4^2$$

Finish by evaluating the remaining power: $4^2=16$.

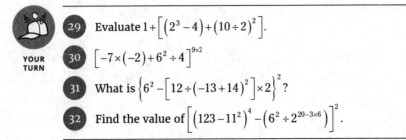

YOUR TURN

29 Evaluate $1+\left[\left(2^3-4\right)+\left(10\div2\right)^2\right]$.

30 $\left[-7\times\left(-2\right)+6^2\div4\right]^{9\times2}$

31 What is $\left\{6^2-\left[12\div\left(-13+14\right)^2\right]\times2\right\}^2$?

32 Find the value of $\left[\left(123-11^2\right)^4-\left(6^2\div2^{20-3\times6}\right)\right]^2$.

Practice Questions Answers and Explanations

(1) **7.** Add and subtract from left to right:

$$9 - 3 + 8 - 7$$
$$= 6 + 8 - 7$$
$$= 14 - 7 = 7$$

(2) **–2.**

$$11 - 5 - 2 + 6 - 12$$
$$= 6 - 2 + 6 - 12$$
$$= 4 + 6 - 12$$
$$= 10 - 12 = -2$$

(3) **–8.**

$$17 - 11 - (-4) + (-10) - 8$$
$$= 6 - (-4) + (-10) - 8$$
$$= 10 + (-10) - 8$$
$$= 0 - 8 = -8$$

(4) **–21.**

$$-7 + (-3) - (-11) + 8(-10) + (-20)$$
$$= -10 - (-11) + 8(-10) + (-20)$$
$$= 1 + 8 - 10 + (-20)$$
$$= 9 - 10 + (-20)$$
$$= -1 + (-20) = -21$$

(5) **5.** Divide and multiply from left to right:

$$18 \div 6 \times 10 \div 6$$
$$= 3 \times 10 \div 6$$
$$= 30 \div 6 = 5$$

(6) **–4.**

$$20 \div 4 \times 8 \div 5 \div (-2)$$
$$= 5 \times 8 \div 5 \div (-2)$$
$$= 40 \div 5 \div (-2)$$
$$= 8 \div (-2) = -4$$

7 **−42.**

$$12 \div (-3) \times (-9) \div 6 \times (-7)$$
$$= -4 \times (-9) \div 6 \times (-7)$$
$$= 36 \div 6 \times (-7)$$
$$= 6 \times (-7) = -42$$

8 **30.**

$$-90 \div 9 \times (-8) \div (-10) \div 4 \times (-15)$$
$$= (-10) \times (-8) \div (-10) \div 4 \times (-15)$$
$$= 80 \div (-10) \div 4 \times (-15)$$
$$= (-8) \div 4 \times (-15)$$
$$= -2 \times (-15)$$
$$= 30$$

9 **7.** Start by underlining and evaluating all multiplication and division from left to right:

$$8 - \underline{3 \times 4} \div 6 \div 1$$
$$= 8 - \underline{12 \div 6} + 1$$
$$= 8 - \underline{2} + 1$$

Now evaluate all addition and subtraction from left to right:

$$= 6 + 1 = 7$$

10 **38.** Start by underlining and evaluating all multiplication and division from left to right:

$$\underline{10 \times 5} - \underline{(-3) \times 8 \div (-2)}$$
$$= 50 - \underline{(-3) \times 8} \div (-2)$$
$$= 50 - \underline{(-24) \div (-2)}$$
$$= 50 - 12$$

Now evaluate the subtraction:

$$= 38$$

11 **−53.** Start by underlining and evaluating all multiplication and division from left to right:

$$-19 - \underline{7 \times 3} + \underline{(-20) \div 4} - 8$$
$$= -19 - 21 + \underline{(-20) \div 4} - 8$$
$$= -19 - 21 + (-5) - 8$$

Then evaluate all addition and subtraction from left to right:

$$= -40 + -5 - 8 = -45 - 8 = -53$$

(12) **40.** Start by underlining and evaluating all multiplication and division from left to right:

$$\underline{60 \div (-10)} - (-2) + \underline{11 \times 8 \div 2}$$
$$= -6 - (-2) + \underline{11 \times 8} \div 2$$
$$= -6 - (-2) + \underline{88 \div 2}$$
$$= -6 - (-2) + 44$$

Now evaluate all addition and subtraction from left to right:

$$= -4 + 44 = 40$$

(13) **7.** First, evaluate all powers:

$$3^2 - 2^3 \div 2^2 = 9 - 8 \div 4$$

Next, evaluate the division:

$$= 9 - 2$$

Finally, evaluate the subtraction:

$$9 - 2 = 7$$

(14) **37.** Evaluate all powers:

$$5^2 - 4^2 - (-7) \times 2^2 = 25 - 16 - (-7) \times 4$$

Evaluate the multiplication:

$$= 25 - 16 - (-28)$$

Finally, evaluate the subtraction from left to right:

$$= 9 - (-28) = 37$$

(15) **8.** Evaluate all powers:

$$70^1 - 3^4 \div (-9) \times (-7) + 123^0 = 70 - 81 \div (-9) \times (-7) + 1$$

Next, evaluate the multiplication and division from left to right:

$$= 70 - (-9) \times -7 + 1 = 70 - 63 + 1$$

Evaluate the addition and subtraction from left to right:

$$= 7 + 1 = 8$$

(16) **2.** First, evaluate all powers:

$$11^2 - 2^7 + 3^5 \div 3^3 = 121 - 128 + 243 \div 27$$

Evaluate the division:

$$= 121 - 128 + 9$$

Finally, evaluate the addition and subtraction from left to right:

$$= -7 + 9 = 2$$

(17) **20.** Start by evaluating what's inside the first set of parentheses:

$$4 \times (3+4) - (16 \div 2)$$
$$= 4 \times 7 - (16 \div 2)$$

Next, evaluate the contents of the second set of parentheses:

$$4 \times 7 - 8$$

Evaluate the multiplication and then the subtraction:

$$= 28 - 8 = 20$$

(18) **19.** Inside the first set of parentheses, evaluate the division first and then the addition:

$$\left[(5 + (-8) \div 2) \right] + (3 \times 6)$$
$$= \left[5 + (-4) \right] + (3 \times 6)$$
$$= 1 + (3 \times 6)$$

Next, evaluate the contents of the second set of parentheses:

$$= 1 + 18$$

Finish up by evaluating the addition:

$$1 + 18 = 19$$

(19) **7.** Begin by focusing on the first set of parentheses, handling all multiplication and division from left to right:

$$(4 + 12 \div 6 \times 7) - (3 + 8)$$
$$= (4 + 2 \times 7) - (3 + 8)$$
$$= (4 + 14) - (3 + 8)$$

Now do the addition inside the first set of parentheses:

$$= 18 - (3 + 8)$$

Next, evaluate the contents of the second set of parentheses:

$$= 18 - 11$$

Finish up by evaluating the subtraction:

$$18 - 11 = 7$$

20 –25. Evaluate the first set of parentheses, then the second, and then the third:

$$\left(2\times-5\right)-\left(10-7\right)\times\left[13+\left(-8\right)\right]$$
$$=-10-\left(10-7\right)\times\left[13+\left(-8\right)\right]$$
$$=-10-3\times\left[13+\left(-8\right)\right]$$
$$=-10-3\times5$$

Next, do multiplication and then finish up with the subtraction:

$$=-10-15=-25$$

21 12. Focusing on the contents of the first set of parentheses, evaluate the power and then the subtraction:

$$\left(6^2-12\right)\div\left(16\div2^3\right)$$
$$=\left(36-12\right)\div\left(16\div2^3\right)$$
$$=24\div\left(16\div2^3\right)$$

Next, work inside the second set of parentheses, evaluating the power first and then the division:

$$=24\div\left(16\div8\right)=24\div2$$

Finish by evaluating the division:

$$=24\div2=12$$

22 24. Focusing on the contents of the parentheses, evaluate the power first, then the multiplication, and then the addition:

$$-10-\left[2+3^2\times\left(-4\right)\right]=-10-\left[2+9\times\left(-4\right)\right]=-10-\left[2+\left(-36\right)\right]=-10\cdot$$

Finish by evaluating the subtraction:

$$-10-\left(-34\right)=24$$

23 17. Focusing *inside* the parentheses, evaluate the power first, then the division, and then the addition:

$$7^2-\left[\left(3+3^2\div\left(-9\right)\right)\right]^5$$
$$=7^2-\left[\left(3+9\div-9\right)\right]^5$$
$$=7^2-\left[\left(3+\left(-1\right)\right)\right]^5$$
$$=7^2-2^5$$

Next, evaluate both powers in order:

$$=49-2^5=49-32$$

To finish, evaluate the subtraction:

$$49 - 32 = 17$$

 64. Focusing inside the first set of parentheses, evaluate the power first, then the multiplication, and then the subtraction:

$$\left(10 - 1^{14} \times 8\right)^{4 \div 4 + 5} = \left(10 - 1 \times 8\right)^{4 \div 4 + 5} = \left(10 - 8\right)^{4 \div 4 + 5} = 2^{4 \div 4 + 5}$$

Next, handle the expression in the exponent, evaluating the division first and then the addition:

$$2^{1+5} = 2^6$$

To finish, evaluate the power:

$$2^6 = 64$$

40. First evaluate the inner set of parentheses:

$$7 + \left\{\left[(10 - 6) \times 5\right] + 13\right\} = 7 + \left\{\left[4 \times 5\right] + 13\right\}$$

Move outward to the next set of parentheses:

$$= 7 + \left\{20 + 13\right\}$$

Next, handle the remaining set of parentheses:

$$= 7 + 33$$

To finish, evaluate the addition:

$$7 + 33 = 40$$

41. Start by focusing on the first set of parentheses. This set contains two inner sets of parentheses, so evaluate these two sets from left to right:

$$\left[(2 + 3) - (30 \div 6)\right] + (-1 + 7 \times 6)$$
$$= \left[(5) - (30 \div 6)\right] + (-1 + 7 \times 6)$$
$$= \left[5 - 5\right] + (-1 + 7 \times 6)$$

Now, the expression has two separate sets of parentheses, so evaluate the first set:

$$= 0 + (-1 + 7 \times 6)$$

Handle the remaining set of parentheses, evaluating the multiplication first and then the addition:

$$= 0 + (-1 + 42) = 0 + 41$$

To finish, evaluate the addition:

$$0 + 41 = 41$$

(27) **5.** Start by evaluating the inner set of parentheses:

$$-4+\left\{\left[-9\times(5-8)\right]\div 3\right\}=-4+\left\{\left[-9\times-3\right]\div 3\right\}$$

Move outward to the next set of parentheses:

$$=-4+\left[27\div 3\right]$$

Next, handle the remaining set of parentheses:

$$=-4+9$$

Finally, evaluate the addition:

$$-4+9=5$$

(28) **−1.** Focus on the inner set of parentheses, $(12-3\times 2)$. Evaluate the multiplication first and then the subtraction:

$$\left\{(4-6)\times\left[18\div(12-3\times 2)\right]\right\}-(-5)$$
$$=\left\{(4-6)\times\left[18\div(12-6)\right]\right\}-(-5)$$
$$=\left\{(4-6)\times\left[18\div 6\right]\right\}-(-5)$$

Now the expression is an outer set of parentheses with two inner sets. Evaluate these two inner sets of parentheses from left to right:

$$=\left\{-2\times\left[18\div 6\right]\right\}-(-5)=\left\{-2\times 3\right\}-(-5)$$

Next, evaluate the final set of parentheses:

$$=-6-(-5)$$

Finish by evaluating the subtraction:

$$-6-(-5)=-1$$

(29) **30.** Start by focusing on the first of the two inner sets of parentheses, $\left(2^{3}-4\right)$. Evaluate the power first and then the subtraction:

$$1+\left[\left(2^{3}-4\right)+(10\div 2)^{2}\right]=1+\left[(8-4)+(10\div 2)^{2}\right]=1+\left[4+(10\div 2)^{2}\right]$$

Continue by focusing on the remaining inner set of parentheses:

$$=1+\left[4+5^{2}\right]$$

Next, evaluate what's inside the last set of parentheses, evaluating the power first and then the addition:

$$=1+\left[4+25\right]=1+29$$

Finish by adding the remaining numbers:

$$1+29=30$$

(30) **23.** Start with the first set of parentheses. Evaluate the power first, then the multiplication and division from left to right, and then the addition:

$$\left[\left(-7\times(-2)+6^2\div4\right)\right]^{9\times2-17}$$
$$=\left[\left(-7\times(-2)+36\div4\right)\right]^{9\times2-17}$$
$$=\left(14+36\div4\right)^{9\times2-17}$$
$$=\left(14+9\right)^{9\times2-17}$$
$$=23^{9\times2-17}$$

Next, work on the exponent, evaluating the multiplication first and then the subtraction:

$$=23^{18-17}=23^1$$

Finish by evaluating the power:

$$23^1=23$$

(31) **144.** Start by evaluating the inner set of parentheses $(-13+14)$:

$$\left\{6^2-\left[12\div(-13+14)^2\right]\times2\right\}^2$$
$$=\left\{6^2-\left[12\div1^2\right]\times2\right\}^2$$

Move outward to the next set of parentheses, $\left[12\div1^2\right]$, evaluating the power and then the division:

$$=\left\{6^2-\left[12\div1\right]\times2\right\}^2$$
$$=\left\{6^2-12\times2\right\}^2$$

Next, work on the remaining set of parentheses, evaluating the power, then the multiplication, and then the subtraction:

$$=\left\{36-12\times2\right\}^2$$
$$=\left\{36-24\right\}^2$$
$$=12^2$$

Finish by evaluating the power:

$$12^2=144$$

49. Start by working on the exponent, $20 - 3 \times 6$, evaluating the multiplication and then the subtraction:

$$\left[\left(123 - 11^2\right)^4 - \left(6^2 \div 2^{20-3\times6}\right)\right]^2$$

$$= \left[\left(123 - 11^2\right)^4 - \left(6^2 \div 2^{20-18}\right)\right]^2$$

$$= \left[\left(123 - 11^2\right)^4 - \left(6^2 \div 2^2\right)\right]^2$$

The result is an expression with two inner sets of parentheses. Focus on the first of these two sets, evaluating the power and then the subtraction:

$$= \left[\left(123 - 121\right)^4 - \left(6^2 \div 2^2\right)\right]^2$$

Work on the remaining inner set of parentheses, evaluating the two powers from left to right and then the division:

$$= \left[2^4 - \left(36 \div 2^2\right)\right]^2$$

$$= \left[2^4 - \left(36 \div 4\right)\right]^2$$

$$= \left[2^4 - 9\right]^2$$

Now evaluate what's left inside the parentheses, evaluating the power and then the subtraction:

$$= \left[16 - 9\right]^2$$

$$= 7^2$$

Finish by evaluating the power: $7^2 = 49$.

Feeling ready to tackle the chapter quiz? The 12 questions in the next section pull together all the topics from this chapter.

Whaddya Know? Chapter 6 Quiz

This quiz includes 12 questions that cover all the topics in this chapter. When you're done, flip to the following section to find solutions and explanations.

Simplify the expressions using the Order of Operations.

1. $7 \times 6 \div 3 \times 5 =$

2. $3 \times \left\{ 3^3 - \left(2^3 - 5\right) - 6 \times 4 \right\} =$

3. $8 - 7 - (-2) + 3 =$

4. $5 \times 8 - 27 \div 3 + 6 =$

5. $\left[18 - \left(6^2 - 3^3\right) \div 5^0 \right] + 7 \times \left[2^2 + 5 - 3 \right] =$

6. $600 \div 100 - 5^2 + 37^0 =$

7. $\left[\left(15 + 1^5\right) \div 8 \right]^{7-5} =$

8. $2 \times \left(4 + 8 \div 2\right) + \left(9 - 3\right) =$

9. $24 \div 6 + 2 \times 3 =$

10. $\left(6^2 + 4\right) \times \left(9 - 3^2\right) =$

11. $4^2 - 3^2 =$

12. $\left(8^2 + 6^2\right)^2 =$

Answers to Chapter 6 Quiz

(1) 70. Perform the multiplication and division in order, moving from left to right.

$$7 \times 6 \div 3 \times 5$$
$$= 42 \div 3 \times 5$$
$$= 14 \times 5$$
$$= 70$$

(2) 0. Working from inside the braces outward, first cube the 2 and subtract the 5.

$$3 \times \left\{ 3^3 - \left(2^3 - 5 \right) - 6 \times 4 \right\}$$
$$= 3 \times \left\{ 3^3 - (8 - 5) - 6 \times 4 \right\}$$
$$= 3 \times \left\{ 3^3 - (3) - 6 \times 4 \right\}$$

Next, in the braces, cube the 3 and multiply the 6 and 4.

$$= 3 \times \left\{ 27 - (3) - 24 \right\}$$

Now subtract the 3 and then the 24.

$$= 3 \times \left\{ 24 - 24 \right\} = 3 \times \left\{ 0 \right\}$$

Finally, multiply.

$$= 3 \times 0 = 0$$

(3) 6. First perform the subtraction on the number in the parentheses.

$$8 - 7 - (-2) + 3 = 8 - 7 + 2 + 3$$

Subtract and add in order, moving from left to right.

$$= 1 + 2 + 3$$
$$= 3 + 3$$
$$= 6$$

(4) 37. First perform the multiplication and division.

$$5 \times 8 - 27 \div 3 + 6$$
$$= 40 - 27 \div 3 + 6$$
$$= 40 - 9 + 6$$

Now subtract and add.

$$40 - 9 + 6$$
$$= 31 + 6 = 37$$

⑤ **51.** Working in the first brackets, square the two numbers in the parentheses and then subtract the results.

$$\left[18-\left(6^2-3^3\right)\div 5^0\right]+7\times\left[2^2+5-3\right]$$
$$=\left[18-(36-27)\div 5^0\right]+7\times\left[2^2+5-3\right]$$
$$=\left[18-(9)\div 5^0\right]+7\times\left[2^2+5-3\right]$$

Next, in the first brackets, raise the 5 to the power. Perform the division and then the subtraction.

$$\left[18-(9)\div 5^0\right]+7\times\left[2^2+5-3\right]$$
$$=\left[18-(9)\div 1\right]+7\times\left[2^2+5-3\right]$$
$$=\left[18-9\right]+7\times\left[2^2+5-3\right]$$
$$=9+7\times\left[2^2+5-3\right]$$

Working in the second brackets, first square the 2. Then perform the addition and subtraction.

$$9+7\times\left[2^2+5-3\right]$$
$$=9+7\times\left[4+5-3\right]$$
$$=9+7\times\left[9-3\right]=9+7\times 6$$

Multiply the 7 and 6. Then add the result to 9.

$$9+7\times 6$$
$$=9+42=51$$

⑥ **–18.** First find the two powers.

$$600\div 100-5^2+37^0$$
$$=600\div 100-25+1$$

Next, perform the division.

$$600\div 100-25+1$$
$$=6-25+1$$

Finally, subtract and add.

$$6-25+1=-19+1=-18$$

⑦ **4.** Perform the power in the parentheses, and simplify the power of the bracket.

$$\left[\left(15+1^5\right)\div 8\right]^{7-5}$$
$$=\left[(15+1)\div 8\right]^{7-5}$$
$$=\left[(15+1)\div 8\right]^{2}$$

Add the two numbers in the parentheses and then perform the division.

$$\left[(15+1) \div 8\right]^2$$
$$=\left[(16) \div 8\right]^2$$
$$=2^2$$

Square the 2 to get the final result of 4.

(8) **22.** In the first parentheses, perform the division. In the second parentheses, subtract.

$$2 \times (4 + 8 \div 2) + (9 - 3)$$
$$= 2 \times (4 + 4) + 6$$

Add the two numbers in the parentheses, then multiply the result by the 2.

$$2 \times (4 + 4) + 6$$
$$= 2 \times 8 + 6$$
$$= 16 + 6$$

The sum of 22 is the final result.

(9) **10.** First perform the division and multiplication. Then add the two results.

$$24 \div 6 + 2 \times 3$$
$$= 4 + 6$$
$$= 10$$

(10) **0.** Raise the two numbers to their powers.

$$\left(6^2 + 4\right) \times \left(9 - 3^2\right)$$
$$= (36 + 4) \times (9 - 9)$$

Perform the addition and subtraction in the parentheses. Then multiply the two results.

$$(36 + 4) \times (9 - 9)$$
$$= 40 \times 0 = 0$$

(11) **7.** First raise the numbers to the powers. Then subtract the results.

$$4^2 - 3^2$$
$$= 16 - 9 = 7$$

(12) **10,000.** Raise the two numbers in the parentheses to the powers.

$$\left(8^2 + 6^2\right)^2$$
$$= (64 + 36)^2$$

Add the two numbers together before squaring the result.

$$(64 + 36)^2$$
$$= 100^2 = 10,000$$

IN THIS CHAPTER

» **Knowing the four steps to solving a word problem**

» **Jotting down simple word equations that condense the important information**

» **Writing more-complex word equations**

» **Plugging numbers into the word equations to solve the problem**

» **Attacking more-complex word problems with confidence**

Chapter **7**

Turning Words into Numbers: Basic Math Word Problems

The very mention of word problems — or story problems, as they're sometimes called — is enough to send a cold shiver of terror into the bones of the average math student. Many would rather swim across a moat full of hungry crocodiles than "figure out how many bushels of corn Farmer Brown picked" or "help Aunt Sylvia decide how many cookies to bake." But word problems help you understand the logic behind setting up equations in real-life situations, making math actually useful — even if the scenarios in the word problems you practice on are pretty far-fetched.

In this chapter, I dispel a few myths about word problems. Then I show you how to solve a word problem in four simple steps. After you understand the basics, I show you how to solve more-complex problems. Some of these problems have longer numbers to calculate, and others may have more complicated stories. In either case, you can see how to work through them step by step.

Dispelling Two Myths about Word Problems

Here are two common myths about word problems:

>> Word problems are always hard.

>> Word problems are only for school — after that, you don't need them.

Both of these ideas are untrue. But they're so common that I want to address them head-on.

Word problems aren't always hard

Word problems don't have to be hard. For example, here's a word problem that you may have run into in first grade:

> Adam had 4 apples. Then Brenda gave him 5 more apples. How many apples does Adam have now?

You can probably do the math in your head, but when you were starting out in math, you may have written it down:

$$4 + 5 = 9$$

Finally, if you had one of those teachers who made you write out your answer in complete sentences, you wrote "Adam has 9 apples." (Of course, if you were the class clown, you probably wrote, "Adam doesn't have any apples because he ate them all.")

Word problems seem hard when they get too complex to solve in your head and you don't have a system for solving them. In this chapter, I give you a system and show you how to apply it to problems of increasing difficulty.

Word problems are useful

In the real world, math rarely comes in the form of equations. It comes in the form of situations that are very similar to word problems.

Whenever you paint a room, prepare a budget, bake a double batch of oatmeal cookies, estimate the cost of a vacation, buy wood to build a shelf, do your taxes, or weigh the pros and cons of buying a car versus leasing one, you need math. And the math skill you need most is understanding how to turn the *situation* you're facing into numbers that you calculate.

Word problems give you practice turning situations — or stories — into numbers.

Solving Basic Word Problems

Generally, solving a word problem involves four steps:

1. **Read through the problem and set up a *word equation* — that is, an equation that contains words as well as numbers.**

2. **Plug in numbers in place of words wherever possible to set up a regular math equation.**

3. **Use math to solve the equation.**

4. **Answer the question the problem asks.**

Most of this book is about Step 3. This chapter is all about Steps 1 and 2. I show you how to break down a word problem sentence by sentence, jot down the information you need to solve the problem, and then substitute numbers for words to set up an equation.

When you know how to turn a word problem into an equation, the hard part is done. Then you can use the rest of what you find in this book to figure out how to do Step 3 — solve the equation. From there, Step 4 is usually pretty easy, though at the end of each example, I make sure you understand how to do it.

Turning word problems into word equations

The first step to solving a word problem is reading it and putting the information you find into a useful form. In this section, I show you how to squeeze the juice out of a word problem and leave the pits behind!

Jotting down information as word equations

Most word problems give you information about numbers, telling you exactly how much, how many, how fast, how big, and so forth. Here are some examples:

Nunu is spinning 17 plates.

The width of the house is 80 feet.

If the local train is going 25 miles per hour

You need this information to solve the problem. And paper is cheap, so don't be afraid to use it. (If you're concerned about trees, write on the back of all that junk mail you get.) Have a piece of scrap paper handy and jot down a few notes as you read through a word problem.

For example, here's how you can jot down "Nunu is spinning 17 plates":

Nunu $= 17$

Here's how to note that "the width of the house is 80 feet":

Width = 80

The third example tells you, "If the local train is going 25 miles per hour" so you can jot down the following:

Local = 25

REMEMBER

Don't let the word *if* confuse you. When a problem says, "If so-and-so were true" and then asks you a question, assume that it *is* true and use this information to answer the question.

When you jot down information this way, you're really turning words into a more useful form called a *word equation.* A word equation has an equals sign like a math equation, but it contains both words and numbers.

EXAMPLE

Q. Write a word equation for the statement, "James bought 36 bagels."

A. James = 36

YOUR TURN

 1 Write a word equation for the statement, "Ralphie took 45 minutes to clean his room."

2 Write a word equation for the statement, "Mavis has saved $2,000 for a new car."

3 Write a word equation for the statement, "Ms. Archer lives 80 miles from her workplace."

 4 Write a word equation for the statement, "Pinewood Drive is 700 feet in length."

Writing relationships: Turning more-complex statements into word equations

When you start doing word problems, you notice that certain words and phrases show up over and over again. For example,

Bobo is spinning five fewer plates than Nunu.

The height of a house is half as long as its width.

The express train is moving three times faster than the local train.

You've probably seen statements such as these in word problems since you were first doing math. Statements like these look like English, but they're really math, so spotting them is important. You can represent each of these types of statements as word equations that also use Big Four operations. Look again at the first example:

Bobo is spinning five fewer plates than Nunu.

You don't know the number of plates that either Bobo or Nunu is spinning. But you know that these two numbers are related.

You can express this relationship like this:

$$Bobo + 5 = Nunu$$

This word equation is shorter than the statement it came from. And as you see in the next section, word equations are easy to turn into the math you need to solve the problem.

Here's another example:

The height of a house is half as long as its width.

You don't know the width or height of the house, but you know that these numbers are connected.

You can express this relationship between the width and height of the house as the following word equation:

$$Height = width \div 2$$

With the same type of thinking, you can express "The express train is moving three times faster than the local train" as this word equation:

$$Express = 3 \times local$$

REMEMBER

As you can see, each of the examples allows you to set up a word equation using one of the Big Four operations — adding, subtracting, multiplying, and dividing.

EXAMPLE

Q. Given the statement, "Abigail has six fewer dresses than Gwendolyn," which girl has the greater number of dresses?

A. **Gwendolyn has more dresses than Abigail.**

Q. Given the word equation Myra = Samuel ÷ 2, does Myra or Samuel have the greater number of objects?

A. **Samuel has more objects than Myra.**

Q. Write a word equation for the statement, "The obelisk is 51 feet shorter than the radio tower."

A. **Obelisk = Radio Tower − 51; obelisk + 51 = radio tower; radio tower − obelisk = 51**

YOUR TURN

5 In each of the following statements, which person has the greater number of objects?

(a) David received five more balloons than Jessica.

(b) Fred has ten fewer action figures than Wally.

(c) Larry owns half as many antique coins as Mei Ling.

(d) Ronaldo has collected four times as many butterflies as Patricia.

6 In each of the following word equations, which person has the greater number of objects?

(a) Scott = Violet + 3

(b) Carly − 15 = Noelle

(c) Mr. Grace × 4 = Ms. Pardee

(d) Aunt Eleanor = Aunt Maddie ÷ 6

7 Consider the statement, "Kathleen owns 22 more comic books than Ulysses." Which of the following word equations is a correct translation of this statement? (More than one answer may be correct.)

(a) Kathleen + 22 = Ulysses

(b) Kathleen = 22 + Ulysses

(c) Kathleen − 22 = Ulysses

(d) Kathleen − Ulysses = 22

8 Write a word equation for each of the following statements:

(a) Kyle owns 9 more goldfish than Drew.

(b) Geoff bought half as many bags of seeds as Shari.

(c) The hare runs five times as fast as the tortoise.

(d) Peggy Sue saved $150 less than Lisa Marie.

Figuring out what the problem's asking

The end of a word problem usually contains the question you need to answer to solve the problem. You can use word equations to clarify this question so you know right from the start what you're looking for.

For example, you can write the question, "Altogether, how many plates are Bobo and Nunu spinning?" as

Bobo + Nunu = ?

You can write the question, "How tall is the house?" as

Height = ?

Finally, you can rephrase the question, "What's the difference in speed between the express train and the local train?" in this way:

Express − local = ?

Q. For a word problem about the cost of groceries, write a word equation for the question, "What was the total cost of the chips, the dip, and the salsa?"

A. **Chips + Dip + Salsa = ?**

9 For a word problem about three children baking cookies for a bake sale, write a word equation for the question, "How many cookies did Greg bake?"

10 For a word problem about runners preparing for a marathon, write a word equation for the question, "How many more miles did Sarah run as compared with Vicki?"

11 For a word problem about saving money, write a word equation for the question, "How much money do Melissa and Tyrone have altogether?"

12 For a word problem about the cost of a new car, write a word equation for the question, "If the salesman gave Yolanda a $700 discount off the sticker price, how much did she actually pay for the car?"

Plugging in numbers for words

After you've written out a bunch of word equations, you have the facts you need in a form you can use. You can often solve the problem by plugging numbers from one word equation into another. In this section, I show you how to use the word equations you built in the last section to solve three problems.

Send in the clowns

Some problems involve simple addition or subtraction. Here's an example:

> Bobo is spinning five fewer plates than Nunu. (Bobo dropped a few.) Nunu is spinning 17 plates. Altogether, how many plates are Bobo and Nunu spinning?

Here's what you have already, just from reading the problem:

Nunu = 17
Bobo + 5 = Nunu

Plugging in the information gives you the following:

Bobo + 5 = 17

If you see how many plates Bobo is spinning, feel free to jump ahead. If not, here's how you rewrite the addition equation as a subtraction equation (see Chapter 4 for details):

Bobo = 17 − 5 = 12

The problem wants you to find out how many plates the two clowns are spinning together. So you need to find out the following:

Bobo + Nunu = ?

Just plug in the numbers, substituting 12 for Bobo and 17 for Nunu:

$12 + 17 = 29$

So Bobo and Nunu are spinning 29 plates.

Our house in the middle of our street

At times, a problem notes relationships that require you to use multiplication or division. Here's an example:

> The height of a house is half as long as its width, and the width of the house is 80 feet. How tall is the house?

You already have a head start from what you determined earlier:

Width = 80
Height = width ÷ 2

You can plug in information as follows, substituting 80 for the word *width*:

height = $80 ÷ 2 = 40$

So you know that the height of the house is 40 feet.

I hear the train a-comin'

Pay careful attention to what the question is asking. You may have to set up more than one equation. Here's an example:

> The express train is moving three times faster than the local train. If the local train is going 25 miles per hour, what's the difference in speed between the express train and the local train?

Here's what you have so far:

Local = 25
Express = $3 × $ local

Plug in the information you need:

Express = $3 × 25 = 75$

In this problem, the question at the end asks you to find the difference in speed between the express train and the local train. Finding the difference between two numbers is subtraction, so here's what you want to find:

Express − local = ?

You can get what you need to know by plugging in the information you've already found:

$75 - 25 = 50$

Therefore, the difference in speed between the express train and the local train is 50 miles per hour.

EXAMPLE

Q. The Panthers have won 4 fewer baseball games this season than the Cobras, and the Cobras have won twice as many games as the Fireballs. If the Cobras have won 28 games, how many more games did the Panthers win than the Fireballs?

(a) Write a word equation for the statement, "The Panthers have won 4 fewer baseball games this season than the Cobras."

(b) Write a word equation for the statement, "The Cobras have won twice as many games as the Fireballs."

(c) Write a word equation for the statement, "The Cobras have won 28 games."

(d) Plug in the information from (c) into the word equations in (a) and (b).

(e) Solve these equations.

(f) Write a word equation for the question, "How many more games did the Panthers win than the Fireballs?"

(g) Solve this equation.

(h) Write the answer to the question in a complete sentence.

A.

(a) Panthers = Cobras − 4

(b) Cobras = Fireballs × 2

(c) Cobras = 28

(d) Panthers = 28 − 4; 28 = Fireballs × 2

(e) Panthers = 24; Fireballs = 14

(f) Panthers − Fireballs = ?

(g) $24 - 14 = 10$

(h) The Panthers have won 10 more games than the Fireballs.

Q. Together, Andre and Toby sold 35 boxes of candy to support their school play. Toby sold twice as many boxes as Samantha. If Samantha sold 12 boxes, how many boxes did Andre and Samantha sell together?

- **(a)** How many boxes did Samantha sell?
- **(b)** How many boxes did Toby sell?
- **(c)** How many boxes did Andre sell?
- **(d)** What is the problem asking? Write the answer in a complete sentence.

A.

- **(a)** Samantha sold 12 boxes.
- **(b)** Toby sold 24 boxes. $(12 \times 2 = 24)$
- **(c)** Andre sold 11 boxes. $(35 - 24 = 11)$
- **(d)** Together, Andre and Samantha sold 23 boxes. $(11 + 12 = 23)$

---

13 Mason Street has five times as many houses as London Avenue. And London Avenue has 40 fewer houses than River Road. How many more houses are on Mason Street than London Avenue, assuming that River Road has 56 houses?

- **(a)** Write a word equation for the statement, "London Avenue has 40 fewer houses than River Road."
- **(b)** Write a word equation for the statement, "River Road has 56 houses."
- **(c)** Plug the information from (b) into the word equation in (a) and solve. How many houses are on London Avenue?
- **(d)** Write a word equation for the statement, "Mason Street has five times as many houses as London Avenue."
- **(e)** Plug the information from (c) into the word equation in (d) and solve. How many houses are on Mason Street?
- **(f)** What is the problem asking? Write the answer in a complete sentence.

14 Last week, Dominic worked twice as many hours as Raymond, and Carter worked 6 fewer hours than Dominic. If Carter worked 36 hours, how many hours did the three men work altogether?

- **(a)** How many hours did Carter work?
- **(b)** How many hours did Dominic work?
- **(c)** How many hours did Raymond work?
- **(d)** What is the question asking? Write the answer in a complete sentence.

15 At rush hour, the bus from Center City to Hackettstown takes three times longer to reach its destination than the train, which takes 35 minutes less than a car. If a car takes 55 minutes, how much longer is the trip by bus than by car?

- **(a)** How long does the trip by car take?
- **(b)** How long does the trip by train take?
- **(c)** How long does the trip by bus take?
- **(d)** What is the problem asking? Write the answer in a complete sentence.

Solving More-Complex Word Problems

The skills I show you previously in the section, "Solving Basic Word Problems," are important for solving any word problem because they streamline the process and make it simpler. What's more, you can use those same skills to find your way through more-complex problems. Problems become more complex when

>> The calculations become harder. (For example, instead of a dress costing $30, it costs $29.95.)

>> The amount of information in the problem increases. (For example, instead of two clowns, you have five.)

Don't let problems like these scare you. In this section, I show you how to use your new problem-solving skills to solve more-difficult word problems.

When numbers get serious

A lot of problems that look tough aren't much more difficult than the problems I show you in the previous sections. For example, consider this problem:

> Aunt Effie has $732.84 hidden in her pillowcase, and Aunt Jezebel has $234.19 less than Aunt Effie has. How much money do the two women have altogether?

One question you may have is how these women ever get any sleep with all that change clinking around under their heads. But moving on to the math, even though the numbers are larger, the principle is still the same as in problems in the earlier sections. Start reading from the beginning: "Aunt Effie has $732.84." This text is just information to jot down as a simple word equation:

Effie = $732.84

Continuing, you read, "Aunt Jezebel has *$234.19 less than* Aunt Effie has." It's another statement you can write as a word equation:

Jezebel = Effie – $234.19

Now you can plug in the number $732.84 where you see Aunt Effie's name in the equation:

Jezebel = $732.84 – $234.19

So far, the big numbers haven't been any trouble. At this point, though, you probably need to stop to do the subtraction:

$$\begin{array}{r} \$732.84 \\ -\$234.19 \\ \hline \$498.65 \end{array}$$

Now you can jot this information down, as always:

Jezebel = $498.65

The question at the end of the problem asks you to find out how much money the two women have altogether. Here's how to represent this question as an equation:

Effie + Jezebel = ?

You can plug information into this equation:

$732.84 + $498.65 = ?

Again, because the numbers are large, you probably have to stop to do the math:

$$\begin{array}{r} \$732.84 \\ +\$498.65 \\ \hline \$1231.49 \end{array}$$

So altogether, Aunt Effie and Aunt Jezebel have $1,231.49.

As you can see, the procedure for solving this problem is basically the same as for the simpler problems in the earlier sections. The only difference is that you have to stop to do some addition and subtraction.

EXAMPLE

Q. Daisy Mae is a circus elephant that weighs 12,587 pounds. Together, she and her twin sister, Luci Jo, weigh 25,322 pounds. How much more does Luci Jo weigh than her sister?

(a) How much does Daisy Mae weigh?

(b) How much does Luci Jo weigh?

(c) What is the problem asking? Write the answer in a complete sentence.

A.

(a) Daisy Mae weighs 12,587 lbs.

(b) Luci Jo weighs 12,735 lbs. (25,322 − 12,587 = 12,735)

(c) Luci Jo weighs 148 lbs. more than Daisy Mae. (12,735 − 12,587 = 148)

YOUR TURN

16 Amelia bought lunch for herself and her two daughters at a restaurant. When the check arrived, it said that she owed $35.65. She asked the server to add on a $7.50 tip, and gave him a $50.00 bill. How much money should the server return to Amelia?

(a) How much was the check for?

(b) How much did Amelia want to leave as a tip?

(c) How much money did Amelia want to pay, including the check and the tip?

(d) What is the problem asking? Write the answer in a complete sentence.

17   The washing machine that Todd wanted to buy costs $58.75 more than the dryer. But he found that buying both items together would give him a $75 discount. If the dryer costs $392.50, how much did Todd end up spending for both items?

    (a) What is the price of the dryer?

    (b) What is the price of the washing machine?

    (c) What would be the combined price of both items if bought separately?

    (d) What is the problem asking? Write the answer in a complete sentence.

18   Last month, Josh spent exactly 5 times more on rent than he spent on electricity and his Internet connection combined. If his Internet bill was $86 and he paid half that amount for electricity, how much did he spend on all three combined?

    (a) How much did Josh spend on the Internet?

    (b) How much did he spend on electricity?

    (c) How much did he spend on Internet and electricity combined?

    (d) How much did he spend on rent?

    (e) What is the problem asking? Write the answer in a complete sentence.

19   Three brothers are all keeping track of how many pushups they can do in one week. Yesterday, Tyler had completed 405 more than Stephen, and Stephen had completed exactly half as many as Andrew. But today, only Stephen had time to do more pushups, so he did 475, bringing his total up to 1,140. How many pushups have the three boys completed altogether?

    (a) How many pushups has Stephen completed?

    (b) How many pushups had Stephen completed yesterday?

    (c) How many pushups has Tyler completed?

    (d) How many pushups has Andrew completed?

    (e) What is the problem asking? Write the answer in a complete sentence.

Too much information

When the going gets tough, knowing the system for writing word equations really becomes helpful. Here's a word problem that's designed to scare you off — but with your new skills, you're ready for it:

> Four women collected money to save the endangered Salt Creek tiger beetle. Keisha collected $160, Brie collected $50 more than Keisha, Amy collected twice as much as Brie, and together Amy and Sophia collected $700. How much money did the four women collect altogether?

If you try to do this problem all in your head, you'll probably get confused. Instead, take it line by line and just jot down word equations as I discuss earlier in this chapter.

First, "Keisha collected $160." So jot down the following:

    Keisha = 160

Next, "Brie collected $50 dollars more than Keisha," so write

$$\text{Brie} = \text{Keisha} + 50$$

After that, "Amy collected twice as much as Brie":

$$\text{Amy} = \text{Brie} \times 2$$

Finally, "together, Amy and Sophia collected $700":

$$\text{Amy} + \text{Sophia} = 700$$

That's all the information the problem gives you, so now you can start working with it. Keisha collected $160, so you can plug in 160 anywhere you find Keisha's name:

$$\text{Brie} = 160 + 50 = 210$$

Now you know how much Brie collected, so you can plug this information into the next equation:

$$\text{Amy} = 210 \times 2 = 420$$

This equation tells you how much Amy collected, so you can plug this number into the last equation:

$$420 + \text{Sophia} = 700$$

To solve this problem, change it from addition to subtraction using inverse operations, as I show you in Chapter 5:

$$\text{Sophia} = 700 - 420 = 280$$

Now that you know how much money each woman collected, you can answer the question at the end of the problem:

$$\text{Keisha} + \text{Brie} + \text{Amy} + \text{Sophia} = ?$$

You can plug in this information easily:

$$160 + 210 + 420 + 280 = 1{,}070$$

So you can conclude that the four women collected $1,070 altogether.

Here's one final example putting together everything from this chapter:

> On a recent shopping trip, Travis bought six shirts for $19.95 each and two pairs of pants for $34.60 each. He then bought a jacket that cost $37.08 less than he paid for both pairs of pants. If he paid the cashier with three $100 bills, how much change did he receive?

On the first read-through, you may wonder how Travis found a store that prices jackets that way. Believe me — it was quite a challenge. Anyway, back to the problem. You can jot down the following word equations:

$$\text{Shirts} = \$19.95 \times 6$$
$$\text{Pants} = \$34.60 \times 2$$
$$\text{Jacket} = \text{pants} - \$37.08$$

The numbers in this problem are probably longer than you can solve in your head, so they require some attention:

$$\begin{array}{r} \$19.95 \\ \times \quad 6 \\ \hline \$119.70 \end{array} \qquad \begin{array}{r} \$34.60 \\ \times \quad 2 \\ \hline \$69.20 \end{array}$$

With this done, you can fill in some more information:

Shirts = $119.70
Pants = $69.20
Jacket = pants − $37.08

Now you can plug in $69.20 for *pants:*

Jacket = $69.20 − $37.08

Again, because the numbers are long, you need to solve this equation separately:

$$\begin{array}{r} \$69.20 \\ -\$37.08 \\ \hline \$32.12 \end{array}$$

This equation gives you the price of the jacket:

Jacket = $32.12

Now that you have the price of the shirts, pants, and jacket, you can find out how much Travis spent:

Amount Travis spent = $119.70 + $69.20 + $32.12

Again, you have another equation to solve:

$$\begin{array}{r} \$119.70 \\ \$69.20 \\ +\$32.12 \\ \hline \$221.02 \end{array}$$

So you can jot down the following:

Amount Travis spent = $221.02

The problem is asking you to find out how much change Travis received from $300, so jot this down:

Change = $300 − amount Travis spent

You can plug in the amount that Travis spent:

Change = $300 − $221.02

And do just one more equation:

$$
\begin{array}{r}
\$300.00 \\
-\$221.02 \\
\hline
\$78.98
\end{array}
$$

So you can jot down the answer:

Change = $78.98

Therefore, Travis received $78.98 in change.

EXAMPLE

Q. Emily loves counting money from her parents' change bowl. Yesterday, she counted 3 times as many pennies as nickels, 46 more nickels than dimes, and 7 times more dimes than quarters. If she counted 285 pennies, how many more pennies were in the bowl than quarters?

(a) How many pennies were in the bowl?

(b) How many nickels?

(c) How many dimes?

(d) How many quarters?

(e) What is the problem asking? Write the answer in a complete sentence.

A.

(a) 285 pennies.

(b) 95 nickels. $(285 \div 3 = 95)$

(c) 49 dimes. $(95 - 46 = 49)$

(d) 7 quarters. $(49 \div 7 = 7)$

(e) There were 278 more pennies in the bowl than quarters. $(285 - 7 = 278)$

YOUR TURN

20 At Seaborne High School, each of the 355 students cast a vote for one of four candidates for the president of student council. Marjorie received exactly one-fifth of the vote, which was 24 fewer votes than Leo, and Leo received 38 more votes than Glen. If Britney won the election by capturing all the rest of the votes, how many votes did she receive?

(a) How many votes were cast in total?

(b) How many votes did Marjorie receive?

(c) How many votes did Leo receive?

(d) How many votes did Glen receive?

(e) How many votes did Marjorie, Leo, and Glen receive together?

(f) What is the problem asking? Write the answer in a complete sentence.

21 If the Amaranth Building were three times taller, it would be exactly 24 feet taller than the Quinoa Hotel. And if the Quinoa Hotel were half as tall, it would be just 19 feet taller than the Teff Tower. Assuming that the Teff Tower is 74 feet, how much taller is the Quinoa Hotel than the Amaranth Building?

(a) How tall is the Teff Tower?

(b) How tall is the Quinoa Hotel?

(c) How tall is the Amaranth Building?

(d) What is the problem asking? Write the answer in a complete sentence.

22 Evan baked four types of cookies for a bake sale. He made twice as many chocolate chip cookies as oatmeal cookies, and a total of 120 mint julep cookies and peanut butter cookies combined. He made 23 more oatmeal cookies than mint julep cookies, and 66 peanut butter cookies. How many cookies did he make altogether?

(a) How many peanut butter cookies did Evan make?

(b) How many mint julep cookies did he make?

(c) How many oatmeal cookies did he make?

(d) How many chocolate chip cookies did he make?

(e) What is the problem asking? Write the answer in a complete sentence.

23 A school has 16 times as many students as teachers, and 5 more teachers than administrators. The number of male students is 7 times the number of non-teaching staff workers, and there are 10 fewer male students than female students. If there are exactly 101 female students in the school, what is the total number of teacher, staff workers, and administrators combined?

(a) How many female students are there?

(b) How many male students are there?

(c) What is the total number of students?

(d) How many staff workers are there?

(e) How many teachers are there?

(f) How many administrators are there?

(g) What is the problem asking? Write the answer in a complete sentence.

24 Five members of an extended family all have birthdays on the same day. On their most recent birthday, Brad's age was three times Sara's, and Sara was 16 years younger than Allen. When you add Allen's and Brad's ages together, the result was Eve's age. And Virginia is exactly four times older than Allen. If Virginia had just turned 84 years old, how old is Eve?

25 The express train starting at Watmore and ending in Yarborough passes through Cramerville, Middle City, and Danbury, in that order. The leg from Watmore to Cramerville takes half as many minutes as the leg from Cramerville to Middle City. And the trip from Watmore to Middle City takes three times longer than the trip from Danbury to Yarborough. If the trip from Middle City to Danbury takes 18 minutes and the trip from Middle City to Yarborough takes 33 minutes, how long does the trip from Cramerville to Danbury take?

26 If five donuts cost $3.25, two donuts and two small cups of coffee cost $3.00, and a cup of coffee and a toasted bagel with cream cheese costs $2.50, what is the cost of one donut, two toasted bagels with cream cheese, and three cups of coffee?

Practice Questions Answers and Explanations

(1) Ralphie = 45.

(2) Mavis = 2,000.

(3) Ms. Archer = 80.

(4) Pinewood Drive = 700.

(5)

    (a) David has more balloons than Jessica.

    (b) Wally has more action figures than Fred.

    (c) Mei Ling has more antique coins than Larry.

    (d) Ronaldo has more butterflies than Patricia.

(6)

    (a) Scott has more items than Violet.

    (b) Carly has more items than Noelle.

    (c) Ms. Pardee has more items than Mr. Grace.

    (d) Aunt Maddie has more items than Aunt Eleanor.

(7)

    (a) Wrong

    (b) Right

    (c) Right

    (d) Right

(8)

    (a) $\text{Kyle} = \text{Drew} + 9$; $\text{Kyle} - 9 = \text{Drew}$; $\text{Kyle} - \text{Drew} = 9$

    (b) $\text{Geoff} = \text{Shari} \div 2$; $\text{Geoff} \times 2 = \text{Shari}$; $\text{Shari} \div \text{Geoff} = 2$

    (c) $\text{Hare} = \text{Tortoise} \times 5$; $\text{Hare} \div 5 = \text{Tortoise}$; $\text{Hare} \div \text{Tortoise} = 5$

    (d) $\text{Peggy Sue} = \text{Lisa Marie} - 150$; $\text{Peggy Sue} + 150 = \text{Lisa Marie}$; $\text{Lisa Marie} - \text{Peggy Sue} = 150$

(9) Greg = ?

(10) Sarah − Vicki = ?

(11) Melissa + Tyrone = ?

(12) Sticker price − 700 = ?

(13)

   (a) London Avenue = River Road − 40.

   (b) River Road = 56.

   (c) London Avenue = 56 − 40 = 16.

   (d) Mason Street = London Avenue × 5.

   (e) Mason Street = 16 × 5 = 80.

   (f) Mason Street has 64 more houses than London Avenue (Mason Street − London Avenue = 80 − 16 = 64.)

(14)

   (a) Carter worked 36 hours.

   (b) Dominic worked 42 hours. $(36 + 6 = 42)$

   (c) Raymond worked 21 hours. $(42 \div 2 = 21)$

   (d) The three men worked a total of 99 hours. $(36 + 42 + 21 = 99)$

(15)

   (a) The trip by car takes 55 minutes.

   (b) The trip by train takes 20 minutes. $(55 - 35 = 20)$

   (c) The trip by bus takes 60 minutes. $(20 \times 3 = 60)$

   (d) The trip by bus takes 5 minutes longer than the trip by car. $(60 - 55 = 5)$

(16)

   (a) $35.65.

   (b) $7.50.

   (c) $43.15. $(\$35.65 + \$7.50 = \$43.15)$

   (d) $6.85 $(\$50.00 - \$43.15 = \$6.85)$. The server should return $6.85 to Amelia.

(17)

   (a) $392.50.

   (b) $451.25. $(\$392.50 + \$58.75 = \$451.25)$

   (c) $843.75. $(\$392.50 + \$451.25 = \$843.75)$

   (d) $768.75 $(\$843.75 - \$75.00 = \$768.75)$. Todd spent $768.75 for both items.

(18)

   (a) $86.

   (b) $43. $(\$86 \div 2 = \$43)$

   (c) $129. $(\$86 + \$43 = \$129)$

   (d) $645. $(\$129 \times 5 = \$645)$

   (e) Josh spent a total of $774 for rent, electricity, and the Internet. $(\$645 + \$86 + 43 = \$774)$

(19)

(a) 1,140

(b) 665 $(1,140 - 475 = 665)$

(c) 1,070 $(665 + 405 = 1,070)$

(d) 1,330 $(665 \times 2 = 1,330)$

(e) Altogether, the three boys have completed 3,540 pushups. $(1,140 + 1,070 + 1,330 = 3,540)$

(20)

(a) A total of 355 were cast.

(b) Marjorie received 71 votes. $(355 \div 5 = 71)$

(c) Leo received 95 votes. $(71 + 24 = 95)$

(d) Glen received 57 votes. $(95 - 38 = 57)$

(e) Together, Marjorie, Leo, and Glen received 223 votes. $(71 + 95 + 57 = 223)$

(f) Britney won the election with 132 votes. $(355 - 223 = 132)$

(21)

(a) 74 feet.

(b) 186 feet. $([74 + 19] \times 2 = 186)$

(c) 70 feet. $([186 + 24] \div 3 = 70)$

(d) The Quinoa Hotel is 116 feet taller than the Amaranth Building. $(186 - 70 = 116)$

(22)

(a) 66

(b) 54 $(120 - 66 = 54)$

(c) 77 $(54 + 23 = 77)$

(d) 154 $(77 \times 2 = 154)$

(e) Evan made a total of 351 cookies. $(66 + 54 + 77 + 154 = 351)$

(23)

(a) 101

(b) 91 $(101 - 10 = 91)$

(c) 192 $(101 + 91 = 192)$

(d) 13 $(91 \div 7 = 13)$

(e) 12 $(192 \div 16 = 12)$

(f) 7 $(12 - 5 = 7)$

(g) There are a total of 32 teachers, staff workers, and administrators. $(13 + 12 + 7 = 32)$

(24) **36 years old.** Virginia was 84, so Allen was 21 $(84 \div 4 = 21)$. Sara was 5 $(21 - 16 = 5)$, so Brad was 15 $(5 \times 3 = 15)$. Therefore, Eve was 36 $(21 + 15)$.

(25) **48 minutes.** The trip from Middle City to Danbury takes 18 minutes and the trip from Middle City to Yarborough takes 33 minutes, so the trip from Danbury to Yarborough takes 15 minutes $(33 - 18 = 15)$. So the trip from Watmore to Middle City takes 45 minutes $(15 \times 3 = 45)$. These 45 minutes consist of two legs, the first of which (from Watmore to Cramerville) takes half as long as the second leg (from Cramerville to Middle City). So, the trip from Watmore to Cramerville takes 15 minutes, and the trip from Cramerville to Middle City takes 30 minutes. Therefore, the trip from Cramerville to Danbury takes 48 minutes.

(26) **$6.50.** Five donuts cost $3.25, so each donut costs $.65 $(\$3.25 \div 5 = \$0.65)$; thus, two donuts cost $1.30 $(\$.65 \times 2 = \$1.30)$. The combined cost of two donuts and two coffees is $3.00, so two coffees cost $1.70 ($3.00 − $1.30 = $1.70); therefore, one coffee costs $.85 ($1.70 ÷ 2 = $.85). The combined cost of a coffee and a bagel is $2.50, so a bagel costs $1.65 $(\$2.50 - \$.85 = \$1.65)$. Therefore, here's how you calculate the cost of a donut, two bagels, and three coffees:

$$\$.65 + (\$1.65 \times 2) + (\$.85 \times 3) = \$.65 + \$3.30 + \$2.55 = \$6.50$$

Now that you've got some practice behind you, check out the following quiz, which tests you on everything in the chapter.

Whaddya Know? Chapter 7 Quiz

Ready for the quiz? The following 12 questions focus on the word-problem skills covered in this chapter. In the next section, you'll find a detailed explanation of each answer.

1 Given the word equation, Jack's speed − 3 = Jill's speed, which person was moving slower?

2 Josh has twice as many baseball cards as Evan, and Scott has 67 more cards than Josh. Who has the least number of cards?

For each of the following problems, write your word equation and then find the answer.

3 You have $20 to spend. You pick out socks that cost $6.59 and a shirt that costs $11.39. Both prices include tax. After paying for the socks and shirt, how much money do you have left?

4 The Anderson triplets are each 12 years old. What will the sum of their ages be in 11 years?

5 Johnny Cash was born in 1932 and died in 2003. How old was he when he died?

6 A square corral has a fence all the way around that measures a total of 312 feet. How long is each side of the corral?

7 A cake recipe calls for $\frac{1}{2}$ cup of white sugar and $\frac{3}{4}$ cup of brown sugar. If you double the recipe, what is the total amount of sugar needed?

8 On your trip to Chicago, you left at 6:00 a.m. and drove until 8:00 a.m., when you stopped for coffee. Then you drove from 8:15 a.m. until 10:30 a.m., when you stopped to eat. Back in the car, you drove from 11:15 a.m. until 2:00 p.m. You took another break and then finished your trip by driving from 2:30 p.m. until you arrived at your destination at 4:00 p.m. How long were you driving?

9 George is on a diet. The first week he lost 7 pounds, and the second week he lost 2 pounds. The third week he gained back 3 pounds, but the fourth week he lost 4 pounds. The fifth week he lost 1 pound, and the sixth week he lost 2 pounds. If he weighed 175 pounds when he started, then what does he weigh now?

10 If Catherine made 7 more free throws than Katie and Katie made 9 free throws, then how many free throws did Catherine make?

11 When Jon, Jim, and Jane were comparing the candy they received on Halloween, they saw that Jon had twice as many candy bars as Jim. But Jane had 16 more candy bars than Jon. If Jane had 48 candy bars, how many candy bars did the three children have altogether?

12 Helen saved up $285 to buy new clothes, and Genna saved up one-third as much. How much more does Helen have to spend than Genna?

13 Janet has 4 cats, 3 dogs, and 10 rabbits. She wants to double her number of cats, triple her number of dogs, and give away half of her rabbits. How many pets will she have after accomplishing this task?

14 A rectangular cardboard box is twice as long as it is wide and half as high as it is wide. If the width of the box is 12 inches, then what is the volume? (Recall: Volume = Length × Width × Height.)

Answers to Chapter 7 Quiz

(1) **Jill.** Jack's speed is being reduced by 3 to make it equal to Jill's speed. He had to slow down, so he was going faster than Jill.

(2) **Evan.** Josh's $= 2 \times$ Evan's and Scott's $= 67 +$ Josh's. So, in order: Scott has more than Josh and Josh has more than Evan.

(3) **\$2.02. Word equation:** $\$20 - (\text{socks} + \text{shirt}) =$ what's left. Adding the prices of the two items together: $\$6.59 + \$11.39 = \$17.98$. Subtract that sum from \$20 to get: $\$20 - \$17.98 = \$2.02$.

(4) **69. Word equation:** $(\text{age now} + 11) \times 3 =$ total. Add 11 to the current age of 12, and then multiply the result by 3. $12 + 11 = 23$ and $23 \times 3 = 69$.

(5) **71 years old. Word equation:** Year died $-$ year born $=$ age. Subtract $2003 - 1932 = 71$.

(6) **78 feet. Word equation:** Total around $\div 4 =$ one side. A square has four equal sides, so divide 312 by 4: $312 \div 4 = 78$.

(7) **2 ½ cups. Word equation:** $(\text{white sugar} + \text{brown sugar}) \times 2 =$ total. Add the two amounts of sugar together, and then multiply the result by 2: $\frac{1}{2} + \frac{3}{4} = \frac{2}{4} + \frac{3}{4} = \frac{5}{4}$. Multiplying by 2, $\frac{5}{4} \times \frac{2}{1} = \frac{5}{\cancel{4}_2} \times \frac{\cancel{2}^1}{1} = \frac{5}{2} = 2\frac{1}{2}$.

(8) **8 ½ hours. Word equation:** $(\text{finish 1} - \text{start 1}) + (\text{finish 2} - \text{start 2}) + (\text{finish 3} - \text{start 3}) + (\text{finish 4} - \text{start 4}) =$ total. Determine the length of each time you drove:

> 8:00 a.m. − 6:00 a.m. = 2 hours; 10:30 a.m. − 8:15 a.m. = 2 hours and 15 minutes; 2:00 p.m. − 11:15 a.m. = 2 hours and 45 minutes; 4:00 p.m. − 2:30 p.m. = 1 hour and 30 minutes. Add the hours and minutes to get 7 hours and 90 minutes. But 90 minutes is 1 hour and 30 minutes, so the total is 8 hours and 30 minutes, or 8 ½ hours.

(9) **162 pounds. Word equation:** $175 +$ total of all changes $=$ new weight. The amounts lost and gained are: $-7, -2, +3, -4, -1, -2$. Totaling the amounts: $-7 - 2 + 3 - 4 - 1 - 2 = -13$. So, subtract 13 from 175 and you have: $175 - 13 = 162$.

(10) **16 free throws. Word equation:** $7 +$ Katie's $=$ Catherine's. Add 7 to the 9 free throws that Katie made: $7 + 9 = 16$.

(11) **96 bars. Word equation:** Jane's $= 48$ and Jane's $=$ Jon's $+ 16$, so Jon's $= 32$. And Jon's $= 2 \times$ Jim's, so Jim's $= 16$. Thus, Jane's $+$ Jon's $+$ Jim's $= 48 + 32 + 16 = 96$.

(12) **\$190. Word equations:** Genna's $= \frac{1}{3} \times$ Helen's, and $285 -$ Genna's $=$ how much more. First find how much one-third of Helen's savings amount to: $285 \times \frac{1}{3} = 95$. Subtract that from the amount Helen saved: $285 - 95 = 190$.

(13) **22 pets. Word equation:** $2 \times \text{cats} + 3 \times \text{dogs} + \frac{1}{2} \times \text{rabbits} =$ new total. Inserting the numbers, $(2 \times 4) + (3 \times 3) + \left(\frac{1}{2} \times 10\right) = 8 + 9 + 5 = 22$.

(14) **1,728 cubic inches. Word equations:** Length $\times$ width $\times$ height $=$ volume; $(2 \times \text{width}) \times \text{width} \times \left(\frac{1}{2} \times \text{width}\right) =$ volume. Twice the width of 12 is 24 inches. And half the width of 12 is 6 inches. So, multiplying, $24 \times 12 \times 6 = 1,728$.

Chapter **8**

Divisibility and Prime Numbers

When one number is *divisible* by another, you can divide the first number by the second number without getting a remainder (see Chapter 3 for details on division). In this chapter, I explore divisibility from a variety of angles.

To start, I show you a bunch of handy tricks for discovering whether one number is divisible by another without actually doing the division. (In fact, you don't find long division anywhere in this chapter!) After that, I talk about prime numbers and composite numbers (which I introduce briefly in Chapter 1).

This discussion, plus what follows in Chapter 10, can help make your encounter with fractions in Unit 4 a lot friendlier.

Knowing the Divisibility Tricks

As you begin to work with fractions in Unit 4, the question of whether one number is divisible by another comes up a lot. In this section, I give you a bunch of time-saving tricks for finding out whether one number is divisible by another without actually making you do the division.

Counting everyone in: Numbers you can divide everything by

Every number is divisible by 1. As you can see, when you divide any number by 1, the answer is the number itself, with no remainder:

$$2 \div 1 = 2$$
$$17 \div 1 = 17$$
$$431 \div 1 = 431$$

Similarly, every number (except 0) is divisible by itself. Clearly, when you divide any number by itself, the answer is 1:

$$5 \div 5 = 1$$
$$28 \div 28 = 1$$
$$873 \div 873 = 1$$

WARNING

You can't divide any number by 0. Mathematicians say that dividing by 0 is *undefined.*

In the end: Looking at the final digits

You can tell whether a number is divisible by 2, 5, 10, 100, or 1,000 simply by looking at how the number ends — no calculations required.

Divisible by 2

Every even number — that is, every number that ends in 2,4,6,8, or 0 — is divisible by 2. For example, the following bolded numbers are divisible by 2:

$$\textbf{6} \div 2 = 3$$
$$\textbf{22} \div 2 = 11$$
$$\textbf{538} \div 2 = 269$$
$$\textbf{6,790} \div 2 = 3,395$$
$$\textbf{77,144} \div 2 = 38,572$$
$$\textbf{212,116} \div 2 = 106,058$$

Divisible by 5

Every number that ends in either 5 or 0 is divisible by 5. The following bolded numbers are divisible by 5:

$$\mathbf{15} \div 5 = 3$$
$$\mathbf{625} \div 5 = 125$$
$$\mathbf{6,970} \div 5 = 1,394$$
$$\mathbf{44,440} \div 5 = 8,888$$
$$\mathbf{511,725} \div 5 = 102,345$$
$$\mathbf{9,876,630} \div 5 = 1,975,326$$

Divisible by 10, 100, or 1,000

Every number that ends in 0 is divisible by 10. The following bolded numbers are divisible by 10:

$$\mathbf{20} \div 10 = 2$$
$$\mathbf{170} \div 10 = 17$$
$$\mathbf{56,720} \div 10 = 5,672$$

Every number that ends in 00 is divisible by 100:

$$\mathbf{300} \div 100 = 3$$
$$\mathbf{8,300} \div 100 = 83$$
$$\mathbf{634,900} \div 100 = 6,349$$

And every number that ends in 000 is divisible by 1,000:

$$\mathbf{6,000} \div 1,000 = 6$$
$$\mathbf{99,000} \div 1,000 = 99$$
$$\mathbf{1,234,000} \div 1,000 = 1,234$$

In general, every number that ends with a string of 0s is divisible by the number you get when you write 1 followed by that many 0s. For example,

900,000 is divisible by 100,000.

235,000,000 is divisible by 1,000,000.

820,000,000,000 is divisible by 10,000,000,000.

Q. Is 348 divisible by 2?

A. Yes, because it's an even number $\left(\text{check:} 348 \div 2 = 174 \right)$.

Q. Is 551 divisible by 5?

A. No, because it ends in 1, not 5 or 0 $\left(\text{check:} 551 \div 5 = 110 \text{ r } 1 \right)$.

Q. Is 1,620 divisible by 10?

A. **Yes,** because it ends in 0 $\left(\text{check: } 1,620 \div 10 = 162\right)$.

Q. Is 3,050 divisible by 100?

A. **No,** because it ends in 50, not 00 $\left(\text{check: } 3,050 \div 100 = 30 \text{ r } 50\right)$.

YOUR
TURN

1 Which of the following numbers are divisible by 2?

    (a) 37

    (b) 82

    (c) 111

    (d) 75,316

2 Which of the following numbers are divisible by 5?

    (a) 75

    (b) 103

    (c) 230

    (d) 9,995

3 Which of the following numbers are divisible by 10?

    (a) 40

    (b) 105

    (c) 200

    (d) 60,001

4 Which of the following numbers are divisible by 100?

    (a) 660

    (b) 900

    (c) 10,200

    (d) 500,080

Count it up: Checking divisibility by adding and subtracting digits

Sometimes you can check divisibility by adding up all or some of the digits in a number. The sum of a number's digits is called its *digital root*. Finding the digital root of a number is easy, and it's handy to know.

REMEMBER

To find the digital root of a number, just add up the digits and repeat this process until you get a one-digit number. Here are some examples:

The digital root of 24 is 6 because $2 + 4 = 6$.

The digital root of 143 is 8 because $1 + 4 + 3 = 8$.

The digital root of 51,111 is 9 because $5 + 1 + 1 + 1 + 1 = 9$.

Sometimes you need to do this process more than once. Here's how to find the digital root of the number 87,482. You have to repeat the process three times, but eventually you find that the digital root of 87,482 is 2:

$$8 + 7 + 4 + 8 + 2 = 29$$
$$2 + 9 = 11$$
$$1 + 1 = 2$$

Read on to find out how sums of digits can help you check for divisibility by 3, 9, or 11.

Divisible by 3

Every number whose digital root is 3, 6, or 9 is divisible by 3.

REMEMBER First, find the digital root of a number by adding its digits until you get a single-digit number. Here are the digital roots of 18, 51, and 975:

18	$1 + 8 = 9$
51	$5 + 1 = 6$
975	$9 + 7 + 5 = 21$; $2 + 1 = 3$

With the numbers 18 and 51, adding the digits leads immediately to digital roots 9 and 6, respectively. With 975, when you add up the digits, you first get 21, so you then add up the digits in 21 to get the digital root 3. Thus, these three numbers are all divisible by 3. If you do the actual division, you find that $18 \div 3 = 6$, $51 \div 3 = 17$, and $975 \div 3 = 325$, so the method checks out.

However, when the digital root of a number is anything other than 3, 6, *or* 9, the number *isn't* divisible by 3:

1,037	$1 + 0 + 3 + 7 = 11$; $1 + 1 = 2$

Because the digital root of 1,037 is 2, 1,037 *isn't* divisible by 3. If you try to divide by 3, you end up with 345 r 2.

Divisible by 9

Every number whose digital root is 9 is divisible by 9.

REMEMBER

To test whether a number is divisible by 9, find its digital root by adding up its digits until you get a one-digit number. Here are some examples:

36	$3+6=9$
243	$2+4+3=9$
7,587	$7+5+8+7=27;\ 2+7=9$

With the numbers 36 and 243, adding the digits leads immediately to digital roots of 9 in both cases. With 7,587, however, when you add up the digits, you get 27, so you then add up the digits in 27 to get the digital root 9. Thus, all three of these numbers are divisible by 9. You can verify this by doing the division:

$$36 \div 9 = 4 \qquad 243 \div 9 = 27 \qquad 7,857 \div 9 = 873$$

However, when the digital root of a number is anything other than 9, the number isn't divisible by 9. Here's an example:

706	$7+0+6=13;\ 1+3=4$

Because the digital root of 706 is 4, 706 *isn't* divisible by 9. If you try to divide 706 by 9, you get 78 r 4.

Divisible by 11

Two-digit numbers that are divisible by 11 are hard to miss because they simply repeat the same digit twice. Here are all the numbers less than 100 that are divisible by 11:

$$11 \quad 22 \quad 33 \quad 44 \quad 55 \quad 66 \quad 77 \quad 88 \quad 99$$

TIP

For numbers between 100 and 200, use this rule: Every three-digit number whose first and third digits add up to its second digit is divisible by 11. For example, suppose you want to decide whether the number 154 is divisible by 11. Just add the first and third digits:

$$1+4=5$$

Because these two numbers add up to the second digit, 5, the number 154 is divisible by 11. If you divide, you get $154 \div 11 = 14$, a whole number.

Now suppose you want to figure out whether 136 is divisible by 11. Add the first and third digits:

$$1+6=7$$

Because the first and third digits add up to 7 instead of 3, the number 136 isn't divisible by 11. You can find that $136 \div 11 = 12$ r 4.

TIP

For numbers of any length, the rule is slightly more complicated, but it's still often easier than doing long division. To find out when a number is divisible by 11, place plus and negative signs alternatively in front of every digit, then calculate the result. If this result is divisible by 11 (including 0), the number is divisible by 11; otherwise, the number isn't divisible by 11.

For example, suppose you want to discover whether the number 15,983 is divisible by 11. To start out, place plus and negative signs in front of alternate digits (every other digit):

$$+1-5+9-8+3=0$$

Because the result is 0, the number 15,983 is divisible by 11. If you check the division, $15,983 \div 11 = 1,453$.

Now suppose you want to find out whether 9,181,909 is divisible by 11. Again, place plus and negative signs in front of alternate digits and calculate the result:

$$+9-1+8-1+9-0+9=33$$

Because 33 is divisible by 11, the number 9,181,909 is also divisible by 11. The actual answer is

$$9,181,909 \div 11 = 834,719$$

Q. Which of the following numbers are divisible by 3?

(a) 31

(b) 54

(c) 768

(d) 2,809

A. Add the digits to determine each number's digital root; if the digital root is 3, 6, or 9, the number is divisible by 3:

(a) **No,** because $3+1=4$ $\left(\text{check: } 31 \div 3 = 10 \text{ r } 1\right)$.

(b) **Yes,** because $5+4=9$ $\left(\text{check: } 54 \div 3 = 18\right)$.

(c) **Yes,** because $7+6+8=21$ and $2+1=3$ $\left(\text{check: } 768 \div 3 = 256\right)$.

(d) **No,** because $2+8+0+9=19$, $1+9=10$, and $1+0=1$ $\left(\text{check: } 2,809 \div 3 = 936 \text{ r } 1\right)$.

Q. Which of the following numbers are divisible by 11?

(a) 71

(b) 154

(c) 528

(d) 28,094

A. Place + and − signs between the numbers and determine whether the result is 0 or a multiple of 11:

(a) **No,** because $+7-1=6$ $\left(\text{check: } 71 \div 11 = 6 \text{ r } 5\right)$.

(b) **Yes,** because $+1-5+4=0$ $\left(\text{check: } 154 \div 11 = 14\right)$.

(c) **Yes,** because $5-2+8=11$ $\left(\text{check: } 528 \div 11 = 48\right)$.

(d) **Yes,** because $+2-8+0-9+4=-11$ $\left(\text{check: } 28,094 \div 11 = 2,554\right)$.

5 Which of the following numbers are divisible by 3?

   (a) 81

   (b) 304

   (c) 986

   (d) 1,027

6 Which of the following numbers are divisible by 3?

   (a) 20,103

   (b) 541,836

   (c) 2,345,678

   (d) 4,444,444

7 Which of the following numbers are divisible by 9?

   (a) 107

   (b) 765

   (c) 9,876

   (d) 1,111,248

8 Which of the following numbers are divisible by 11?

   (a) 42

   (b) 187

   (c) 726

   (d) 1,969

Less is more: Checking divisibility by subtracting

The tests for divisibility by 4 and 8 are a little unusual, but with practice you may find them useful from time to time.

Divisible by 4

To find out whether a number is divisible by 4:

REMEMBER **1. Drop all but the last two digits.**

      2. Keep subtracting 20 until the result is 20 or less.

If the result is divisible by 4 (that is, 4, 8, 12, 16, or 20), then the number you started with is also divisible by 4.

Divisible by 8

To test a number for divisibility by 8:

REMEMBER

1. **Drop all but the last three digits.**

2. **Keep subtracting 200 until the result is 200 or less.**

3. **Keep subtracting 40 until the result is 40 or less.**

If the result is divisible by 8 (that is, 8, 16, 24, 32, or 40), the original number is also divisible by 8.

Odd numbers are *never* divisible by 4 or 8.

Q. Is 856 divisible by 4?

A. **Yes.** Drop all but the last two digits. Then keep subtracting 20 until you know whether the number is divisible by 4: $856 \rightarrow 56 \rightarrow 36 \rightarrow 16$, and 16 is divisible by 4 (check: $856 \div 4 = 214$).

Q. Is 1,492 divisible by 8?

A. **No.** Drop all but the last three digits. Then keep subtracting 200 and then 40 until you know whether the number is divisible by 8: $1,492 \rightarrow 492 \rightarrow 292 \rightarrow 92 \rightarrow 52 \rightarrow 12$, and 12 is not divisible by 8 (check: $1,492 \div 8 = 186 \text{ r } 4$).

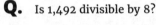

(9) Which of the following numbers are divisible by 4?

   (a) 82

   (b) 756

   (c) 4,463

   (d) 789,508

(10) Which of the following numbers are divisible by 8?

   (a) 112

   (b) 386

   (c) 5,458

   (d) 39,216

Cross-checking: Using multiple tests

To test for divisibility by 6, 12, 15, and 18, use a pair of divisibility tests explained earlier in this chapter:

» **By 6.** Test for divisibility by 2 and 3: Any even number whose digital root is 3, 6, or 9 is divisible by 6; otherwise, the number isn't.

» **By 12.** Test for divisibility by 3 and 4: Any number whose digital root is 3, 6, or 9 *and* that can be decreased to a number that's divisible by 4 by repeatedly subtracting 20 is divisible by 12; all other numbers aren't.

>> **By 15.** Test for divisibility by 3 and 5: Any number that ends in 5 or 0 and has a digital root of 3, 6, or 9 is divisible by 15; all other numbers aren't.

>> **By 18.** Test for divisibility by 2 and 9: Any even number whose digital root is 9 is divisible by 18; other numbers aren't.

EXAMPLE

Q. Is 702 divisible by 6?

A. **Yes,** because 702 is even and $7 + 0 + 2 = 9$ (check: $702 \div 6 = 117$).

Q. Is 624 divisible by 12?

A. **Yes,** because $6 + 2 + 4 = 12$ and $1 + 2 = 3$, and $624 \rightarrow 24 \rightarrow 4$, which is divisible by 4 (check: $624 \div 12 = 52$).

Q. Is 2,160 divisible by 15?

A. **Yes,** because 2,160 ends in 0 and $2 + 1 + 6 + 0 = 9$ (check: $2{,}160 \div 15 = 144$).

Q. Is 8,142 divisible by 18?

A. **No,** because 8,142 is even, but $8 + 1 + 4 + 2 = 15$ and $1 + 5 = 6$ (check: $8{,}142 \div 18 = 452 \text{ r } 6$).

YOUR TURN

11 Is 178 divisible by 6?

12 Is 338 divisible by 12?

13 Is 505 divisible by 15?

14 Is 1,656 divisible by 18?

Identifying Prime and Composite Numbers

In the earlier section, "Counting everyone in: Numbers you can divide everything by," I show you that every number (except 0 and 1) is divisible by at least two numbers: 1 and itself. In this section, I explore prime numbers and composite numbers (which I introduce you to in Chapter 1).

In Chapter 9, you need to know how to tell prime numbers from composite to break a number down into its prime factors. This tactic is important when you begin working with fractions.

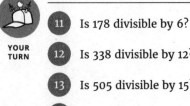

REMEMBER

A *prime number* is divisible by exactly two positive whole numbers: 1 and the number itself. A *composite number* is divisible by at least three numbers.

For example, 2 is a prime number because when you divide it by any number but 1 and 2, you get a remainder. So there's only one way to multiply two counting numbers and get 2 as a product:

$$1 \times 2 = 2$$

Similarly, 3 is prime because when you divide by any number but 1 or 3, you get a remainder. So the only way to multiply two numbers together and get 3 as a product is the following:

$$1 \times 3 = 3$$

On the other hand, 4 is a composite number because it's divisible by three numbers: 1, 2, and 4. In this case, you have two ways to multiply two counting numbers and get a product of 4:

$$1 \times 4 = 4$$
$$2 \times 2 = 4$$

But 5 is a prime number because it's divisible only by 1 and 5. Here's the only way to multiply two counting numbers and get 5 as a product:

$$1 \times 5 = 5$$

And 6 is a composite number because it's divisible by 1, 2, 3, and 6. Here are two ways to multiply two counting numbers and get a product of 6:

$$1 \times 6 = 6$$
$$2 \times 3 = 6$$

REMEMBER

Every counting number except 1 is either prime or composite. The reason 1 is neither is that it's divisible by only one number, which is 1.

Here's a list of the prime numbers that are less than 30:

2, 3, 5, 7, 11, 13, 17, 19, 23, 29

TIP

When testing to see whether a number is prime or composite, perform divisibility tests in the following order (from easiest to hardest): 2, 5, 3, 11, 7, and 13. If you find that a number is divisible by one of these, you know that it's composite and you don't have to perform the remaining tests. Here's how you know which tests to perform:

» If a number less than 121 isn't divisible by 2, 3, 5, or 7, it's prime; otherwise, it's composite.

» If a number less than 289 isn't divisible by 2, 3, 5, 7, 11, or 13, it's prime; otherwise, it's composite.

Remember that 2 is the only prime number that's even. The next three odd numbers are prime — 3, 5, and 7. To keep the list going, think "lucky 7, lucky 11, unlucky 13" — they're all prime.

EXAMPLE

Q. For each of the following numbers, tell which is prime and which is composite.

(a) 185

(b) 243

(c) 253

(d) 263

A. Check divisibility to identify prime and composite numbers:

   (a) **185 is composite.** The number 185 ends in 5, so it's divisible by 5.

   (b) **243 is composite.** The number 243 ends in an odd number, so it isn't divisible by 2. It doesn't end in 5 or 0, so it isn't divisible by 5. Its digital root is 9 (because $2+4+3=9$), so it's divisible by 3. The math shows you that $243 \div 3 = 81$.

   (c) **253 is composite.** The number 253 ends in an odd number, so it isn't divisible by 2. It doesn't end in 5 or 0, so it isn't divisible by 5. Its digital root is 1 (because $2+5+3=10$ and $1+0=1$), so it isn't divisible by 3. But it is divisible by 11, because it passes the + and – test $(+2-5+3=0)$. If you do the math, you find that $253 = 11 \times 23$.

   (d) **263 is prime.** The number 263 ends in an odd number, so it isn't divisible by 2. It doesn't end in 5 or 0, so it isn't divisible by 5. Its digital root is 2 (because $2+6+3=11$ and $1+1=2$), so it isn't divisible by 3. It isn't divisible by 11, because it fails the + and – test $(+2-6+3=-1$, which isn't 0 or divisible by 11$)$. It isn't divisible by 7, because $263/7 = 37r4$. And it isn't divisible by 13, because $263 \div 13 = 20r3$.

---

 15 Which of the following numbers are prime, and which are composite?

YOUR TURN

   (a) 3

   (b) 9

   (c) 11

   (d) 14

16 Of the following numbers, tell which are prime and which are composite.

   (a) 65

   (b) 73

   (c) 111

   (d) 172

 17 Find out whether each of these numbers is prime or composite.

   (a) 23

   (b) 51

   (c) 91

   (d) 113

18 Figure out which of the following are prime numbers and which are composite numbers.

   (a) 143

   (b) 169

   (c) 187

   (d) 283

Practice Questions Answers and Explanations

1

(a) **No,** because it's odd $(\text{check: } 37 \div 2 = 18 \text{ r } 1)$.

(b) **Yes,** because it's even $(\text{check: } 82 \div 2 = 41)$.

(c) **No,** because it's odd $(\text{check: } 111 \div 2 = 55 \text{ r } 1)$.

(d) **Yes,** because it's even $(\text{check: } 75,316 \div 2 = 37,658)$.

2

(a) **Yes,** because it ends in 5 $(\text{check: } 75 \div 5 = 15)$.

(b) **No,** because it ends in 3, not 0 or 5 $(\text{check: } 103 \div 5 = 20 \text{ r } 3)$.

(c) **Yes,** because it ends in 0 $(\text{check: } 230 \div 5 = 46)$.

(d) **Yes,** because it ends in 5 $(\text{check: } 9,995 \div 5 = 1,999)$.

3

(a) **Yes,** because it ends in 0 $(\text{check: } 40 \div 10 = 4)$.

(b) **No,** because it ends in 5, not 0 $(\text{check: } 105 \div 10 = 10 \text{ r } 5)$.

(c) **Yes,** because it ends in 0 $(\text{check: } 200 \div 10 = 20)$.

(d) **No,** because it ends in 1, not 0 $(\text{check: } 60,001 \div 10 = 6,000 \text{ r } 1)$.

4

(a) **No,** because it ends in 60, not 00 $(\text{check: } 660 \div 100 = 6 \text{ r } 60)$.

(b) **Yes,** because it ends in 00 $(\text{check: } 900 \div 100 = 9)$.

(c) **Yes,** because it ends in 00 $(\text{check: } 10,200 \div 100 = 102)$.

(d) **No,** because it ends in 80, not 00 $(\text{check: } 500,080 \div 100 = 5,000 \text{ r } 80)$.

5 *Note:* The digital root is 3, 6, or 9 for numbers divisible by 3.

(a) **Yes,** because $8 + 1 = 9$ $(\text{check: } 81 \div 3 = 27)$.

(b) **No,** because $3 + 0 + 4 = 7$ $(\text{check: } 304 \div 3 = 101 \text{ r } 1)$.

(c) **No,** because $9 + 8 + 6 = 23$ and $2 + 3 = 5$ $(\text{check: } 986 \div 3 = 328 \text{ r } 2)$.

(d) **No,** because $1 + 2 + 7 = 10$ and $1 + 0 = 1$ $(\text{check: } 1,027 \div 3 = 342 \text{ r } 1)$.

6 *Note:* The digital root is 3, 6, or 9 for numbers divisible by 3.

(a) **Yes,** because $2 + 0 + 1 + 0 + 3 = 6$ $(\text{check: } 20,103 \div 3 = 6,701)$.

(b) **Yes,** because $5 + 4 + 1 + 8 + 3 + 6 = 27$ and $2 + 7 = 9$ $(\text{check: } 541,836 \div 3 = 180,612)$.

(c) **No,** because $2 + 3 + 4 + 5 + 6 + 7 + 8 = 35$ and $3 + 5 = 8$ $(\text{check: } 2,345,678 \div 3 = 781,892 \text{ r } 2)$.

(d) **No,** because $4 + 4 + 4 + 4 + 4 + 4 + 4 = 28$, $2 + 8 = 10$, and
$1 + 0 = 1$ $(\text{check: } 4,444,444 \div 3 = 1,481,481 \text{ r } 1)$.

(7) *Note:* The digital root is 9 for numbers divisible by 9.

(a) **No,** because $1+0+7=8$ (check: $107 \div 9 = 11$ r 8).

(b) **Yes,** because $7+6+5=18$ and $1+8=9$ (check: $765 \div 9 = 85$).

(c) **No,** because $9+8+7+6=30$ and $3+0=3$ (check: $9,876 \div 9 = 1,097$ r 3).

(d) **Yes,** because $1+1+1+1+2+4+8=18$ and $1+8=9$ (check: $1,111,248 \div 9 = 123,472$).

(8) *Note:* Answers add up to 0 or a multiple of 11 for numbers divisible by 11.

(a) **No,** because $+4-2=2$ (check: $42 \div 11 = 3$ r 9).

(b) **Yes,** because $+1-8+7=0$ (check: $187 \div 11 = 17$).

(c) **Yes,** because $+7-2+6=11$ (check: $726 \div 11 = 66$).

(d) **Yes,** because $+1-9+6-9=-11$ (check: $1,969 \div 11 = 179$).

(9) Drop all but the last two digits. Then keep subtracting 20 until you know whether the number is divisible by 4. (***Remember:*** If the number is odd, it's *not* divisible by 4.)

(a) **No,** because $82 \rightarrow 62 \rightarrow 42 \rightarrow 22 \rightarrow 2$, and 2 is not divisible by 4 (check: $82 \div 4 = 20$ r 2).

(b) **Yes,** because $756 \rightarrow 56 \rightarrow 36 \rightarrow 16$, and 16 is divisible by 4 (check: $756 \div 4 = 189$).

(c) **No,** because 4,463 is an odd number (check: $4,463 \div 4 = 1,115$ r 3).

(d) **Yes,** because $789,508 \rightarrow 08$, and 8 is divisible by 4 (check: $789,508 \div 4 = 197,377$).

(10) Drop all but the last three digits. Then keep subtracting 200, then 40, until you know whether the number is divisible by 8. (***Remember:*** If the number is odd, it's *not* divisible by 8.)

(a) **Yes,** because $112 \rightarrow 72 \rightarrow 32$, and 32 is divisible by 8 (check: $112 \div 8 = 14$).

(b) **No,** because $386 \rightarrow 186 \rightarrow 146 \rightarrow 106 \rightarrow 66 \rightarrow 26$, and 26 is not divisible by 8 (check: $386 \div 8 = 48$ r 2).

(c) **No,** because $5,458 \rightarrow 458 \rightarrow 258 \rightarrow 58 \rightarrow 18$, and 18 is not divisible by 8 (check: $5,458 \div 8 = 682$ r 2).

(d) **Yes,** because $39,216 \rightarrow 216 \rightarrow 16$, and 16 is divisible by 8 (check: $39,216 \div 8 = 4,902$).

(11) **No,** because 178 is even, but $1+7+8=16$ and $1+6=7$ (check: $178 \div 6 = 29$ r 4).

(12) **No,** because $3+3+8=14$ and $1+4=5$ (check: $338 \div 12 = 28$ r 2).

(13) **No,** because 505 ends in 5, but $5+0+5=10$ and $1+0=1$ (check: $505 \div 15 = 33$ r 10).

(14) **Yes,** because 1,656 is even, and $1+6+5+6=18$ and $1+8=9$ (check: $1,656 \div 18 = 92$).

(15)

(a) **3 is prime.** The only factors of 3 are 1 and 3.

(b) **9 is composite.** The factors of 9 are 1, 3, and 9.

(c) **11 is prime.** The only factors of 11 are 1 and 11.

(d) **14 is composite.** As an even number, 14 is also divisible by 2 and therefore can't be prime.

(16)

    (a) 65 is composite. Because 65 ends in 5, it's divisible by 5.

    (b) 73 is prime. The number 73 isn't even, doesn't end in 5 or 0, and isn't a multiple of 7.

    (c) 111 is composite. The digital root of 111 is $1+1+1=3$, so it's divisible by 3 $\left(\text{check: } 111 \div 3 = 37\right)$.

    (d) 172 is composite. The number 172 is even, so it's divisible by 2.

(17)

    (a) 23 is prime. The number 23 isn't even, doesn't end in 5 or 0, has a digital root of 5, and isn't a multiple of 7.

    (b) 51 is composite. The digital root of 51 is 6, so it's a multiple of 3 $\left(\text{check: } 51 \div 3 = 17\right)$.

    (c) 91 is composite. The number 91 is a multiple of $7: 7 \times 13 = 91$.

    (d) 113 is prime. The number 113 is odd, doesn't end in 5 or 0, and has a digital root of 5, so it's not divisible by 2, 5, or 3. It's also not a multiple of $7: 113 \div 7 = 16$ r 1.

(18)

    (a) 143 is composite. $+1-4+3=0$, so 143 is divisible by 11.

    (b) 169 is composite. You can evenly divide 13 into 169 to get 13.

    (c) 187 is composite. $+1-8+7=0$, so 187 is a multiple of 11.

    (d) 283 is prime. The number 283 is odd, doesn't end in 5 or 0, and has a digital root of 4; therefore, it's not divisible by 2, 5, or 3. It's not divisible by 11, because $+2-8+3=-3$, which isn't a multiple of 11. It also isn't divisible by 7 $\left(\text{because } 283 \div 7 = 40 \text{ r } 3\right)$ or 13 $\left(\text{because } 283 \div 13 = 21 \text{ r } 10\right)$.

By now, you should be ready for the chapter quiz in the next section, covering the gamut of divisibility topics in this chapter.

Whaddya Know? Chapter 8 Quiz

Ready for the quiz? The 14 questions that follow test you on divisibility by all the numbers covered in this chapter, as well as prime and composite numbers. When you're done, you can find the solutions and explanations in the next section.

 1   The number 1,110 is divisible by six of these numbers: 2, 3, 4, 5, 6, 8, 9, 10, 11, 12, 15, 18. Which six are they?

 2   The number 5,036 is divisible by two of these numbers: 2, 3, 4, 5, 6, 8, 9, 10, 11, 12, 15, 18. Which two are they?

 3   The number 615 is divisible by three of these numbers: 2, 3, 4, 5, 6, 8, 9, 10, 11, 12, 15, 18. Which three are they?

 4   The number 2,860 is divisible by five of these numbers: 2, 3, 4, 5, 6, 8, 9, 10, 11, 12, 15, 18. Which five are they?

 5   The number 9,408 is divisible by six of these numbers: 2, 3, 4, 5, 6, 8, 9, 10, 11, 12, 15, 18. Which six are they?

 6   The number 7,011 is divisible by two of these numbers: 2, 3, 4, 5, 6, 8, 9, 10, 11, 12, 15, 18. Which two are they?

 7   The number 336 is divisible by six of these numbers: 2, 3, 4, 5, 6, 8, 9, 10, 11, 12, 15, 18. Which six are they?

 8   The number 9,000,000 is divisible by ten of these numbers: 2, 3, 4, 5, 6, 9, 10, 11, 12, 15, 100. Which ten are they (or which one does NOT divide it)?

 9   The number 1,350 is divisible by eight of these numbers: 2, 3, 4, 5, 6, 8, 9, 10, 11, 12, 15, 18. Which eight are they?

 10   The number 1,950 is divisible by six of these numbers: 2, 3, 4, 5, 6, 8, 9, 10, 11, 12, 15, 18. Which six are they?

11   The number 2,662 is divisible by two of these numbers: 2, 3, 4, 5, 6, 8, 9, 10, 11, 12, 15, 18. Which two are they?

12   The number 616 is divisible by four of these numbers: 2, 3, 4, 5, 6, 8, 9, 10, 11, 12, 15, 18. Which four are they?

13   One of the following numbers is prime: 27, 47, 91, 121, 201. Which one is prime?

14   One of the following numbers is composite: 13, 31, 43, 51, 67. Which one is composite?

Answers to Chapter 8 Quiz

(1) **2, 3, 5, 6, 10, 15.** The number ends in 0, so you can include 2, 5, and 10. The sum of the digits is 3, so you can include 3. And since it's divisible by 2 and 3, it's divisible by 6. Being divisible by 3 and 5 makes it divisible by 15. Placing signs alternately, you get $+1-1+1-0=1$, so it's not divisible by 11.

(2) **2, 4.** The last three digits form the number 36, which is divisible by 2 and 4. The sum of the digits is 14, so it's not divisible by 3 or any of its multiples. Placing signs alternately, you get $+5-0+3-6=2$, so it's not divisible by 11.

(3) **3, 5, 15.** This is an odd number, so none of the even values work. The sum of the digits is 12, so it's divisible by 3 also include that because it ends with 5 or 0, it's divisible by 5 and because it's divisible by 3 and 5 it's divisible by 15.

(4) **2, 4, 5, 10, 11.** The last two digits form the number 60, which is divisible by 4. Placing signs alternately, you get $+2-8+6-0=0$, so it's divisible by 11.

(5) **2, 3, 4, 6, 8, 12.** The last three digits form the number 408, which is divisible by 8. Placing signs alternately, you get $+9-4+0-8=-3$, so it's not divisible by 11.

(6) **3, 9.** The sum of the digits is 9, and this is an odd number.

(7) **2, 3, 4, 6, 8, 12.** The sum of the digits is 12. The number 8 divides 336 evenly.

(8) **2, 3, 4, 5, 6, 9, 10, 12, 15, 100.** The only number it isn't divisible by is 11. (It's also divisible by 8 and 16, but they weren't on the list to choose from.)

(9) **2, 3, 5, 6, 9, 10, 15, 18.** Since it's divisible by 3 and 5, it's divisible by 15. And since it's divisible by 2 and 9, it's divisible by 18.

(10) **2, 3, 5, 6, 10, 15.** The sum of the digits is 15, and it ends in 0.

(11) **2, 11.** The number is even. Placing signs alternately, you get $+2-6+6-2=0$, so it's divisible by 11.

(12) **2, 4, 8, 11.** The last two digits form the number 16, which is divisible by 4. Placing signs alternately, you get $+6-1+6=11$, so it's divisible by 11.

(13) **47.** The number 27 is divisible by 3; 91 is divisible by 7; 121 is divisible by 11; and 201 is divisible by 3.

(14) **51.** The number 51 is divisible by 3 and 17. The other numbers are prime.

Chapter **9**

Divided Attention: Factors and Multiples

In Chapter 1, I introduce you to sequences of numbers based on the multiplication table. In this chapter, I tell you about two important ways to think about these sequences: as *factors* and as *multiples.* Factors and multiples are really two sides of the same coin. Here I show you what you need to know about these two important concepts.

For starters, I discuss how factors and multiples are connected to multiplication and division. Then I show you how to find all the factor pairs of a number and how to decompose (split up) any number into its prime factors. To finish up on factors, I show you how to find the greatest common factor (GCF) of any set of numbers. After that, I tackle multiples, showing you how to generate the multiples of a number and then use this skill to find the least common multiple (LCM) of a set of numbers.

Knowing Six Ways to Say the Same Thing

In this section, I introduce you to factors and multiples, and I show you how these two important concepts are connected. As I discuss in Chapter 5, multiplication and division are inverse operations. For example, the following equation is true:

$$5 \times 4 = 20$$

So this equation using the inverse operation is also true:

$$20 \div 4 = 5$$

You may have noticed that, in math, you tend to run into the same ideas over and over again. For example, mathematicians have six different ways to talk about this relationship.

The following three statements all focus on the relationship between 5 and 20 from the perspective of multiplication:

» 5 *multiplied* by some number is 20.

» 5 is a *factor* of 20.

» 20 is a *multiple* of 5.

In two of the examples, you can see this relationship reflected in the words *multiplied* and *multiple*. For the remaining example, keep in mind that two factors are multiplied to equal a product.

Similarly, the following three statements all focus on the relationship between 5 and 20 from the perspective of division:

» 20 *divided* by some number is 5.

» 20 is *divisible by* 5.

» 5 is a *divisor* of 20.

Why do mathematicians need all these words for the same thing? Maybe for the same reason that Inuit people need a bunch of words for snow (my two favorites are *qanik* [falling snow] and *qinu* [slushy sea ice] — what are yours?). In any case, in this chapter, I focus on the words *factor* and *multiple*. When you understand the concepts, which word you choose doesn't matter a whole lot.

Understanding Factors and Multiples

In the preceding section, I connect the concept of divisibility (see Chapter 8) with factors and multiples.

For example, 12 is divisible by 3 because 12 ÷ 3 = 4, with no remainder. You can also describe this relationship between 12 and 3 using the words *factor* and *multiple*. When you're working with positive numbers, the factor is always the *smaller number* and the multiple is the *bigger number*. For example, 12 is divisible by 3, so

>> The number 3 is a *factor* of 12.

>> The number 12 is a *multiple* of 3.

EXAMPLE

Q. The number 40 is divisible by 5, so which number is the factor and which is the multiple?

A. The number **5 is the factor and 40 is the multiple,** because 5 is smaller and 40 is larger.

Q. Which two of the following statements means the same thing as "18 is a multiple of 6"?

(a) 6 is a factor of 18.

(b) 18 is divisible by 6.

(c) 6 is divisible by 18.

(d) 18 is a factor of 6.

A. **Choices** *a* **and** *b*. The number 6 is the factor and 18 is the multiple, because 6 is smaller than 18, so *a* is correct. And 18 ÷ 6 = 3, so 18 is divisible by 6; therefore, *b* is correct.

YOUR TURN

1 Which of the following statements are true, and which are false?

(a) 5 is a factor of 15.

(b) 9 is a multiple of 3.

(c) 11 is a factor of 12.

(d) 7 is a multiple of 14.

2 Which two of these statements mean the same thing as "18 is divisible by 6"?

(a) 18 is a factor of 6.

(b) 18 is a multiple of 6.

(c) 6 is a factor of 18.

(d) 6 is a multiple of 18.

3 Which two of these statements mean the same thing as "10 is a factor of 50"?

(a) 10 is divisible by 50.

(b) 10 is a multiple of 50.

(c) 50 is divisible by 10.

(d) 50 is a multiple of 10.

4 Which of the following statements are true, and which are false?

(a) 3 is a factor of 42.

(b) 11 is a multiple of 121.

(c) 88 is a multiple of 9.

(d) 11 is a factor of 121.

Finding Fabulous Factors

In this section, I introduce you to factors. First, I show you how to find out whether one number is a factor of another. Then I show you how to list all the factor pairs of a number. After that, I introduce the key idea of a number's prime factors. This information all leads up to an essential skill: finding the greatest common factor (GCF) of a set of numbers.

Deciding when one number is a factor of another

REMEMBER

You can easily tell whether a number is a factor of a second number: Just divide the second number by the first. If it divides evenly (with no remainder), the number is a factor; otherwise, it's not a factor.

For example, suppose you want to know whether 7 is a factor of 56. Here's how you find out:

$$56 \div 7 = 8$$

Because 7 divides 56 without leaving a remainder, 7 is a factor of 56.

And here's how you find out whether 4 is a factor of 34:

$$34 \div 4 = 8 \text{ r } 2$$

Because 4 divides 34 with a remainder of 2, 4 isn't a factor of 34.

This method works no matter how large the numbers are.

Understanding factor pairs

REMEMBER

A factor pair of a number is any pair of two numbers that, when multiplied together, equal that number. For example, 35 has two factor pairs — 1×35 and 5×7 — because

$$1 \times 35 = 35$$
$$5 \times 7 = 35$$

Similarly, 24 has four factor pairs — 1×24, 2×12, 3×8, and 4×6 — because

$$1 \times 24 = 24$$
$$2 \times 12 = 24$$
$$3 \times 8 = 24$$
$$4 \times 6 = 24$$

TIP

Every positive integer has at least one factor pair: 1 times the number itself. For example:

$$1 \times 2 = 2 \qquad 1 \times 11 = 11 \qquad 1 \times 43 = 43$$

When a number greater than 1 has only one factor pair, it's a prime number (see Chapter 8 for more on prime numbers).

Generating a Number's Factors

When one number is divisible by a second number, that second number is a *factor* of the first. For example, 10 is divisible by 2, so 2 is a factor of 10.

REMEMBER

A good way to find all the factors of a number is by finding all of that number's factor pairs. A *factor pair* of a number is any pair of two numbers that, when multiplied together, equal that number. For example, 30 has four factor pairs — 1×30, 2×15, 3×10, and 5×6 — because

$$1 \times 30 = 30$$
$$2 \times 15 = 30$$
$$3 \times 10 = 30$$
$$5 \times 6 = 30$$

TIP

Here's how to find *all* the factor pairs of a number:

1. **Begin the list with 1 times the number itself.**

2. **Try to find a factor pair that includes 2 — that is, see whether the number is divisible by 2 (for tricks on testing for divisibility, see Chapter 8). If it is, list the factor pair that includes 2.**

3. **Test the number 3 in the same way.**

4. **Continue testing numbers until you find no more factor pairs.**

EXAMPLE

Q. Write down all the factor pairs of 18.

A. 1×18, 2×9, 3×6. According to Step 1, begin with 1×18:

$$1 \times 18$$

The number 18 is even, so it's divisible by 2. And $18 \div 2 = 9$, so the next factor pair is 2×9:

1×18
2×9

The digital root of 18 is 9 (because $1 + 8 = 9$), so 18 is divisible by 3. And $18 \div 3 = 6$, so the next factor pair is 3×6:

1×18
2×9
3×6

The number 18 isn't divisible by 4, because $18 \div 4 = 4r2$. And 18 isn't divisible by 5, because it doesn't end with 5 or 0. This list of factor pairs is complete because there are no more numbers between 3 and 6, the last factor pair on the list.

YOUR TURN

5 Find all the factor pairs of 12.

6 Write down all the factor pairs of 28.

7 Figure out all the factor pairs of 40.

8 Find all the factor pairs of 66.

Decomposing a Number into Its Prime Factors

Every number is the product of a unique set of *prime factors,* a group of prime numbers (including repeats) that, when multiplied together, equals that number. This section shows you how to find those prime factors for a given number, a process called *decomposition.*

TIP

An easy way to decompose a number is to make a factorization tree. Here's how:

1. **Find two numbers that multiply to equal the original number; write them as numbers that branch off the original one.**

    Knowing the multiplication table can often help you here.

2. **If either number is prime, circle it and end that branch.**

3. **Continue branching off non-prime numbers into two factors; whenever a branch reaches a prime number, circle it and close the branch.**

    When every branch ends in a circled number, you're finished — just gather up the circled numbers.

Q. Decompose the number 48 into its prime factors.

EXAMPLE **A.** $48 = 2 \times 2 \times 2 \times 2 \times 3$. Begin making a factorization tree by finding two numbers that multiply to equal 48:

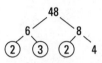

Continue making branches of the tree by doing the same for 6 and 8:

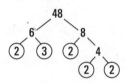

Circle the prime numbers and close those branches. At this point, the only open branch is 4. Break it down into 2 and 2:

Every branch ends in a circled number, so you're finished. The prime factors are 2, 2, 2, 2, and 3.

 YOUR TURN

9 Decompose 18 into its prime factors.

10 Decompose 42 into its prime factors.

11 Decompose 81 into its prime factors.

12 Decompose 120 into its prime factors.

Finding the Greatest Common Factor

The *greatest common factor* (GCF) of a set of numbers is the largest number that's a factor of every number in that set. Finding the GCF is helpful when you want to reduce a fraction to its lowest terms (see Chapter 10).

You can find the GCF in two ways. The first option is to list all the factor pairs of the numbers and choose the largest factor that appears in both (or all) the lists. (For info on finding factor pairs, see the earlier section, "Generating a Number's Factors.")

The other method uses prime factors, which I discuss in the preceding section. Here's how to find the GCF:

TIP

1. **Decompose the numbers into their prime factors.**

2. **Underline the factors that all the original numbers have in common.**

3. **Multiply the underlined numbers to get the GCF.**

Q. Find the greatest common factor of 12 and 20.

EXAMPLE **A.** 4. Write down all the factor pairs of 12 and 20:

Factor pairs of 12: 1×12, 2×6, 3×4

Factor pairs of 20: 1×20, 2×10, 4×5

The number 4 is the greatest number that appears in both lists of factor pairs, so it's the GCF.

Q. Find the greatest common factor of 24, 36, and 42.

A. 6. Decompose all three numbers down to their prime factors:

$24 = 2 \times 2 \times 2 \times 3$
$36 = 2 \times 2 \times 3 \times 3$
$42 = 2 \times 3 \times 7$

Underline all factors that are common to all three numbers:

$24 = \underline{2} \times 2 \times 2 \times \underline{3}$
$36 = \underline{2} \times 2 \times \underline{3} \times 3$
$42 = \underline{2} \times \underline{3} \times 7$

Multiply those underlined numbers to get your answer:

$2 \times 3 = 6$

YOUR
TURN

13 Find the greatest common factor of 10 and 22.

14 What's the GCF of 8 and 32?

15 Find the GCF of 30 and 45.

16 Figure out the GCF of 27 and 72.

17 Find the GCF of 15, 20, and 35.

18 Figure out the GCF of 44, 56, and 72.

Generating the Multiples of a Number

Generating the multiples of a number is easier than generating the factors: Just multiply the number by 1, 2, 3, and so forth. But unlike the factors of a number — which are always less than the number itself — in the positive numbers, the multiples of a number are greater than or equal to that number. Therefore, you can never write down all the multiples of any positive number.

Q. Find the first six (positive) multiples of 4.

A. **4, 8, 12, 16, 20, 24.** Write down the number 4 and keep adding 4 to it until you've written down six numbers.

Q. List the first six multiples of 12.

A. **12, 24, 36, 48, 60, 72.** Write down 12 and keep adding 12 to it until you've written down six numbers.

19 Write the first six multiples of 5.

20 Generate the first six multiples of 7.

21 List the first ten multiples of 8.

22 Write the first six multiples of 15.

Finding the Least Common Multiple

The *least common multiple* (LCM) of a set of numbers is the smallest number that's a multiple of every number in that set. For small numbers, you can simply list the first several multiples of each number until you get a match.

When you're finding the LCM of two numbers, you may want to list the multiples of the larger number first, stopping when the number of multiples you've written down equals the smaller number. Then list the multiples of the smaller number and look for a match.

However, you may have to write down a lot of multiples with this method, and the disadvantage becomes even greater when you're trying to find the LCM of large numbers. I recommend a method that uses prime factors when you're facing big numbers or more than two numbers. Here's how:

1. **Write down the prime decompositions of all the numbers.**

    See the earlier section, "Decomposing a Number into Its Prime Factors," for details.

2. **For each prime number you find, underline the *most repeated occurrences* of each.**

    In other words, compare the decompositions. If one breakdown contains two 2s and another contains three 2s, you'd underline the three 2s. If one decomposition contains one 7 and the rest don't have any, you'd underline the 7.

3. **Multiply the underlined numbers to get the LCM.**

Q. Find the LCM of 6 and 8.

EXAMPLE **A.** 24. Because 8 is the larger number, write down six multiples of 8:

Multiples of 8: 8, 16, 24, 32, 40, 48

Now, write down multiples of 6 until you find a matching number:

Multiples of 6: 6, 12, 18, 24

Q. Find the LCM of 12, 15, and 18.

A. 180. Begin by writing the prime decompositions of all three numbers. Then, for each prime number you find, underline the most repeated occurrences of each:

$12 = \underline{2} \times \underline{2} \times 3$
$15 = 3 \times \underline{5}$
$18 = 2 \times \underline{3} \times \underline{3}$

Notice that 2 appears in the decomposition of 12 most often (twice), so I underline both of those 2s. Similarly, 3 appears in the decomposition of 18 most often (twice), and 5 appears in the decomposition of 15 most often (once). Now, multiply all the underlined numbers:

$2 \times 2 \times 3 \times 3 \times 5 = 180$

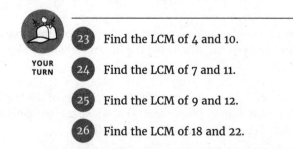

YOUR TURN

23 Find the LCM of 4 and 10.

24 Find the LCM of 7 and 11.

25 Find the LCM of 9 and 12.

26 Find the LCM of 18 and 22.

Practice Questions Answers and Explanations

1

    (a) True: $5 \times 3 = 15$.

    (b) True: $3 \times 3 = 9$.

    (c) False: You can't multiply 11 by any whole number to get 12.

    (d) False: The number 7 is a factor of 14.

2 **Choices *b* and *c*.** You're looking for something that says that the smaller number, 6, is a factor of the larger number (choice *c*) or one that says the larger number, 18, is a multiple of the smaller number (choice *b*).

3 **Choices *c* and *d*.** Factors are numbers you multiply to get larger ones, so you can say 50 is divisible by 10 (choice *c*). Multiples are the larger numbers, the products you get when you multiply two factors; you can say 50 is a multiple of 10 (choice *d*) because $10 \times 5 = 50$.

4

    (a) True: $3 \times 14 = 42$.

    (b) False: The number 11 is a factor of 121.

    (c) False: You can't multiply 9 by any whole numbers to get 88 because $88 \div 9 = 9r7$.

    (d) True: $11 \times 11 = 121$.

5 1×12, **2×6, and 3×4.** The first factor pair is 1×12. And 12 is divisible by $2 \left(12 \div 2 = 6\right)$, so the next factor pair is 2×6. Because 12 is divisible by $3 \left(12 \div 3 = 4\right)$, the next factor pair is 3×4.

6 **1×28, 2×14, and 4×7.** The first factor pair is 1×28. And 28 is divisible by $2 \left(28 \div 2 = 14\right)$, so the next factor pair is 2×14. Although 28 isn't divisible by 3, it's divisible by $4 \left(28 \div 4 = 7\right)$, so the next factor pair is 4×7. Finally, 28 isn't divisible by 5 or 6.

7 **1×40, 2×20, 4×10, and 5×8.** The first factor pair is 1×40. Because 40 is divisible by $2 \left(40 \div 2 = 20\right)$, the next factor pair is 2×20. Although 40 isn't divisible by 3, it's divisible by $4 \left(40 \div 4 = 10\right)$, so the next factor pair is 4×10. And 40 is divisible by $5 \left(40 \div 5 = 8\right)$, so the next factor pair is 5×8. Finally, 40 isn't divisible by 6 or 7.

8 **1×66, 2×33, 3×22, and 6×11.** The first factor pair is 1×66. The number 66 is divisible by $2 \left(66 \div 2 = 33\right)$, so the next factor pair is 2×33. It's also divisible by $3 \left(66 \div 3 = 22\right)$, so the next factor pair is 3×22. Although 66 isn't divisible by 4 or 5, it's divisible by $6 \left(66 \div 6 = 11\right)$, so the next factor pair is 6×11. Finally, 66 isn't divisible by 7,8,9, *or* 10.

9 **$18 = 2 \times 3 \times 3$.** Here's one possible factoring tree:

10 **$42 = 2 \times 3 \times 7$.** Here's one possible factoring tree:

11 $81 = 3 \times 3 \times 3 \times 3$. Here's one possible factoring tree:

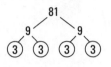

12 $120 = 2 \times 2 \times 2 \times 3 \times 5$. Here's one possible factoring tree:

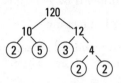

13 **2.** Write down all the factor pairs of 10 and 22:

  10: $1 \times 10, 2 \times 5$
  22: $1 \times 22, 2 \times 11$

The number 2 is the greatest number that appears on both lists.

14 **8.** Write down all the factor pairs of 8 and 32:

  8: $1 \times 8, 2 \times 4$
  32: $1 \times 32, 2 \times 16, 4 \times 8$

The greatest number that appears on both lists is 8.

15 **15.** Write down all the factor pairs of 30 and 45:

  30: $1 \times 30, 2 \times 15, 3 \times 10, 5 \times 6$
  45: $1 \times 45, 3 \times 15, 5 \times 9$

The greatest number that appears on both lists is 15.

16 **9.** Decompose 27 and 72 into their prime factors and underline every factor that's common to both:

  $27 = \underline{3} \times \underline{3} \times 3$
  $72 = 2 \times 2 \times 2 \times \underline{3} \times \underline{3}$

Multiply those underlined numbers to get your answer: $3 \times 3 = 9$.

17 **5.** Decompose the three numbers into their prime factors and underline every factor that's common to all three:

  $15 = 3 \times \underline{5}$
  $20 = 2 \times 2 \times \underline{5}$
  $35 = \underline{5} \times 7$

The only factor common to all three numbers is 5.

(18) **4.** Decompose all three numbers to their prime factors and underline each factor that's common to all three:

$$44 = \underline{2} \times \underline{2} \times 11$$
$$56 = \underline{2} \times \underline{2} \times 2 \times 7$$
$$72 = \underline{2} \times \underline{2} \times 2 \times 3 \times 3$$

Multiply those underlined numbers to get your answer: $2 \times 2 = 4$.

(19) **5, 10, 15, 20, 25, 30**

Write down the number 5 and keep adding 5 to it until you've written six numbers.

(20) **7, 14, 21, 28, 35, 42**

(21) **8, 16, 24, 32, 40, 48, 56, 64, 72, 80**

(22) **15, 30, 45, 60, 75, 90**

(23) **20.** Write down four multiples of 10:

Multiples of $10 : 10, \mathbf{20}, 30, 40$

Next, generate multiples of 4 until you find a matching number:

Multiples of $4 : 4, 8, 12, 16, \mathbf{20}$

(24) **77.** Write down seven multiples of 11:

Multiples of $11 : 11, 22, 33, 44, 55, 66, \mathbf{77}$

Next, generate multiples of 7 until you find a matching number:

Multiples of $7 : 7, 14, 21, 28, 35, 42, 49, 56, 63, 70, \mathbf{77}$

(25) **36.** Write down nine multiples of 12:

Multiples of $12 : 12, 24, \mathbf{36}, 48, 60, 72, 84, 96, 108$

Next, generate multiples of 9 until you find a matching number:

Multiples of $9 : 9, 18, 27, \mathbf{36}$

(26) **198.** First, decompose both numbers into their prime factors. Then underline the most frequent occurrences of each prime number:

$$18 = \underline{2} \times \underline{3} \times \underline{3}$$
$$22 = 2 \times \underline{11}$$

The factor 2 appears only once in any decomposition, so underline a 2. The number 3 appears twice in the decomposition of 18, so underline both of these. The number 11 appears only once, in the decomposition of 22, so underline it. Now, multiply all the underlined numbers:

$$2 \times 3 \times 3 \times 11 = 198$$

How did you do? When you're feeling ready, try out the quiz in the next section.

Whaddya Know? Chapter 9 Quiz

Quiz time! Answer the 12 questions in this section, which focus on the topics covered in this chapter. Then check out the next section for the answers and detailed explanations.

1. Write all the factor pairs of the number 60.

2. Write the prime factorization of 45.

3. What are the first five multiples of 11?

4. What is the GCF of 36 and 48?

5. What is the LCM of 40 and 45?

6. What are the first six multiples of 12?

7. Write the prime factorization of 143.

8. What is the LCM of 8, 15, and 20?

9. What is the prime factorization of 120?

10. What is the GCF of 24 and 51?

11. Write all the factor pairs of the number 200.

12. What is the LCM of 18, 24, and 30?

Answers to Chapter 9 Quiz

(1) $1 \times 60, 2 \times 30, 3 \times 20, 4 \times 15, 5 \times 12, 6 \times 10$

(2) $45 = 3 \times 3 \times 5$

(3) **11, 22, 33, 44, 55**

(4) **12.** Writing the prime factorization of the numbers and underlining the shared factors, you have $36 = \underline{2} \times \underline{2} \times \underline{3} \times 3$ and $48 = \underline{2} \times \underline{2} \times 2 \times 2 \times \underline{3}$. So the shared factors are: $\underline{2} \times \underline{2} \times \underline{3} = 12$.

(5) **360.** Writing the prime factorization of the numbers and underlining the most repeated occurrence of any particular factor, you have $40 = \underline{2} \times \underline{2} \times \underline{2} \times \underline{5}$ and $45 = \underline{3} \times \underline{3} \times 5$. So the LCM is the product of the factors:

$$2 \times 2 \times 2 \times 3 \times 3 \times 5 = 360$$

(6) **12, 24, 36, 48, 60, 72**

(7) **143 $= 11 \times 13$. Both 11 and 13 are prime.**

(8) **120.** Writing the prime factorizations of the numbers and underlining the most repeated occurrence of any particular factor, you have $8 = \underline{2} \times \underline{2} \times \underline{2}$, $15 = \underline{3} \times \underline{5}$, and $20 = 2 \times 2 \times 5$, so the LCM is the product $2 \times 2 \times 2 \times 3 \times 5 = 120$.

(9) **$120 = 2 \times 2 \times 2 \times 3 \times 5$**

(10) **3.** Writing the prime factorization of the numbers and underlining the shared factors, you have $24 = 2 \times 2 \times 2 \times \underline{3}$ and $51 = \underline{3} \times 17$. So the only shared factor is 3.

(11) $1 \times 200, 2 \times 100, 4 \times 50, 5 \times 40, 8 \times 25, 10 \times 20$

(12) **360.** Writing the prime factorization of the numbers and underlining the most repeated occurrence of any particular factor, you have $18 = 2 \times \underline{3} \times \underline{3}$, $24 = \underline{2} \times \underline{2} \times \underline{2} \times 3$, and $30 = 2 \times 3 \times \underline{5}$. So the LCM is the product of the factors: $2 \times 2 \times 2 \times 3 \times 3 \times 5 = 360$.

Fractions

4

In This Unit . . .

» Knowing the numerator from the denominator

» Understanding proper fractions, improper fractions, and mixed numbers

» Increasing and reducing the terms of fractions

» Converting between improper fractions and mixed numbers

» Using cross-multiplication to compare fractions

Chapter **10**

Understanding Fractions

Suppose that today is your birthday and your friends are throwing you a surprise party. After opening all your presents, you finish blowing out the candles on your cake, but now you have a problem: Eight of you want some cake, but you have only *one cake.* Several solutions are proposed:

>> You can all go into the kitchen and bake seven more cakes.

>> Instead of eating cake, everyone can eat celery sticks.

>> Because it's your birthday, you can eat the *whole* cake and everyone else can eat celery sticks. (That idea was yours.)

>> You can cut the cake into eight equal slices so that everyone can enjoy it.

After careful consideration, you choose the last option. With that decision, you've opened the door to the exciting world of fractions. Fractions represent parts of a thing that can be cut into pieces. In this chapter, I give you some basic information about fractions that you need to know, including the three basic types of fractions: proper fractions, improper fractions, and mixed numbers.

I move on to increasing and reducing the terms of fractions, which you need when you begin applying the Big Four operations to fractions in Chapter 11. I also show you how to convert between improper fractions and mixed numbers. Finally, I show you how to compare fractions using cross-multiplication. By the time you're done with this chapter, you'll see how fractions really can be a piece of cake!

Slicing a Cake into Fractions

Here's a simple fact: When you cut a cake into two equal pieces, each piece is half of the cake. As a fraction, you write that as $\frac{1}{2}$. In Figure 10-1, the shaded piece is half of the cake.

FIGURE 10-1:
Two halves of a cake.

© John Wiley & Sons, Inc.

Every fraction is made up of two numbers separated by a line, or a fraction bar. The line can be either diagonal or horizontal — so you can write this fraction in either of the following two ways:

$$\frac{1}{2} \qquad\qquad 1/2$$

The number above the line is called the numerator. The numerator tells you how many pieces you have. In this case, you have one dark-shaded piece of cake, so the numerator is 1.

The number below the line is called the denominator. The denominator tells you how many equal pieces the whole cake has been cut into. In this case, the denominator is 2.

Similarly, when you cut a cake into three equal slices, each piece is a third of the cake (see Figure 10-2).

This time, the shaded piece is one-third — $\frac{1}{3}$ — of the cake. Again, the numerator tells you how many pieces you have, and the denominator tells you how many equal pieces the whole cake has been cut up into (see Figure 10-3).

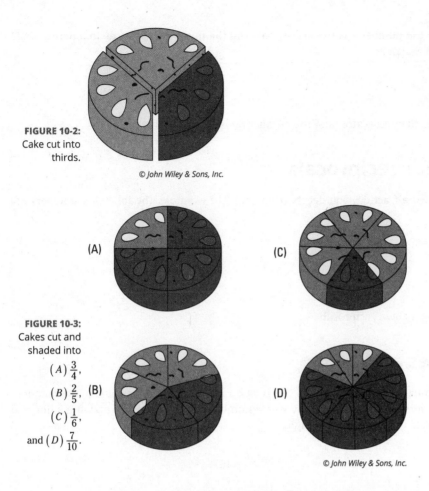

FIGURE 10-2:
Cake cut into thirds.

FIGURE 10-3:
Cakes cut and shaded into
$(A) \frac{3}{4}$,
$(B) \frac{2}{5}$,
$(C) \frac{1}{6}$,
and $(D) \frac{7}{10}$.

In each case, the numerator tells you how many pieces are shaded, and the denominator tells how many pieces there are altogether.

The fraction bar can also mean a division sign. In other words, $\frac{3}{4}$ signifies $3 \div 4$. If you take three cakes and divide them among four people, each person gets $\frac{3}{4}$ of a cake.

Knowing the Fraction Facts of Life

Fractions have their own special vocabulary and a few important properties that are worth knowing right from the start. When you know them, you find working with fractions a lot easier.

Telling the numerator from the denominator

The top number in a fraction is called the *numerator,* and the bottom number is called the *denominator.* For example, look at the following fraction:

$$\frac{3}{4}$$

In this example, the number 3 is the numerator, and the number 4 is the denominator. Similarly, look at this fraction:

$$\frac{55}{89}$$

The number 55 is the numerator, and the number 89 is the denominator.

Flipping for reciprocals

When you flip over a fraction, you get its reciprocal. For example, the following numbers are reciprocals:

$\frac{2}{3}$ and $\frac{3}{2}$

$\frac{11}{14}$ and $\frac{14}{11}$

The fraction $\frac{19}{19}$ is its own reciprocal.

Using ones and zeros

When the denominator (bottom number) of a fraction is 1, the fraction is equal to the numerator by itself. Conversely, you can turn any whole number into a fraction by drawing a line and placing the number 1 under it. For example,

$$\frac{2}{1} = 2 \qquad\qquad \frac{9}{1} = 9 \qquad\qquad \frac{157}{1} = 157$$

REMEMBER

When the numerator and denominator match, the fraction equals 1. After all, if you cut a cake into eight pieces and you keep all eight of them, you have the entire cake. Here are some fractions that equal 1:

$$\frac{8}{8} = 1 \qquad\qquad \frac{11}{11} = 1 \qquad\qquad \frac{365}{365} = 1$$

When the numerator of a fraction is 0, the fraction is equal to 0. For example,

$$\frac{0}{1} = 0 \qquad\qquad \frac{0}{12} = 0 \qquad\qquad \frac{0}{113} = 0$$

WARNING

The denominator of a fraction can never be 0. Fractions with 0 in the denominator are *undefined* — that is, they have no mathematical meaning.

Remember from earlier in this chapter that placing a number in the denominator is similar to cutting a cake into that number of pieces. You can cut a cake into two, or ten, or even a million pieces. You can even cut it into one piece (that is, don't cut it at all). But you can't cut a cake into zero pieces. For this reason, putting 0 in the denominator — much like lighting an entire book of matches on fire — is something you should never, never do.

Mixing things up

A mixed number is a combination of a whole number and a proper fraction added together. Here are some examples:

$$1\frac{1}{2} \qquad\qquad 5\frac{3}{4} \qquad\qquad 99\frac{44}{100}$$

A mixed number is always equal to the whole number plus the fraction attached to it. So $1\frac{1}{2}$ means $1+\frac{1}{2}$, $5\frac{3}{4}$ means $5+\frac{3}{4}$, and so on.

Knowing proper from improper

When the numerator (top number) is less than the denominator (bottom number), the fraction is less than 1:

$$\frac{1}{2}<1 \qquad\qquad \frac{3}{5}<1 \qquad\qquad \frac{63}{73}<1$$

Fractions like these are called *proper fractions*. Positive proper fractions are always between 0 and 1. However, when the numerator is greater than the denominator, the fraction is greater than 1. Take a look:

$$\frac{3}{2}\geq 1 \qquad\qquad \frac{7}{4}\geq 1 \qquad\qquad \frac{98}{97}\geq 1$$

Any fraction that's greater than or equal to 1 is called an *improper fraction*. Converting an improper fraction to a mixed number is customary, especially when it's the final answer to a problem.

REMEMBER

An improper fraction is always top heavy, as if it's unstable and wants to fall over. To stabilize it, convert it to a mixed number. Proper fractions are always stable.

Later in this chapter, I discuss improper fractions in more detail when I show you how to convert between improper fractions and mixed numbers.

EXAMPLE

Q. For each cake pictured, identify the fraction of the cake that's shaded.

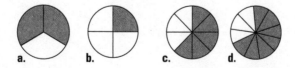

a.   b.   c.   d.

A. Put the number of shaded slices over the number of total slices in each cake:

(a) $\frac{2}{3}$

(b) $\frac{1}{4}$

(c) $\frac{5}{8}$

(d) $\frac{7}{10}$

Q. What's the reciprocal of each of the following fractions?

    (a) $\frac{3}{4}$

    (b) $\frac{6}{11}$

    (c) $\frac{22}{7}$

    (d) $\frac{41}{48}$

A. To find the reciprocal, switch around the numerator and the denominator:

    (a) The reciprocal of $\frac{3}{4}$ is $\frac{4}{3}$.

    (b) The reciprocal of $\frac{6}{11}$ is $\frac{11}{6}$.

    (c) The reciprocal of $\frac{22}{7}$ is $\frac{7}{22}$.

    (d) The reciprocal of $\frac{41}{48}$ is $\frac{48}{41}$.

YOUR
TURN

1 For each cake pictured, identify the fraction of the cake that's shaded.

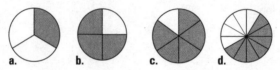

a.   b.   c.   d.

2 Which of the following fractions are proper? Which are improper?

    (a) $\frac{3}{2}$

    (b) $\frac{8}{9}$

    (c) $\frac{20}{23}$

    (d) $\frac{75}{51}$

3 Rewrite each of the following fractions as a whole number:

    (a) $\frac{3}{3}$

    (b) $\frac{10}{1}$

    (c) $\frac{10}{10}$

    (d) $\frac{81}{1}$

4 Find the reciprocal of the following fractions:

    (a) $\frac{5}{7}$

    (b) $\frac{10}{3}$

    (c) $\frac{12}{17}$

    (d) $\frac{80}{91}$

Increasing and Reducing Terms of Fractions

Take a look at these three fractions:

$$\frac{1}{2} \qquad\qquad \frac{2}{4} \qquad\qquad \frac{3}{6}$$

If you cut three cakes (as I do earlier in this chapter) into these three fractions (see Figure 10-4), exactly half of the cake will be shaded, just like in Figure 10-1, no matter how you slice it. (Get it? No matter how you slice it? You may as well laugh at the bad jokes, too — they're free.) The important point here isn't the humor, or the lack of it, but the idea about fractions.

(A)

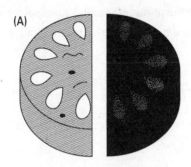

(B)

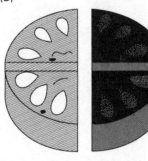

(C)

FIGURE 10-4:
Cakes cut and
shaded into

$(A)\,\frac{1}{2}$,

$(B)\,\frac{2}{4}$,

and $(C)\,\frac{3}{6}$.

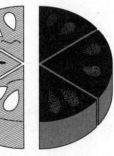

The fractions $\frac{1}{2}, \frac{2}{4}$, and $\frac{3}{6}$ are all equal in value. In fact, you can write a lot of fractions that are also equal to these. As long as the numerator is exactly half the denominator, the fractions are all equal to

$$\frac{1}{2}$$

— for example:

$$\frac{11}{22} \qquad\qquad \frac{100}{200} \qquad\qquad \frac{1,000,000}{2,000,000}$$

These fractions are equal to $\frac{1}{2}$, but their terms (the numerator and denominator) are different. In this section, I show you how to both increase and reduce the terms of a fraction without changing its value.

Increasing the terms of fractions

REMEMBER

To increase the terms of a fraction by a certain number, multiply both the numerator and the denominator by that number.

For example, to increase the terms of the fraction $\frac{3}{4}$ by 2, multiply both the numerator and the denominator by 2:

$$\frac{3}{4} = \frac{3 \times 2}{4 \times 2} = \frac{6}{8}$$

Similarly, to increase the terms of the fraction $\frac{5}{11}$ by 7, multiply both the numerator and the denominator by 7:

$$\frac{5}{11} = \frac{5 \times 7}{11 \times 7} = \frac{35}{77}$$

REMEMBER

Increasing the terms of a fraction doesn't change its value. Because you're multiplying the numerator and denominator by the same number, you're essentially multiplying the fraction by a fraction that equals 1.

One key point to know is how to increase the terms of a fraction so that the denominator becomes a preset number. Here's how you do it:

1. **Divide the new denominator by the old denominator.**

    To keep the fractions equal, you have to multiply the numerator and denominator of the old fraction by the same number. This first step tells you what the old denominator was multiplied by to get the new one.

    For example, suppose you want to raise the terms of the fraction $\frac{4}{7}$ so that the denominator is 35. You're trying to fill in the question mark here:

    $$\frac{4}{7} = \frac{?}{35}$$

Divide 35 by 7, which tells you that the denominator was multiplied by 5.

2. **Multiply this result by the old numerator to get the new numerator.**

   You now know how the two denominators are related. The numerators need to have the same relationship, so multiply the old numerator by the number you found in Step 1.

   Multiply 5 by 4, which gives you 20. So here's the answer:

$$\frac{4}{7} = \frac{4 \times 5}{7 \times 5} = \frac{20}{35}$$

Reducing fractions to lowest terms (simplifying fractions)

Reducing fractions (also called simplifying fractions) is similar to increasing fractions, except that it involves division rather than multiplication. But because you can't always divide, reducing takes a bit more finesse.

In practice, reducing fractions is similar to factoring numbers. For this reason, if you're not up on factoring, you may want to review this topic in Chapter 9.

In this section, I show you the formal way to reduce fractions, which works in all cases. Then I show you a more informal way you can use when you're more comfortable.

Reducing fractions the formal way

Reducing (simplifying) fractions the formal way relies on understanding how to break down a number into its prime factors. I discuss this in detail in Chapter 9, so if you're shaky on this concept, you may want to review it first.

Here's how to reduce a fraction:

1. **Break down both the numerator (top number) and the denominator (bottom number) into their prime factors.**

   For example, suppose you want to reduce the fraction $\frac{12}{30}$. Break down both 12 and 30 into their prime factors:

$$\frac{12}{30} = \frac{2 \times 2 \times 3}{2 \times 3 \times 5}$$

2. **Cross out any common factors.**

   As you can see, I cross out a 2 and a 3 because they're common factors — that is, they appear in both the numerator and the denominator:

$$\frac{12}{30} = \frac{\cancel{2} \times 2 \times \cancel{3}}{\cancel{2} \times \cancel{3} \times 5}$$

3. **Multiply the remaining numbers to get the reduced numerator and denominator.**

You can see now that the fraction $\frac{12}{30}$ reduces to $\frac{2}{5}$:

$$\frac{12}{30} = \frac{\cancel{2} \times 2 \times \cancel{3}}{\cancel{2} \times \cancel{3} \times 5} = \frac{2}{5}$$

As another example, here's how you reduce the fraction $\frac{32}{100}$:

$$\frac{32}{100} = \frac{\cancel{2} \times \cancel{2} \times 2 \times 2 \times 2}{\cancel{2} \times \cancel{2} \times 5 \times 5} = \frac{8}{25}$$

This time, cross out two 2s from both the top and the bottom as common factors. The remaining 2s on top and the 5s on the bottom aren't common factors. So the fraction $\frac{32}{100}$ reduces to $\frac{8}{25}$.

Reducing fractions the informal way

Here's an easier way to reduce fractions when you get comfortable with the concept:

1. **If the numerator (top number) and denominator (bottom number) are both divisible by 2 — that is, if they're both even — divide both by 2.**

For example, suppose you want to reduce the fraction $\frac{36}{60}$. The numerator and the denominator are both even, so divide them both by 2:

$$\frac{36}{60} = \frac{18}{30}$$

2. **Repeat Step 1 until the numerator or denominator (or both) is no longer divisible by 2.**

In the resulting fraction, both numbers are still even, so repeat the first step again:

$$\frac{18}{30} = \frac{9}{15}$$

3. **Repeat Step 1 using the number 3, and then 5, and then 7, continuing testing prime numbers until you're sure that the numerator and denominator have no common factors.**

Now, the numerator and the denominator are both divisible by 3 (see Chapter 8 for easy ways to tell if one number is divisible by another), so divide both by 3:

$$\frac{9}{15} = \frac{3}{5}$$

Neither the numerator nor the denominator is divisible by 3, so this step is complete. At this point, you can move on to test for divisibility by 5, 7, and so on, but you really don't need to. The numerator is 3, and it obviously isn't divisible by any larger number, so you know that the fraction $\frac{36}{60}$ reduces to $\frac{3}{5}$.

Q. Increase the terms of the fraction $\frac{4}{5}$ to a new fraction whose denominator is 15.

A. $\frac{12}{15}$. To start out, write the problem as follows:

$$\frac{4}{5} = \frac{?}{15}$$

The question mark stands for the numerator of the new fraction, which you want to fill in. Now divide the larger denominator (15) by the smaller denominator (5):

$$15 \div 5 = 3$$

Multiply this result by the numerator:

$$3 \times 4 = 12$$

Finally, take this number and use it to replace the question mark:

$$\frac{4}{5} = \frac{12}{15}$$

Q. Reduce the fraction $\frac{18}{42}$ to lowest terms.

A. $\frac{3}{7}$. The numerator and denominator aren't too large, so use the informal way: To start out, try to find a small number that the numerator and denominator are both divisible by. In this case, notice that the numerator and denominator are both divisible by 2, so divide both by 2:

$$\frac{18}{42} = \frac{18 \div 2}{42 \div 2} = \frac{9}{21}$$

Next, notice that the numerator and denominator are both divisible by 3 (see Chapter 8 for more on how to tell whether a number is divisible by 3), so divide both by 3:

$$\frac{9}{21} = \frac{9 \div 3}{21 \div 3} = \frac{3}{7}$$

At this point, there's no number (except for 1) that evenly divides both the numerator and denominator, so this is your answer.

Q. Reduce the fraction $\frac{135}{196}$ to lowest terms.

A. $\frac{135}{196}$. The numerator and denominator are both over 100, so use the formal way. First, decompose both the numerator and denominator down to their prime factors:

$$\frac{135}{196} = \frac{3 \times 3 \times 3 \times 5}{2 \times 2 \times 7 \times 7}$$

The numerator and denominator have no common factors, so the fraction is already in lowest terms.

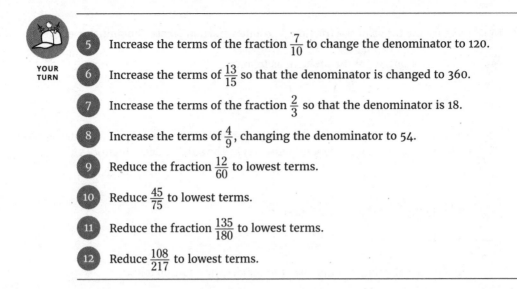

YOUR TURN

5 Increase the terms of the fraction $\frac{7}{10}$ to change the denominator to 120.

6 Increase the terms of $\frac{13}{15}$ so that the denominator is changed to 360.

7 Increase the terms of the fraction $\frac{2}{3}$ so that the denominator is 18.

8 Increase the terms of $\frac{4}{9}$, changing the denominator to 54.

9 Reduce the fraction $\frac{12}{60}$ to lowest terms.

10 Reduce $\frac{45}{75}$ to lowest terms.

11 Reduce the fraction $\frac{135}{180}$ to lowest terms.

12 Reduce $\frac{108}{217}$ to lowest terms.

Converting between Improper Fractions and Mixed Numbers

In the section, "Knowing the Fraction Facts of Life," I tell you that any fraction whose numerator is greater than its denominator is an improper fraction. Improper fractions are useful and easy to work with, but for some reason, people just don't like them. (The word *improper* should've tipped you off.) Teachers especially don't like them, and they really don't like an improper fraction to appear as the answer to a problem. However, they love mixed numbers. One reason they love them is that estimating the approximate size of a mixed number is easy.

For example, if I tell you to put $\frac{31}{3}$ of a gallon of gasoline in my car, you probably find it hard to estimate roughly how much that is: 5 gallons, 10 gallons, 20 gallons?

But if I tell you to get $10\frac{1}{3}$ gallons of gasoline, you know immediately that this amount is a little more than 10 but less than 11 gallons. Although $10\frac{1}{3}$ is the same as $\frac{31}{3}$, knowing the mixed number is a lot more helpful in practice. For this reason, you often have to convert improper fractions to mixed numbers.

Knowing the parts of a mixed number

Every mixed number has both a whole number part and a fractional part. So the three numbers in a mixed number are

>> The whole number

>> The numerator

>> The denominator

For example, in the mixed number $3\frac{1}{2}$, the whole number part is 3 and the fractional part is $\frac{1}{2}$. So this mixed number is made up of three numbers: the whole number (3), the numerator (1), and the denominator (2). Knowing these three parts of a mixed number is helpful for converting back and forth between mixed numbers and improper fractions.

Converting a mixed number to an improper fraction

To convert a mixed number to an improper fraction, follow these steps:

1. **Multiply the denominator of the fractional part by the whole number, and add the result to the numerator.**

   For example, suppose you want to convert the mixed number $5\frac{2}{3}$ to an improper fraction. First, multiply 3 by 5 and add 2:

   $$3 \times 5 + 2 = 17$$

2. **Use this result as your numerator, and place it over the denominator you already have.**

   Place this result over the denominator:

   $$\frac{17}{3}$$

   So the mixed number $5\frac{2}{3}$ equals the improper fraction $\frac{17}{3}$. This method works for all mixed numbers. Furthermore, if you start with the fractional part reduced, the answer is also reduced (see the earlier section, "Increasing and Reducing Terms of Fractions").

EXAMPLE

Q. Convert the mixed number $2\frac{3}{4}$ to an improper fraction.

A. $\frac{11}{4}$. Multiply the whole number (2) by the denominator (4), and then add the numerator (3):
$$2 \times 4 + 3 = 11$$

Use this number as the numerator of your answer, keeping the same denominator:

$$\frac{11}{4}$$

Q. Convert the mixed number $3\frac{5}{7}$ to an improper fraction.

A. $\frac{26}{7}$. Multiply the whole number (3) by the denominator (7), and then add the numerator (5). This time, I do the whole process in one step:

$$3\frac{5}{7} = \frac{3 \times 7 + 5}{7} = \frac{26}{7}$$

Converting an improper fraction to a mixed number

In this section, I show you two ways to change improper fractions to mixed numbers. The first is a quick way to convert relatively small improper fractions using only subtraction. The second way is slower and depends on division, but it works for all improper fractions, no matter how large.

Changing small improper fractions to mixed numbers the quick way

When an improper fraction is between 1 and 2, you can use this quick way to change it to a mixed number using only subtraction:

1. **Write the number 1 and copy the denominator.**

2. **Subtract the numerator minus the denominator to get the numerator.**

For example, suppose you want to convert $\frac{13}{9}$ to a mixed number. First, write the number 1 and then copy the denominator:

$$\frac{13}{9} = 1\frac{}{9}$$

To complete the problem, subtract $13 - 9 = 4$ and place this value in the numerator:

$$\frac{13}{9} = 1\frac{4}{9}$$

So the improper fraction $\frac{13}{9}$ written as a mixed number is $1\frac{4}{9}$.

When an improper fraction is greater than 2, you can repeat this process to find the answer. For example, suppose you want to change $\frac{17}{5}$ to a mixed number. Begin by following the steps that I just outlined:

$$\frac{17}{5} = 1\frac{12}{5}$$

This result isn't finished because $\frac{12}{5}$ is still an improper fraction. To complete the problem, repeat the subtraction process, adding 1 to the whole number each time:

$$\frac{17}{5} = 1\frac{12}{5} = 2\frac{7}{5} = 3\frac{2}{5}$$

So the improper fraction $\frac{17}{5}$ is equivalent to the mixed number $3\frac{2}{5}$.

Q. Convert the improper fraction $\frac{11}{6}$ to a mixed number.

EXAMPLE **A.** $1\frac{5}{6}$. First, write the number 1 and then copy the denominator of 6:

$$\frac{11}{6} = 1\frac{}{6}$$

To complete the problem, subtract $11 - 6 = 5$ and place this value in the numerator:

$$\frac{11}{6} = 1\frac{5}{6}$$

Q. Convert the improper fraction $\frac{39}{8}$ to a mixed number.

A. $4\frac{7}{8}$. Begin by following the steps outlined in the previous example:

$$\frac{39}{8} = 1\frac{31}{8}$$

This result isn't finished because $\frac{31}{8}$ is still an improper fraction. To complete the problem, repeat the subtraction process, adding 1 to the whole number each time:

$$\frac{39}{8} = 1\frac{31}{8} = 2\frac{23}{8} = 3\frac{15}{8} = 4\frac{7}{8}$$

Working with larger improper fractions

In the previous section, I show you a quick way to change improper fractions to mixed numbers by using repeated subtraction. When the value of an improper fraction is greater than 3 or 4, however, this process can be tedious. In these cases, use the following method.

To convert an improper fraction to a mixed number, divide the numerator by the denominator. Then write the mixed number in this way:

>> The quotient (answer) is the whole-number part.

>> The remainder is the numerator.

>> The denominator of the improper fraction is the denominator.

For example, suppose you want to write the improper fraction $\frac{47}{5}$ as a mixed number. First, divide 47 by 5:

$$47 \div 5 = 9 \text{ r } 2$$

Then write the mixed number as follows:

$$9\frac{2}{5}$$

This method works for all improper fractions. And as is true of conversions in the other direction, if you start with a reduced fraction, you don't have to reduce your answer (see the section, "Increasing and Reducing Terms of Fractions").

Q. Convert the improper fraction $\frac{39}{7}$ to a mixed number.

EXAMPLE **A.** $5\frac{4}{7}$. Divide the numerator (39) by the denominator (7):

$$39 \div 7 = 5 \text{ r } 4$$

Build your answer using the quotient (5) as the whole number and the remainder (4) as the numerator, keeping the same denominator (7):

$5\frac{4}{7}$

Q. What is $\frac{137}{12}$ written as a mixed number?

A. $11\frac{5}{12}$. Divide the numerator (137) by the denominator (12):

$137 \div 12 = 11 \text{ r } 5$

Build your answer using the quotient (11) as the whole number and the remainder (5) as the numerator, keeping the same denominator (12):

$11\frac{5}{12}$

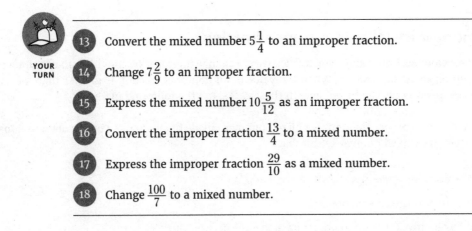

YOUR TURN

13 Convert the mixed number $5\frac{1}{4}$ to an improper fraction.

14 Change $7\frac{2}{9}$ to an improper fraction.

15 Express the mixed number $10\frac{5}{12}$ as an improper fraction.

16 Convert the improper fraction $\frac{13}{4}$ to a mixed number.

17 Express the improper fraction $\frac{29}{10}$ as a mixed number.

18 Change $\frac{100}{7}$ to a mixed number.

Comparing Fractions with Cross-Multiplication

Cross-multiplication is a handy tool for getting a common denominator for two fractions, which is important for many operations involving fractions. In this section, I show you how to cross-multiply to compare a pair of fractions to find out which is greater or less. (In Chapter 11, I show you how to use cross-multiplication to add fractions, and in Chapter 23, I show you how to use it to help solve algebra equations.)

Here's how to cross-multiply two fractions:

1. **Multiply the numerator (top number) of the first fraction by the denominator (bottom number) of the second, writing the answer below the first fraction.**

2. **Multiply the numerator of the second fraction by the denominator of the first, writing the answer below the second fraction.**

The result is that each fraction now has a new number written underneath it. The larger number is below the larger fraction.

You can use cross-multiplication to rewrite a pair of fractions as two new fractions with a common denominator:

1. **Cross-multiply the two fractions to find the numerators of the new fractions.**

2. **Multiply the denominators of the two fractions to find the new denominators.**

When two fractions have the same denominator, the one with the greater numerator is the greater fraction.

Q. Which fraction is greater: $\frac{5}{8}$ or $\frac{6}{11}$?

A. $\frac{5}{8}$. Cross-multiply the two fractions:

$$\frac{5}{8} \diagup\!\!\!\!\diagdown \frac{6}{11}$$
$$55 \qquad 48$$

Because 55 is greater than 48, $\frac{5}{8}$ is greater than $\frac{6}{11}$.

Q. Which of these three fractions is the least: $\frac{3}{4}$, $\frac{7}{10}$, or $\frac{8}{11}$?

A. $\frac{7}{10}$. Cross-multiply the first two fractions: $\frac{3}{4} \diagup\!\!\!\!\diagdown \frac{7}{10}$
$$\qquad\qquad\qquad\qquad\qquad\qquad\qquad\quad 30 \qquad 28$$

Because 28 is less than 30, $\frac{7}{10}$ is less than $\frac{3}{4}$, so you can rule out $\frac{3}{4}$. Now compare $\frac{7}{10}$

and $\frac{8}{11}$ similarly: $\frac{7}{10} \diagup\!\!\!\!\diagdown \frac{8}{11}$
$$\qquad\qquad\qquad\qquad\qquad 77 \qquad 80$$

Because 77 is less than 80, $\frac{7}{10}$ is less than $\frac{8}{11}$. Therefore, $\frac{7}{10}$ is the least of the three fractions.

19 Which is the greater fraction: $\frac{1}{5}$ or $\frac{2}{9}$?

20 Find the lesser fraction: $\frac{3}{7}$ or $\frac{5}{12}$.

21 Among these three fractions, which is greatest: $\frac{1}{10}$, $\frac{2}{21}$, or $\frac{3}{29}$?

22 Figure out which of the following fractions is the least: $\frac{1}{3}$, $\frac{2}{7}$, $\frac{4}{13}$, or $\frac{8}{25}$.

Working with Ratios and Proportions

A *ratio* is a mathematical comparison of two numbers, based on division. For example, suppose you bring 3 shirts and 5 ties with you on a business trip. Here are a few ways to express the ratio of shirts to ties:

3:5   3 to 5 $\frac{3}{5}$

A good way to work with a ratio is to turn it into a fraction. Be sure to keep the order the same: The first number goes on top of the fraction, and the second number goes on the bottom.

You can use a ratio to solve problems by setting up a *proportion equation* — that is, an equation involving two ratios.

Typically, a proportion looks like a word equation, as follows:

$$\frac{\text{Scarves}}{\text{Caps}} = \frac{2}{3}$$

For example, suppose you know that your friend Andrew brought a 2 to 3 ratio of scarves to caps on a ski trip in Chile. If you also know that Andrew brought 8 scarves, you can use this proportion to find out how many caps he brought. Just increase the terms of the fraction $\frac{2}{3}$ so that the numerator becomes 8. I do this in two steps:

$$\frac{\text{Scarves}}{\text{Caps}} = \frac{2 \times 4}{3 \times 4}$$

$$\frac{\text{Scarves}}{\text{Caps}} = \frac{8}{12}$$

As you can see, the ratio 8:12 is equivalent to the ratio 2:3 because the fractions $\frac{2}{3}$ and $\frac{8}{12}$ are equal. Therefore, Andrew brought 12 caps.

EXAMPLE

Q. Clarence has 1 daughter and 4 sons. Set up a proportion equation based on this ratio.

A. $\dfrac{\text{Daughters}}{\text{Sons}} = \dfrac{1}{4}$

Q. An English language school has a 3:7 ratio of European to Asian students. If the school has 28 students from Asia, how many students in the school are from Europe?

A. **12 students.** To begin, set up a proportion based on the ratio of European to Asian students (be sure that the order of the two numbers is the same as the two attributes that they stand for — 3 Europeans and 7 Asians):

$$\frac{\text{Europe}}{\text{Asia}} = \frac{3}{7}$$

Now, increase the terms of the fraction $\frac{3}{7}$ so that the number representing the count of Asian students becomes 28:

$$\frac{\text{Europe}}{\text{Asia}} = \frac{3 \times 4}{7 \times 4} = \frac{12}{28}$$

Therefore, given that the school has 28 Asian students, it has 12 European students.

YOUR TURN

23 A farmers' market sells a 4 to 5 ratio of vegetables to fruit. If it sells 35 different types of fruits, how many different types of vegetables does it sell?

24 An art gallery is currently featuring an exhibition with a 2:7 ratio of sculpture to paintings. If the exhibition includes 18 sculptures, how many paintings does it include?

25 A summer camp has a 7 to 9 ratio of girls to boys. If it has 117 boys, what is the total number of children attending the summer camp?

26 The budget of a small town has a 3:8 ratio of state funding to municipal funding. If the town received $600,000 in municipal funding last year, what was its total budget from both state and municipal sources?

Practice Questions Answers and Explanations

1

   (a) You have 1 shaded slice and 3 slices in total, so it's $\frac{1}{3}$.

   (b) You have 3 shaded slices and 4 slices in total, so it's $\frac{3}{4}$.

   (c) You have 5 shaded slices and 6 slices in total, so it's $\frac{5}{6}$.

   (d) You have 7 shaded slices and 12 slices in total, so it's $\frac{7}{12}$.

2

   (a) The numerator (3) is greater than the denominator (2), so $\frac{3}{2}$ is an **improper fraction.**

   (b) The numerator (8) is less than the denominator (9), so $\frac{8}{9}$ is a **proper fraction.**

   (c) The numerator (20) is less than the denominator (23), so $\frac{20}{23}$ is a **proper fraction.**

   (d) The numerator (75) is greater than the denominator (51), so $\frac{75}{51}$ is an **improper fraction.**

3

   (a) The numerator and denominator are the same, so $\frac{3}{3} = 1$.

   (b) The denominator is 1, so $\frac{10}{1} = 10$.

   (c) The numerator and denominator are the same, so $\frac{10}{10} = 1$.

   (d) The denominator is 1, so $\frac{81}{1} = 81$.

4 Find the reciprocal by switching the numerator and denominator.

   (a) The reciprocal of $\frac{5}{7}$ is $\frac{7}{5}$.

   (b) The reciprocal of $\frac{10}{3}$ is $\frac{3}{10}$.

   (c) The reciprocal of $\frac{12}{17}$ is $\frac{17}{12}$.

   (d) The reciprocal of $\frac{80}{91}$ is $\frac{91}{80}$.

5 $\frac{7}{10} = \frac{84}{120}$. To start out, write the problem as follows:

$$\frac{7}{10} = \frac{?}{120}$$

Divide the larger denominator (120) by the smaller denominator (10) and then multiply this result by the numerator (7):

$$12 \times 7 = 84$$

Take this numerator and use it to replace the question mark; your answer is $\frac{84}{120}$.

6 $\frac{13}{15} = \frac{312}{360}$. To start out, write the problem as follows:

$$\frac{13}{15} = \frac{?}{360}$$

Divide the larger denominator (360) by the smaller denominator (15) and then multiply this result by the numerator (13):

$$24 \times 13 = 312$$

Take this numerator and use it to replace the question mark; your answer is $\frac{312}{360}$.

(7) $\frac{2}{3} = \frac{12}{18}$. To start out, write the problem as follows:

$$\frac{2}{3} = \frac{?}{18}$$

Divide the larger denominator (18) by the smaller denominator (3) and then multiply this result by the numerator (2):

$$6 \times 2 = 12$$

Take this number and use it to replace the question mark; your answer is $\frac{12}{18}$.

(8) $\frac{4}{9} = \frac{24}{54}$. Write the problem as follows:

$$\frac{4}{9} = \frac{?}{54}$$

Divide the larger denominator (54) by the smaller denominator (9) and then multiply this result by the numerator (4):

$$6 \times 4 = 24$$

Take this number and use it to replace the question mark; your answer is $\frac{24}{54}$.

(9) $\frac{12}{60} = \frac{1}{5}$. The numerator (12) and the denominator (60) are both even, so divide both by 2:

$$\frac{12}{60} = \frac{6}{30}$$

They're still both even, so divide both by 2 again:

$$= \frac{3}{15}$$

Now the numerator and denominator are both divisible by 3, so divide both by 3:

$$= \frac{1}{5}$$

(10) $\frac{45}{75} = \frac{3}{5}$. The numerator (45) and the denominator (75) are both divisible by 5, so divide both by 5:

$$\frac{45}{75} = \frac{9}{15}$$

Now the numerator and denominator are both divisible by 3, so divide both by 3:

$$= \frac{3}{5}$$

(11) $\frac{135}{180} = \frac{3}{4}$. The numerator (135) and the denominator (180) are both divisible by 5, so divide both by 5:

$$\frac{135}{180} = \frac{27}{36}$$

Now the numerator and denominator are both divisible by 3, so divide both by 3:

$$= \frac{9}{12}$$

They're still both divisible by 3, so divide both by 3 again:

$$= \frac{3}{4}$$

(12) $\frac{108}{217} = \frac{\mathbf{108}}{\mathbf{217}}$. With a numerator and denominator this large, reduce using the formal way. First, decompose both the numerator and denominator down to their prime factors:

$$\frac{108}{217} = \frac{2 \times 2 \times 3 \times 3 \times 3}{7 \times 31}$$

The numerator and denominator have no common factors, so the fraction is already in lowest terms.

(13) $5\frac{1}{4} = \frac{5 \times 4 + 1}{4} = \frac{\mathbf{21}}{\mathbf{4}}$

(14) $7\frac{2}{9} = \frac{7 \times 9 + 2}{9} = \frac{\mathbf{65}}{\mathbf{9}}$

(15) $10\frac{5}{12} = \frac{10 \times 12 + 5}{12} = \frac{\mathbf{125}}{\mathbf{12}}$

(16) $\frac{13}{4} = 3\frac{1}{4}$. Divide the numerator (13) by the denominator (4): $13 \div 4 = 3$ r 1

Use the repeated subtraction method:

$$\frac{13}{4} = 1\frac{9}{4} = 2\frac{5}{4} = 3\frac{1}{4}$$

(17) $\frac{29}{10} = 2\frac{\mathbf{9}}{\mathbf{10}}$. Divide the numerator (29) by the denominator (10): $29 \div 10 = 2$ r 9

Use the repeated subtraction method:

$$\frac{29}{10} = 1\frac{19}{10} = 2\frac{9}{10}$$

(18) $\frac{100}{7} = 14\frac{2}{7}$. Divide the numerator (100) by the denominator (7): $100 \div 7 = 14$ r 2

Build your answer using the quotient (14) as the whole number and the remainder (2) as the numerator, keeping the same denominator (7): $14\frac{2}{7}$.

(19) $\frac{2}{9}$ **is greater than** $\frac{1}{5}$. Cross-multiply to compare the two fractions:

$$\frac{2}{9} \quad \frac{1}{5}$$
$$10 \quad 9$$

Because 10 is greater than 9, $\frac{2}{9}$ is greater than $\frac{1}{5}$.

(20) $\frac{5}{12}$ **is less than** $\frac{3}{7}$. Cross-multiply to compare the two fractions:

$$\frac{5}{12} \quad \frac{3}{7}$$
$$35 \quad 36$$

Because 35 is less than 36, $\frac{5}{12}$ is less than $\frac{3}{7}$.

(21) $\frac{3}{29}$ **is greater than** $\frac{1}{10}$ **and** $\frac{2}{21}$. Use cross-multiplication to compare the first two fractions:

$$\frac{1}{10} \quad \frac{2}{21}$$
$$21 \quad 20$$

Because 21 is greater than 20, $\frac{1}{10}$ is greater than $\frac{2}{21}$, so you can rule out $\frac{2}{21}$. Next, compare $\frac{1}{10}$ and $\frac{3}{29}$ by cross-multiplying.

$$\frac{1}{10} \quad \frac{3}{29}$$
$$30 \quad 29$$

Because 30 is greater than 29, $\frac{3}{29}$ is greater than $\frac{1}{10}$. Therefore, $\frac{3}{29}$ is the greatest of the three fractions.

(22) $\frac{2}{7}$ **is less than** $\frac{1}{3}$, $\frac{4}{13}$, **and** $\frac{8}{25}$. Cross-multiply to compare the first two fractions:

$$\frac{1}{3} \quad \frac{2}{7}$$
$$7 \quad 6$$

Because 6 is less than 7, $\frac{2}{7}$ is less than $\frac{1}{3}$, so you can rule out $\frac{1}{3}$. Next, compare $\frac{2}{7}$ and $\frac{4}{13}$:

$$\frac{2}{7} \quad \frac{4}{13}$$
$$26 \quad 28$$

Because 26 is less than 28, $\frac{2}{7}$ is less than $\frac{4}{13}$, so you can rule out $\frac{4}{13}$.

Finally, compare $\frac{2}{7}$ and $\frac{8}{25}$:

$$\frac{2}{7} \quad \frac{8}{25}$$
$$50 \quad 56$$

Because 50 is less than 56, $\frac{2}{7}$ is less than $\frac{8}{25}$. Therefore, $\frac{2}{7}$ is the lowest of the four.

23) **The market sells 35 varieties of fruit and 28 varieties of vegetables.**

To begin, set up a proportion based on the ratio of vegetables to fruit:

$$\frac{\text{Vegetables}}{\text{Fruit}} = \frac{4}{5}$$

Now, increase the terms of the fraction $\frac{4}{5}$ so that the number representing fruit becomes 35:

$$\frac{\text{Vegetables}}{\text{Fruit}} = \frac{4 \times 7}{5 \times 7} = \frac{28}{35}$$

Therefore, given that the farmers' market has 35 varieties of fruit, it has 28 varieties of vegetables.

24) **63.** To begin, set up a proportion based on the ratio of sculptures to paintings:

$$\frac{\text{Sculptures}}{\text{Paintings}} = \frac{2}{7}$$

Now, increase the terms of the fraction $\frac{2}{7}$ so that the number representing sculptures becomes 18:

$$\frac{\text{Sculptures}}{\text{Paintings}} = \frac{2 \times 9}{7 \times 9} = \frac{18}{63}$$

Therefore, the gallery has 63 paintings.

25) **208.** To begin, set up a proportion based on the ratio of girls to boys:

$$\frac{\text{Girls}}{\text{Boys}} = \frac{7}{9}$$

Now, you want to increase the terms of the fraction $\frac{7}{9}$ so that the number representing boys becomes 117. (To do this, notice that $117 \div 9 = 13$, so $9 \times 13 = 117$):

$$\frac{\text{Girls}}{\text{Boys}} = \frac{7 \times 13}{9 \times 13} = \frac{91}{117}$$

Therefore, the camp has 91 girls and 117 boys, so the total number of children is 208.

(26) **$825,000.** To begin, set up a proportion based on the ratio of state to municipal funding:

$$\frac{\text{State}}{\text{Municipal}} = \frac{3}{8}$$

Now, you want to increase the terms of the fraction $\frac{3}{8}$ so that the number representing municipal funding becomes 600,000. (To do this, notice that $600,000 \div 8 = 75,000$, so $8 \times 75,000 = 600,000$):

$$\frac{\text{State}}{\text{Municipal}} = \frac{3 \times 75,000}{8 \times 75,000} = \frac{225,000}{600,000}$$

Therefore, the town's budget includes $225,000 in state funding and $600,000 in municipal funding, so the total budget is $825,000.

When you're ready to test your skills, take the chapter quiz in the next section.

Whaddya Know? Chapter 10 Quiz

Ready for a quiz? The 12 questions in this section cover a variety of topics related to fractions and ratios. When you're done, answers and complete explanations await you in the section that follows.

 1 Identify what fraction of the pie is shaded in, and write your answer in lowest terms.

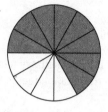

2 Write $\frac{3}{7}$ as an equivalent fraction whose denominator is 35.

3 Change $\frac{21}{8}$ to a mixed number.

4 Which fraction is greater: $\frac{5}{8}$ or $\frac{2}{3}$?

5 Reduce to lowest terms: $\frac{72}{108}$

6 What is the reciprocal of $6\frac{2}{5}$?

7 Write $5\frac{2}{7}$ as an improper fraction.

8 Write $\frac{5}{9}$ as an equivalent fraction with a denominator of 36.

9 Reduce to lowest terms: $\frac{270}{243}$

10 What is the reciprocal of $\frac{8}{11}$?

11 The ratio of dogs to cats in the shelter is 5 to 8. If there are 40 cats, then how many dogs are there?

12 Write the fractions $\frac{1}{6}$, $\frac{3}{20}$, and $\frac{2}{9}$ in order from the least to the greatest.

Answers to Chapter 10 Quiz

(1) $\frac{2}{3}$. The pie is divided into 12 equal pieces, and 8 of them are shaded in. Writing the fraction $\frac{8}{12}$ and then reducing, you get $\frac{2}{3}$.

(2) $\frac{15}{35}$. Multiply the numerator and denominator by 5.

$$\frac{3}{7} \times \frac{5}{5} = \frac{15}{35}$$

(3) $2\frac{5}{8}$. The number 8 divides 21 twice with a remainder of 5.

(4) $\frac{2}{3}$. Change both to equivalent fractions with denominators of 24.

$$\frac{5}{8} \times \frac{3}{3} = \frac{15}{24} \text{ and } \frac{2}{3} \times \frac{8}{8} = \frac{16}{24}$$

(5) $\frac{2}{3}$. Write the prime factorizations of the numerator and denominator and reduce.

$$\frac{72}{108} = \frac{2 \times 2 \times 2 \times 3 \times 3}{2 \times 2 \times 3 \times 3 \times 3} = \frac{\cancel{2} \times \cancel{2} \times 2 \times \cancel{3} \times \cancel{3}}{\cancel{2} \times \cancel{2} \times 3 \times \cancel{3} \times \cancel{3}} = \frac{2}{3}$$

(6) $\frac{5}{32}$. Multiply the 6 times the denominator, 5, and add the 2. This is the numerator of the mixed number, giving you the improper fraction $\frac{32}{5}$. Flip the fraction for the reciprocal.

(7) $\frac{37}{7}$. Multiply the 5 times the denominator, 7, and add the 2. This is the numerator of the mixed number, and the denominator stays the same.

(8) $\frac{20}{36}$. Multiply both the numerator and denominator by 4.

$$\frac{5}{9} \times \frac{4}{4} = \frac{20}{36}$$

(9) $\frac{10}{9}$. Write the prime factorizations of the numerator and denominator and reduce.

$$\frac{270}{243} = \frac{2 \times 3 \times 3 \times 3 \times 5}{3 \times 3 \times 3 \times 3 \times 3} = \frac{2 \times \cancel{3} \times \cancel{3} \times \cancel{3} \times 5}{\cancel{3} \times \cancel{3} \times \cancel{3} \times 3 \times 3} = \frac{10}{9}$$

(10) $\frac{11}{8}$. Reverse the positions of the numerator and denominator.

(11) **25.** Write the ratio as a fraction with dogs in the numerator and cats in the denominator. Set the fraction equal to another fraction with 40 in the denominator. Determine what needs to multiply the numerator and denominator of the ratio's fraction to create equivalent fractions.

$\frac{\text{Dogs}}{\text{Cats}} = \frac{5}{8}$ and $\frac{\text{Dogs}}{\text{Cats}} = \frac{}{40}$. Multiply both the numerator and denominator of the ratio's

fraction by 5: $\frac{5}{8} \times \frac{5}{5} = \frac{25}{40}$.

(12) $\frac{3}{20}, \frac{1}{6}, \frac{2}{9}$. Rewrite each fraction with a common denominator of 180.

$$\frac{1}{6} \times \frac{30}{30} = \frac{30}{180} \text{ and } \frac{3}{20} \times \frac{9}{9} = \frac{27}{180} \text{ and } \frac{2}{9} \times \frac{20}{20} = \frac{40}{180}$$

» Canceling factors to make multiplying and dividing fractions easier

» Adding and subtracting fractions that have the same denominator

» Increasing the terms of one fraction to add and subtract fractions with different denominators

» Using cross-multiplication to find a common denominator

Chapter **11**

Fractions and the Big Four Operations

I n this chapter, the focus is on applying the Big Four operations to fractions. I start by show-ing you how to multiply and divide fractions, which isn't much more difficult than multi-plying whole numbers. You also discover how to make fraction multiplication and division problems easier by canceling equivalent factors in the numerator and denominator.

Then, you move on to the trickier task of adding and subtracting fractions. To begin, you add and subtract fractions that have the same denominator. Then I show you how, in some cases, you can add and subtract fractions that have different denominators by increasing the terms of one fraction. Finally, for the most difficult problems, I show you how use cross-multiplication to find a common denominator when adding and subtracting fractions.

Multiplying and Dividing Fractions

One of the odd little ironies of life is that multiplying and dividing fractions is usually easier than adding or subtracting them. For this reason, I show you how to multiply and divide fractions before I show you how to add or subtract them.

In fact, you may find multiplying fractions easier than multiplying whole numbers because the numbers you're working with are usually small. More good news is that dividing fractions is nearly as easy as multiplying them. So I'm not even wishing you good luck — you don't need it!

Multiplying numerators and denominators straight across

Everything in life should be as simple as multiplying fractions. All you need for multiplying fractions is a pen or pencil, something to write on, and a basic knowledge of the multiplication table. (See Chapter 3 for a multiplication refresher.)

REMEMBER

To multiply two fractions, multiply the numerators (the numbers on top) to get the numerator of the answer, then multiply the denominators (the numbers on the bottom) to get the denominator of the answer.

For example, here's how to multiply $\frac{2}{5} \times \frac{3}{7}$:

$$\frac{2}{5} \times \frac{3}{7} = \frac{2 \times 3}{5 \times 7} = \frac{6}{35}$$

TIP

When multiplying fractions, you can often make your job easier by canceling out equal factors in the numerator and denominator. Canceling out equal factors makes the numbers that you're multiplying smaller and easier to work with, and it also saves you the trouble of reducing at the end. Here's how it works:

>> When the numerator of one fraction and the denominator of the other are the same, change both of these numbers to 1s. (See the nearby sidebar for why this works.)

>> When the numerator of one fraction and the denominator of the other are divisible by the same number, factor this number out of both. In other words, divide the numerator and denominator by that common factor. (For more on how to find factors, see Chapter 9.)

For example, suppose you want to multiply the following two numbers: $\frac{5}{13} \times \frac{13}{20}$.

You can make this problem easier by canceling out the number 13, as follows:

$$\frac{5}{13} \times \frac{13}{20} = \frac{5}{\cancel{13}_1} \times \frac{\cancel{13}^1}{20}$$

You can make it even easier by noticing that 20 and 5 are both divisible by 5, so you can also factor out the number 5 before multiplying:

$$= \frac{\cancel{5}^{1}}{\cancel{13}_{1}} \times \frac{\cancel{13}^{1}}{\cancel{20}_{4}}$$

Now, multiply across to complete the problem:

$$= \frac{1}{4}$$

ONE IS THE EASIEST NUMBER

With fractions, the relationship between the numbers, not the actual numbers themselves, is most important. Understanding how to multiply and divide fractions can give you a deeper understanding of why you can increase or decrease the numbers within a fraction without changing the value of the whole fraction.

When you multiply or divide any number by 1, the answer is the same number. This rule also goes for fractions, so

$$\frac{3}{8} \times 1 = \frac{3}{8} \text{ and } \frac{3}{8} \div 1 = \frac{3}{8}$$

$$\frac{5}{13} \times 1 = \frac{5}{13} \text{ and } \frac{5}{13} \div 1 = \frac{5}{13}$$

$$\frac{67}{70} \times 1 = \frac{67}{70} \text{ and } \frac{67}{70} \div 1 = \frac{67}{70}$$

And as I discuss in Chapter 10, when a fraction has the same number in both the numerator and the denominator, its value is 1. In other words, the fractions $\frac{2}{2}$, $\frac{3}{3}$, and $\frac{4}{4}$ are all equal to 1. Look what happens when you multiply the fraction $\frac{3}{4}$ by $\frac{2}{2}$:

$$\frac{3}{4} \times \frac{2}{2} = \frac{3 \times 2}{4 \times 2} = \frac{6}{8}$$

The net effect is that you've increased the terms of the original fraction by 2. But all you've done is multiply the fraction by 1, so the value of the fraction hasn't changed. The fraction $\frac{6}{8}$ is equal to $\frac{3}{4}$.

Similarly, reducing the fraction $\frac{6}{9}$ by a factor of 3 is the same as dividing that fraction by $\frac{3}{3}$ (which is equal to 1):

$$\frac{6}{9} \div \frac{3}{3} = \frac{6 \div 3}{9 \div 3} = \frac{2}{3}$$

So $\frac{6}{9}$ is equal to $\frac{2}{3}$.

Q. Multiply $\frac{2}{5}$ by $\frac{4}{9}$.

EXAMPLE

A. $\frac{8}{45}$. Multiply the two numerators (top numbers) to get the numerator of the answer. Then multiply the two denominators (bottom numbers) to get the denominator of the answer:

$$\frac{2}{5} \times \frac{4}{9} = \frac{2 \times 4}{5 \times 9} = \frac{8}{45}$$

Q. Find $\frac{4}{7} \times \frac{5}{8}$.

A. $\frac{5}{14}$. Before you multiply, notice that the numerator 4 and the denominator 8 are both factors of 4. So, divide both of these numbers by 4 just as you would when reducing a fraction:

$$\frac{4}{7} \times \frac{5}{8} = \frac{\cancel{4}^{1}}{7} \times \frac{5}{\cancel{8}_{2}}$$

At this point, neither numerator has a common factor with either denominator, so you're ready to multiply. Multiply the two numerators to get the numerator of the answer. Then multiply the two denominators to get the denominator of the answer:

$$= \frac{5}{14}$$

Because you canceled all common factors before multiplying, this answer is in lowest terms.

YOUR TURN

1. Multiply $\frac{2}{3}$ by $\frac{7}{9}$.

2. Find $\frac{3}{8} \times \frac{6}{11}$.

3. Multiply $\frac{2}{9}$ by $\frac{3}{10}$.

4. Figure out $\frac{9}{14} \times \frac{8}{15}$.

Doing a flip to divide fractions

Dividing fractions is just as easy as multiplying them. In fact, when you divide fractions, you really turn the problem into multiplication.

REMEMBER

To divide one fraction by another, multiply the first fraction by the reciprocal of the second. (As I discuss in Chapter 10, the *reciprocal* of a fraction is simply that fraction turned upside down.) To remember this rule, use the mnemonic **Keep-Change-Flip**:

Keep the first fraction as it is.

Change multiplication to division.

Flip the second fraction to its reciprocal.

For example, here's how you turn fraction division into multiplication:

$$\frac{1}{3} \div \frac{4}{5} = \frac{1}{3} \times \frac{5}{4}$$

As you can see, keep $\frac{1}{3}$ as it is, change the division sign to a multiplication sign, and flip $\frac{4}{5}$ to its reciprocal of $\frac{5}{4}$. After that, just multiply the fractions as I describe in the section, "Multiplying numerators and denominators straight across":

$$\frac{1}{3} \times \frac{5}{4} = \frac{1 \times 5}{3 \times 4} = \frac{5}{12}$$

REMEMBER

As with multiplication, you can also make the numbers smaller, and also eliminate the need to reduce the result to lowest terms, by canceling out equal factors before multiplying. (See the preceding section.)

EXAMPLE

Q. Divide $\frac{5}{8}$ by $\frac{3}{7}$.

A. $1\frac{11}{24}$. Use Keep-Change-Flip to turn the division to multiplication:

$$\frac{5}{8} \div \frac{3}{7} = \frac{5}{8} \times \frac{7}{3}$$

Solve the problem using fraction multiplication:

$$= \frac{5 \times 7}{8 \times 3} = \frac{35}{24}$$

The answer is an improper fraction (because the numerator is greater than the denominator), so change it to a mixed number. Divide the numerator by the denominator and put the remainder over the denominator:

$$= 1\frac{11}{24}$$

Q. Calculate $\frac{7}{10} \div \frac{2}{5}$.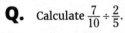

A. $1\frac{3}{4}$. Change the division to multiplication using Keep-Change-Flip:

$$\frac{7}{10} \div \frac{2}{5} = \frac{7}{10} \times \frac{5}{2}$$

Notice that you have a 5 in one of the numerators and a 10 in the other fraction's denominator, so you can cancel out the common factor of 5 before you multiply:

$$= \frac{7}{\cancel{10}_2} \times \frac{\cancel{5}^1}{2} = \frac{7}{4}$$

Because the numerator is greater than the denominator, the fraction is improper, so change it to a mixed number:

$$1\frac{3}{4}$$

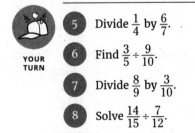

YOUR TURN

5 Divide $\frac{1}{4}$ by $\frac{6}{7}$.

6 Find $\frac{3}{5} \div \frac{9}{10}$.

7 Divide $\frac{8}{9}$ by $\frac{3}{10}$.

8 Solve $\frac{14}{15} \div \frac{7}{12}$.

Adding and Subtracting Fractions with the Same Denominator

When you add and subtract fractions, one important item to notice is whether their denominators (the numbers on the bottom) are the same. If they're the same — woo-hoo! Adding and subtracting fractions that have the same denominator is a walk in the park.

REMEMBER

To add two fractions that have the same denominator (bottom number), add the numerators (top numbers) and leave the denominator unchanged.

For example, consider the following problem:

$$\frac{1}{5} + \frac{2}{5} = \frac{1+2}{5} = \frac{3}{5}$$

As you can see, to add these two fractions, you add the numerators $(1+2)$ and keep the denominator (5).

Why does this work? Chapter 10 tells you that you can think about fractions as pieces of cake. The denominator in this case tells you that the entire cake has been cut into five pieces. So when you add $\frac{1}{5} + \frac{2}{5}$, you're really adding one piece plus two pieces. The answer, of course, is three pieces — that is, $\frac{3}{5}$.

Even if you have to add more than two fractions, as long as the denominators are all the same, you just add the numerators and leave the denominator unchanged:

$$\frac{1}{17} + \frac{3}{17} + \frac{4}{17} + \frac{6}{17} = \frac{1+3+4+6}{17} = \frac{14}{17}$$

Sometimes when you add fractions with the same denominator, you have to reduce the result to lowest terms (to find out more about reducing, flip to Chapter 10). Take this problem, for example:

$$\frac{1}{4} + \frac{1}{4} = \frac{1+1}{4} = \frac{2}{4}$$

The numerator and the denominator are both even, so you know they can be reduced:

$$\frac{2}{4} = \frac{1}{2}$$

In other cases, the sum of two proper fractions is an improper fraction. You get a numerator that's larger than the denominator when the two fractions add up to more than 1, as in this case:

$$\frac{3}{7} + \frac{5}{7} = \frac{8}{7}$$

If you have more work to do with this fraction, leave it as an improper fraction so that it's easier to work with. But if this is your final answer, you may need to turn it into a mixed number (I cover mixed numbers in Chapter 10):

$$\frac{8}{7} = 1\frac{1}{7}$$

As with addition, subtracting fractions with the same denominator is always easy. When the denominators are the same, you can just think of the fractions as pieces of cake.

To subtract one fraction from another when they both have the same denominator (bottom number), subtract the numerator (top number) of the second from the numerator of the first and keep the denominator the same. For example:

$$\frac{3}{5} - \frac{2}{5} = \frac{3-2}{5} = \frac{1}{5}$$

Sometimes, as when you add fractions, you have to reduce:

$$\frac{3}{10} - \frac{1}{10} = \frac{3-1}{10} = \frac{2}{10}$$

Because the numerator and denominator are both even, you can reduce this fraction by a factor of 2:

$$\frac{2}{10} = \frac{1}{5}$$

Unlike with addition, when you subtract one proper fraction from another, you never get an improper fraction.

EXAMPLE

Q. Find $\frac{5}{8} + \frac{7}{8}$.

A. $1\frac{1}{2}$. The denominators are both 8, so add the numerators (5 and 7) to get the new numerator and keep the denominator the same:

$$\frac{5}{8} + \frac{7}{8} = \frac{5+7}{8} = \frac{12}{8}$$

The numerator is greater than the denominator, so the answer is an improper fraction. Change it to a mixed number and then reduce (as I show you in Chapter 10):

$$= 1\frac{2}{4} = 1\frac{1}{2}$$

Q. What is $\frac{11}{12} - \frac{5}{12}$?

A. The denominators are both 12, so subtract the numerators (11 and 5) to get the new numerator and keep the denominator the same:

$$\frac{11}{12} - \frac{5}{12} = \frac{11-5}{12} = \frac{6}{12}$$

The numerator is a factor of the denominator, so reduce it (as I show you in Chapter 10):

$$= \frac{1}{2}$$

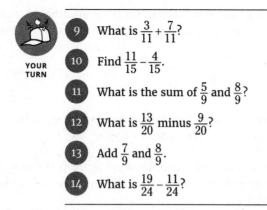

9 What is $\frac{3}{11} + \frac{7}{11}$?

10 Find $\frac{11}{15} - \frac{4}{15}$.

11 What is the sum of $\frac{5}{9}$ and $\frac{8}{9}$?

12 What is $\frac{13}{20}$ minus $\frac{9}{20}$?

13 Add $\frac{7}{9}$ and $\frac{8}{9}$.

14 What is $\frac{19}{24} - \frac{11}{24}$?

Adding and Subtracting Fractions with Different Denominators

When a pair of fractions have different denominators (bottom numbers), a good way to add or subtract them is to rewrite them using the same denominator. In some easy cases, you can do this by increasing the terms of only one fraction. In other more difficult cases, you may need to increase the terms of both fractions.

In this section, I show you when and how to do both.

The easy case: Increasing the terms of one fraction

Sometimes, you can add or subtract a pair of fractions that have different denominators by increasing the terms of just one fraction.

REMEMBER Before you add or subtract two fractions with different denominators, check the denominators to see whether one is a factor of the other (for more on factors, flip to Chapter 9). If so, increase the terms of the fraction with the lower denominator so that its denominator is the same as that of the other fraction.

For example:

$$\frac{2}{9} + \frac{1}{3}$$

Notice here that 3 is a factor of 9. So in this problem, you can increase the terms of $\frac{1}{3}$ to $\frac{3}{9}$, as I show you in Chapter 10:

$$= \frac{2}{9} + \frac{3}{9}$$

The result is a pair of fractions with the same denominator, so you can add them as I show you in the previous section:

$$= \frac{5}{9}$$

So the answer is $\frac{5}{9}$.

You can use the same method to subtract some fractions. For example:

$$\frac{7}{12} - \frac{1}{2}$$

This time, the denominator 2 is a factor of 12, so increase the terms of $\frac{1}{2}$ to $\frac{6}{12}$ and then complete the subtraction:

$$\frac{7}{12} - \frac{6}{12} = \frac{1}{12}$$

So the answer is $\frac{1}{12}$.

EXAMPLE

Q. $\frac{11}{15} + \frac{3}{5} = ?$

A. $1\frac{1}{3}$. Because 5 is a factor of 15, increase the terms of $\frac{3}{5}$ to $\frac{9}{15}$ and then add:

$$= \frac{11}{15} + \frac{9}{15} = \frac{20}{15}$$

The result is an improper fraction, so change this to a mixed number and then reduce the result to lowest terms:

$$= 1\frac{5}{15} = 1\frac{1}{3}$$

So the answer is $1\frac{1}{3}$.

Q. Subtract $\frac{5}{6} - \frac{13}{18}$.

A. $\frac{1}{9}$. Notice that 6 is a factor of 18, so you can increase the terms of $\frac{5}{6}$ to $\frac{15}{18}$ and then add:

$$\frac{5}{6} - \frac{13}{18} = \frac{15}{18} - \frac{13}{18} = \frac{2}{18}$$

To complete the problem, reduce this result to lowest terms:

$$= \frac{1}{9}$$

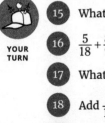

YOUR TURN

15. What is the sum of the fractions $\frac{13}{20}$ and $\frac{3}{10}$?

16. $\frac{5}{18} + \frac{8}{9} = ?$

17. What is $\frac{11}{24}$ plus $\frac{7}{8}$?

18. Add $\frac{47}{100} + \frac{19}{25}$.

19. $\frac{9}{10} - \frac{1}{2} = ?$

20. What is the difference between $\frac{8}{9}$ and $\frac{11}{18}$?

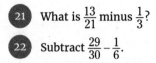

21 What is $\frac{13}{21}$ minus $\frac{1}{3}$?

22 Subtract $\frac{29}{30} - \frac{1}{6}$.

The difficult case: Increasing the terms of both fractions

When adding or subtracting fractions, sometimes you need to increase the terms of both fractions before you add. This situation arises when the smaller denominator isn't a factor of the larger one.

The simplest way to increase the terms of both fractions is to use cross-multiplication, a tool for comparing fractions that I discuss in Chapter 10.

For example, consider this fraction addition problem:

$$\frac{1}{5} + \frac{2}{3}$$

In this case, the denominator 3 isn't a factor of 5, so you can't simply increase the terms of $\frac{1}{5}$. Instead, multiply the two denominators to find a *common denominator* for the two fractions:

$$\frac{1}{5} + \frac{2}{3} = \frac{}{15} + \frac{}{15}$$

Now, cross-multiply to find the two numerators:

$$\frac{1}{5} + \frac{2}{3} = \frac{1 \times 3}{15} + \frac{5 \times 2}{15}$$

To complete the problem, simplify and add the two fractions:

$$\frac{3}{15} + \frac{10}{15} = \frac{13}{15}$$

Thus, the answer is $\frac{13}{15}$.

This process is similar when subtracting fractions. For example:

$$\frac{5}{7} - \frac{1}{2}$$

Begin by multiplying to find a common denominator $(7 \times 2 = 14)$:

$$\frac{5}{7} - \frac{1}{2} = \frac{}{14} - \frac{}{14}$$

Now, cross-multiply $(5 \times 2 = 10$ and $7 \times 1 = 7)$ to find the numerators:

$$\frac{5}{7} - \frac{1}{2} = \frac{10}{14} - \frac{7}{14}$$

To complete the problem, subtract as usual:

$$= \frac{3}{14}$$

Therefore, the answer is $\frac{3}{14}$.

Q. $\frac{5}{6} + \frac{3}{5} = ?$

A. $1\frac{13}{30}$. Cross-multiply to find the increased numerators of the two fractions, and multiply the denominators to find the denominator:

$$\frac{5}{6} + \frac{3}{5} = \frac{25}{30} + \frac{18}{30}$$

Add the results:

$$= \frac{43}{30}$$

The result is an improper fraction, so change it to a mixed number:

$$= 1\frac{13}{30}$$

Q. What is $\frac{3}{4} - \frac{1}{6}$?

A. $1\frac{13}{30}$. Cross-multiply to find the increased numerators of the two fractions, and multiply the denominators to find the denominator:

$$\frac{3}{4} - \frac{1}{6} = \frac{18}{24} - \frac{4}{24}$$

Subtract the results:

$$= \frac{14}{24}$$

Reduce the result to lowest terms:

$$= \frac{7}{12}$$

YOUR TURN

23 What is $\frac{2}{9}$ plus $\frac{3}{4}$?

24 Add $\frac{3}{8} + \frac{6}{7}$.

25 $\frac{5}{12} + \frac{1}{10} = ?$

26 What is the sum of $\frac{5}{6}$ and $\frac{7}{9}$?

27 Subtract $\frac{1}{4} - \frac{1}{10}$.

28 What is the difference between $\frac{5}{7}$ and $\frac{3}{10}$?

29 What is the result when you subtract $\frac{6}{11} - \frac{3}{8}$?

30 $\frac{7}{10} - \frac{7}{12} = ?$

Practice Questions Answers and Explanations

 $\frac{14}{27}$. Multiply both the numerators and the denominators across:

$$\frac{2}{3} \times \frac{7}{9} = \frac{2 \times 7}{3 \times 9} = \frac{14}{27}$$

(2) $\frac{9}{44}$. Begin by canceling a factor of 2 in both the numerator and denominator:

$$\frac{3}{8} \times \frac{6}{11} = \frac{3}{\cancel{8}_4} \times \frac{\cancel{6}^3}{11}$$

Multiply the numerator by the numerator and the denominator by the denominator:

$$= \frac{3 \times 3}{4 \times 11} = \frac{9}{44}$$

 $\frac{1}{15}$. Begin by canceling out common factors. The numerator 2 and the denominator 10 are both even, so divide both by 2:

$$\frac{2}{9} \times \frac{3}{10} = \frac{\cancel{2}^1}{9} \times \frac{3}{\cancel{10}_5}$$

Next, the numerator 3 and the denominator 9 are both divisible by 3, so divide both by 3:

$$= \frac{\cancel{2}^1}{\cancel{9}_3} \times \frac{\cancel{3}^1}{\cancel{10}_5}$$

Now multiply straight across:

$$= \frac{1 \times 1}{3 \times 5} = \frac{1}{15}$$

 $\frac{12}{35}$. Start by canceling out common factors. The numbers 14 and 8 are both divisible by 2, and 9 and 15 are divisible by 3:

$$\frac{9}{14} \times \frac{8}{15} = \frac{\cancel{9}^3}{\cancel{14}_7} \times \frac{\cancel{8}^4}{\cancel{15}_5}$$

Now multiply:

$$= \frac{3 \times 4}{7 \times 5} = \frac{12}{35}$$

(5) $\frac{7}{24}$. First, use Keep-Change-Flip to convert the problem to multiplication:

$$\frac{1}{4} \div \frac{6}{7} = \frac{1}{4} \times \frac{7}{6}$$

Now complete the problem using fraction multiplication:

$$= \frac{1 \times 7}{4 \times 6} = \frac{7}{24}$$

$\textcircled{6}$ $\frac{2}{3}$. Change the problem to multiplication using Keep–Change–Flip:

$$\frac{3}{5} \div \frac{9}{10} = \frac{3}{5} \times \frac{10}{9}$$

Cancel common factors in both the numerator and denominator:

$$= \frac{\cancel{3}^1}{\cancel{5}_1} \times \frac{\cancel{10}^2}{\cancel{9}_3}$$

Multiply across to complete the problem:

$$= \frac{1 \times 2}{1 \times 3} = \frac{2}{3}$$

$\textcircled{7}$ $2\frac{26}{27}$. Change the problem to multiplication using Keep–Change–Flip:

$$\frac{8}{9} \div \frac{3}{10} = \frac{8}{9} \times \frac{10}{3}$$

Complete the problem using fraction multiplication:

$$= \frac{8 \times 10}{9 \times 3} = \frac{80}{27}$$

The numerator is greater than the denominator, so change this improper fraction to a mixed number. I do this in two steps, as I show you in Chapter 10:

$$= 1\frac{53}{27} = 2\frac{26}{27}$$

$\textcircled{8}$ $1\frac{3}{5}$. Use Keep–Change–Flip to turn the problem to multiplication:

$$\frac{14}{15} \div \frac{7}{12} = \frac{14}{15} \times \frac{12}{7}$$

Cancel common factors:

$$= \frac{\cancel{14}^2}{\cancel{15}_5} \times \frac{\cancel{12}^4}{\cancel{7}_1}$$

Multiply across:

$$= \frac{2 \times 4}{5 \times 1} = \frac{8}{5}$$

Convert this improper fraction to a mixed number:

$$= 1\frac{3}{5}$$

$\textcircled{9}$ $\frac{10}{11}$. The denominators are the same, so add numerators and keep the denominator the same:

$$\frac{3}{11} + \frac{7}{11} = \frac{10}{11}$$

(10) $\frac{7}{15}$. The denominators are the same, so subtract the second numerator from the first and keep the denominator the same:

$$\frac{11}{15} - \frac{4}{15} = \frac{7}{15}$$

(11) $1\frac{4}{9}$. The denominators are the same, so add the numerators and keep the denominator the same:

$$\frac{5}{9} + \frac{8}{9} = \frac{13}{9}$$

The result is an improper fraction, so convert it to a mixed number:

$$= 1\frac{4}{9}$$

(12) $\frac{1}{5}$. The denominators are the same, so subtract the first numerator minus the second and keep the denominator the same:

$$\frac{13}{20} - \frac{9}{20} = \frac{4}{20}$$

Both the numerator and denominator are divisible by 4, so reduce the fraction by 4:

$$= \frac{1}{5}$$

(13) $1\frac{2}{3}$. The denominators are the same, so add the numerators:

$$\frac{7}{9} + \frac{8}{9} = \frac{15}{9}$$

The result is an improper fraction, so convert it to a mixed number:

$$= 1\frac{6}{9}$$

Both the numerator and denominator are divisible by 3, so reduce the fraction by a factor of 3:

$$= 1\frac{2}{3}$$

(14) $\frac{1}{3}$. The denominators are the same, so subtract the first numerator minus the second:

$$\frac{19}{24} - \frac{11}{24} = \frac{8}{24}$$

Both the numerator and denominator are divisible by 8, so reduce the fraction by 8:

$$= \frac{1}{3}$$

15 $\frac{19}{20}$. The number 10 is a factor of 20, so increase the fraction $\frac{3}{10}$ to a fraction with 20 in the denominator, then add the fractions:

$$\frac{13}{20} + \frac{3}{10} = \frac{13}{20} + \frac{6}{20} = \frac{19}{20}$$

16 $1\frac{1}{6}$. The number 9 is a factor of 18, so increase the fraction $\frac{8}{9}$ to a fraction with 18 in the denominator, then add the fractions:

$$\frac{5}{18} + \frac{8}{9} = \frac{5}{18} + \frac{16}{18} = \frac{21}{18}$$

The result is an improper fraction, so change it to a mixed number:

$$= 1\frac{3}{18}$$

This fraction can be reduced by a factor of 3:

$$= 1\frac{1}{6}$$

17 $1\frac{1}{3}$. The number 8 is a factor of 24, so increase the fraction $\frac{7}{8}$ to a fraction with 24 in the denominator, then add the fractions:

$$\frac{11}{24} + \frac{7}{8} = \frac{11}{24} + \frac{21}{24} = \frac{32}{24}$$

The result is an improper fraction, so change it to a mixed number:

$$= 1\frac{8}{24}$$

This fraction can be reduced by a factor of 8:

$$= 1\frac{1}{3}$$

18 $1\frac{23}{100}$. The number 25 is a factor of 100, so increase the fraction $\frac{19}{25}$ to a fraction with 100 in the denominator, then add the fractions:

$$\frac{47}{100} + \frac{19}{25} = \frac{47}{100} + \frac{76}{100} = \frac{123}{100}$$

The result is an improper fraction, so change it to a mixed number:

$$= 1\frac{23}{100}$$

19 $\frac{2}{5}$. The number 2 is a factor of 10, so increase the fraction $\frac{1}{2}$ to a fraction with 10 in the denominator, then subtract:

$$\frac{9}{10} - \frac{1}{2} = \frac{9}{10} - \frac{5}{10} = \frac{4}{10}$$

Reduce this fraction by a factor of 2:

$$= \frac{2}{5}$$

(20) $\frac{5}{18}$. The number 9 is a factor of 18, so increase the fraction $\frac{8}{9}$ to a fraction with 18 in the denominator, then subtract:

$$\frac{8}{9} - \frac{11}{18} = \frac{16}{18} - \frac{11}{18} = \frac{5}{18}$$

(21) $\frac{2}{7}$. The number 3 is a factor of 21, so increase the fraction $\frac{1}{3}$ to a fraction with 21 in the denominator, then subtract:

$$\frac{13}{21} - \frac{1}{3} = \frac{13}{21} - \frac{7}{21} = \frac{6}{21}$$

This fraction can be reduced by a factor of 3:

$$= \frac{2}{7}$$

(22) $\frac{4}{5}$. The number 6 is a factor of 30, so increase the fraction $\frac{1}{6}$ to a fraction with 30 in the denominator, then subtract:

$$\frac{29}{30} - \frac{1}{6} = \frac{29}{30} - \frac{5}{30} = \frac{24}{30}$$

This fraction can be reduced by a factor of 6:

$$= \frac{4}{5}$$

(23) $\frac{35}{36}$. Cross-multiply to change the numerators of the fractions, and multiply the denominators to find the common denominator, then add the resulting fractions:

$$\frac{2}{9} + \frac{3}{4} = \frac{8}{36} + \frac{27}{36} = \frac{35}{36}$$

(24) $1\frac{13}{56}$. Cross-multiply to change the numerators of the fractions, and multiply the denominators to find the common denominator, then add the resulting fractions:

$$\frac{3}{8} + \frac{6}{7} = \frac{21}{56} + \frac{48}{56} = \frac{69}{56}$$

The result is an improper fraction, so change it to a mixed number:

$$= 1\frac{13}{56}$$

(25) $\frac{31}{60}$. Cross-multiply to change the numerators of the fractions, and multiply the denominators to find the common denominator, then add the resulting fractions:

$$\frac{5}{12} + \frac{1}{10} = \frac{50}{120} + \frac{12}{120} = \frac{62}{120}$$

This fraction can be reduced by a factor of 2:

$$= \frac{31}{60}$$

26 $1\frac{11}{18}$. Cross-multiply to change the numerators of the fractions, and multiply the denominators to find the common denominator, then add the resulting fractions:

$$\frac{5}{6} + \frac{7}{9} = \frac{45}{54} + \frac{42}{54} = \frac{87}{54}$$

The result is an improper fraction, so change it to a mixed number:

$$= 1\frac{33}{54}$$

This fraction can be reduced by a factor of 3:

$$= 1\frac{11}{18}$$

27 $\frac{3}{20}$. Cross-multiply to change the numerators of the fractions, and multiply the denominators to find the common denominator, then subtract the resulting fractions:

$$\frac{1}{4} - \frac{1}{10} = \frac{10}{40} - \frac{4}{40} = \frac{6}{40}$$

This fraction can be reduced by a factor of 2:

$$= \frac{3}{20}$$

28 $\frac{29}{70}$. Cross-multiply to change the numerators of the fractions, and multiply the denominators to find the common denominator, then subtract the resulting fractions:

$$\frac{5}{7} - \frac{3}{10} = \frac{50}{70} - \frac{21}{70} = \frac{29}{70}$$

29 $\frac{15}{88}$. Cross-multiply to change the numerators of the fractions, and multiply the denominators to find the common denominator, then subtract the resulting fractions:

$$\frac{6}{11} - \frac{3}{8} = \frac{48}{88} - \frac{33}{88} = \frac{15}{88}$$

30 $\frac{7}{60}$. Cross-multiply to change the numerators of the fractions, and multiply the denominators to find the common denominator, then subtract the resulting fractions:

$$\frac{7}{10} - \frac{7}{12} = \frac{84}{120} - \frac{70}{120} = \frac{14}{120}$$

This fraction can be reduced by a factor of 2:

$$= \frac{7}{60}$$

When you're ready, the chapter test in the next section tests your skills in adding, subtracting, multiplying, and dividing fractions.

Whaddya Know? Chapter 11 Quiz

This 12-question quiz tests your knowledge on the topics covered in this chapter. When you're done, check your answers in the next section.

1. Multiply: $\frac{6}{7} \times \frac{5}{8}$

2. Divide: $\frac{14}{9} \div \frac{7}{10}$

3. Add: $\frac{6}{7} + \frac{4}{7}$

4. Subtract: $\frac{7}{8} - \frac{3}{4}$

5. Multiply: $\frac{18}{25} \times \frac{10}{27}$

6. Divide: $\frac{8}{15} \div \frac{24}{35}$

7. Add: $\frac{5}{8} + \frac{5}{6}$

8. Subtract: $\frac{9}{10} - \frac{7}{10}$

9. Multiply: $\frac{9}{4} \times \frac{8}{3}$

10. Add: $\frac{2}{15} + \frac{7}{10}$

11. Divide: $\frac{6}{5} \div \frac{2}{7}$

12. Subtract: $\frac{13}{20} - \frac{2}{5}$

Answers to Chapter 11 Quiz

(1) $\frac{15}{28}$. Reduce the fractions first by dividing by the common factor of 2 before multiplying.

$$\frac{6}{7} \times \frac{5}{8} = \frac{\cancel{6}^{3}}{7} \times \frac{5}{\cancel{8}_{4}} = \frac{15}{28}$$

(2) $2\frac{2}{9}$. Use Keep-Change-Flip to change the division to multiplication.

$$\frac{14}{9} \div \frac{7}{10} = \frac{14}{9} \times \frac{10}{7}$$

Next, reduce the fractions by dividing by the common factor of 7 before multiplying.

$$\frac{14}{9} \times \frac{10}{7} = \frac{\cancel{14}^{2}}{9} \times \frac{10}{\cancel{7}_{1}} = \frac{20}{9}$$

Rewrite the improper fraction as a mixed number.

$$= 2\frac{2}{9}$$

(3) $1\frac{3}{7}$. The denominators are the same, so just add the numerators.

$$\frac{6}{7} + \frac{4}{7} = \frac{10}{7}$$

Now change the improper fraction to a mixed number.

$$= 1\frac{3}{7}$$

(4) $\frac{1}{8}$. The denominators need to be the same. Change the second fraction to one with a denominator of 8 by multiplying both the numerator and the denominator by 2.

$$\frac{7}{8} - \frac{3}{4} = \frac{7}{8} - \frac{6}{8}$$

Now subtract.

$$= \frac{1}{8}$$

(5) $\frac{4}{15}$. First reduce the fractions by dividing by the common factor of 9 and then the common factor of 5.

$$\frac{18}{25} \times \frac{10}{27} = \frac{\cancel{18}^{2}}{\cancel{25}_{5}} \times \frac{\cancel{10}^{2}}{\cancel{27}_{3}}$$

Now multiply the numerators and the denominators.

$$= \frac{4}{15}$$

6 $\frac{7}{9}$. Use Keep–Change–Flip to change the division to multiplication.

$$\frac{8}{15} \div \frac{24}{35} = \frac{8}{15} \times \frac{35}{24}$$

Now reduce the fractions by dividing by the common factor of 8 and then the common factor of 5.

$$= \frac{\cancel{8}^1}{\cancel{15}_3} \times \frac{\cancel{35}^7}{\cancel{24}_3}$$

Finally, multiply.

$$= \frac{7}{9}$$

7 $1\frac{11}{24}$. You need a common denominator, so multiply the 8 and 6 to get 48. Then cross-multiply to find the numerators.

$$\frac{5}{8} + \frac{5}{6} = \frac{5 \times 6}{48} + \frac{5 \times 8}{48} = \frac{30}{48} + \frac{40}{48}$$

Now add the fractions.

$$= \frac{70}{48}$$

Change the improper fraction to a mixed number. Then reduce the fraction by dividing by 2.

$$= 1\frac{22}{48} = 1\frac{11}{24}$$

8 $\frac{1}{5}$. The denominators are the same, so subtract the numerators. Then reduce the fraction by dividing by 2.

$$\frac{9}{10} - \frac{7}{10} = \frac{2}{10} = \frac{1}{5}$$

9 **6.** First reduce the fractions by dividing by the common factor of 3 and then the common factor of 4.

$$\frac{9}{4} \times \frac{8}{3} = \frac{\cancel{9}^3}{\cancel{4}_1} \times \frac{\cancel{8}^2}{\cancel{3}_1}$$

Now multiply.

$$= \frac{6}{1} = 6$$

10 $\frac{5}{6}$. You need a common denominator, so multiply the 15 and 10 to get 150. Then cross-multiply to find the numerators.

$$\frac{2}{15} + \frac{7}{10} = \frac{2 \times 10}{150} + \frac{7 \times 15}{150} = \frac{20}{150} + \frac{105}{150}$$

Perform the addition. Then reduce the answer by dividing by the common factor of 25.

$$= \frac{125}{150} = \frac{\cancel{125}^{5}}{\cancel{150}_{6}} = \frac{5}{6}$$

(11) $4\frac{1}{5}$. Use Keep-Change-Flip to change the division to multiplication.

$$\frac{6}{5} \div \frac{2}{7} = \frac{6}{5} \times \frac{7}{2}$$

Before multiplying, reduce the fractions by dividing by the common factor of 2.

$$\frac{6}{5} \times \frac{7}{2} = \frac{\cancel{6}^{3}}{5} \times \frac{7}{\cancel{2}_{1}}$$

Multiply. Then change the improper fraction to a mixed number.

$$= \frac{21}{5} = 4\frac{1}{5}$$

(12) $\frac{1}{4}$. Multiply the 20 and 5 to get the common denominator of 100. Then cross-multiply to find the numerators.

$$\frac{13}{20} - \frac{2}{5} = \frac{13 \times 5}{100} - \frac{2 \times 20}{100} = \frac{65}{100} - \frac{40}{100}$$

Subtract, and then reduce the fraction by dividing by the common factor of 25.

$$= \frac{25}{100} = \frac{1}{4}$$

Chapter **12**

Mixing Things Up with Mixed Numbers

In this chapter, you apply your fractions skills to Big Four operations on mixed numbers.

First, I show you how to multiply and divide mixed numbers by converting them to improper fractions. Next, you discover how to add mixed numbers that have the same denominator and different denominators. Then, I show you how to add mixed numbers when carrying a 1 becomes necessary.

Then, you move on to subtracting mixed numbers, first with the same denominator and then with different denominators. Finally, I show you how to crack the difficult skill of subtracting mixed numbers when borrowing from the whole-number column becomes necessary.

Multiplying and Dividing Mixed Numbers

The best way to multiply and divide mixed numbers is to convert them to improper fractions and then multiply or divide as usual. Here's how to multiply or divide mixed numbers:

1. **Convert all mixed numbers to improper fractions, as I show you in Chapter 10.**

   For example, suppose you want to multiply $1\frac{3}{5} \times 2\frac{1}{3}$. First, convert $1\frac{3}{5}$ and $2\frac{1}{3}$ to improper fractions:

   $$1\frac{3}{5} = \frac{5 \times 1 + 3}{5} = \frac{8}{5} \qquad\qquad 2\frac{1}{3} = \frac{3 \times 2 + 1}{3} = \frac{7}{3}$$

2. **Multiply these improper fractions, as I explain in Chapter 11.**

   $$\frac{8}{5} \times \frac{7}{3} = \frac{56}{15}$$

3. **This result is also an improper fraction, so convert it back to a mixed number (see Chapter 10).**

   $$\frac{56}{15} = 1\frac{41}{15} = 2\frac{26}{15} = 3\frac{11}{15}$$

   In this case, the answer is already in lowest terms, so you don't have to reduce it.

As a second example, suppose you want to divide $3\frac{2}{3}$ by $1\frac{4}{7}$.

1. **Convert $3\frac{2}{3}$ and $1\frac{4}{7}$ to improper fractions.**

   $$3\frac{2}{3} = \frac{3 \times 3 + 2}{3} = \frac{11}{3} \qquad\qquad 1\frac{4}{7} = \frac{7 \times 1 + 4}{7} = \frac{11}{7}$$

2. **Divide these improper fractions.**

   Begin by using Keep–Change–Flip to change division to multiplication, as I show you in Chapter 11, and then divide:

   $$\frac{11}{3} \div \frac{11}{7} = \frac{11}{3} \times \frac{7}{11}$$

   In this case, before you multiply, you can cancel out factors of 11 in the numerator and denominator:

   $$\frac{\cancel{11}^{1}}{3} \times \frac{7}{\cancel{11}_{1}} = \frac{1 \times 7}{3 \times 1} = \frac{7}{3}$$

3. **If this result is an improper fraction, convert it to a mixed number.**

   $$\frac{7}{3} = 1\frac{4}{3} = 2\frac{1}{3}$$

EXAMPLE

Q. What is $2\frac{1}{5} \times 3\frac{1}{4}$?

A. $7\frac{3}{20}$. First, convert both mixed numbers to improper fractions. Multiply the whole number by the denominator and add the numerator; then place your answer over the original denominator:

$$2\frac{1}{5} = \frac{2 \times 5 + 1}{5} = \frac{11}{5} \qquad\qquad 3\frac{1}{4} = \frac{3 \times 4 + 1}{4} = \frac{13}{4}$$

Now multiply the two fractions:

$$\frac{11}{5} \times \frac{13}{4} = \frac{143}{20}$$

The answer is an improper fraction, so change it back to a mixed number:

$$\left(143 \div 20 = 7 \text{ r } 3\right): 7\frac{3}{20}$$

Q. What is $3\frac{1}{2} \div 1\frac{1}{7}$?

A. $3\frac{1}{16}$. First, convert both mixed numbers to improper fractions:

$$3\frac{1}{2} = \frac{3 \times 2 + 1}{2} = \frac{7}{2} \qquad\qquad 1\frac{1}{7} = \frac{1 \times 7 + 1}{7} = \frac{8}{7}$$

Now divide the two fractions:

$$\frac{7}{2} \div \frac{8}{7} = \frac{7}{2} \times \frac{7}{8} = \frac{49}{16}$$

Because the answer is an improper fraction, convert it to a mixed number:

$$= 1\frac{33}{16} = 2\frac{17}{16} = 3\frac{1}{16}$$

YOUR TURN

1. Multiply $2\frac{1}{3}$ by $1\frac{3}{7}$.

2. Find $2\frac{2}{5} \times 1\frac{5}{6}$.

3. Multiply $4\frac{4}{5}$ by $3\frac{1}{8}$.

4. Calculate $4\frac{1}{2} \div 1\frac{5}{8}$.

5. Divide $2\frac{1}{10}$ by $2\frac{1}{4}$.

6. What is $1\frac{2}{7} \div 6\frac{3}{10}$?

Adding Mixed Numbers

Adding mixed numbers looks a bit like column addition with whole numbers: You stack them one on top of the other, draw a line, and add. In this section, I show you everything you need to know.

Adding mixed numbers that have the same denominator

As with any problem involving fractions, adding is always easier when the denominators are the same.

For example, suppose you want to add $3\frac{1}{6} + 5\frac{1}{6}$. Begin by stacking these mixed numbers as follows:

$$3\frac{1}{6}$$
$$+\,5\frac{1}{6}$$

As you can see, this arrangement is similar to how you add whole numbers, but it includes an extra column for fractions. Start by adding the fractional parts and reduce as necessary:

$$\frac{1}{6} + \frac{1}{6} = \frac{2}{6} = \frac{1}{3}$$

Place this result in the fractions column:

$$3\frac{1}{6}$$
$$+\,5\frac{1}{6}$$
$$\frac{1}{3}$$

Now, add the whole-number parts and place this result in the whole-numbers column:

$$3\frac{1}{6}$$
$$+\,5\frac{1}{6}$$
$$8\frac{1}{3}$$

So the answer is $8\frac{1}{3}$.

Adding mixed numbers that have different denominators

When the fractional parts of a pair of mixed numbers have different denominators, use one of the two methods shown in Chapter 10 to change them to a common denominator. For example:

$$2\frac{3}{5}$$
$$+\,7\frac{1}{4}$$

In this problem, $\frac{3}{5}$ and $\frac{1}{4}$, the fractional parts, have different denominators, so rewrite them using a common denominator, as I show you in Chapter 11:

$$\frac{3}{5}+\frac{1}{4}=\frac{12}{20}+\frac{5}{20}$$

Rewrite each of the mixed numbers using this common denominator:

$$2\frac{3}{5}=2\frac{12}{20}$$
$$+\,7\frac{1}{4}=7\frac{5}{20}$$

Now, add the fractional parts and the whole-number parts as shown in the previous section:

$$2\frac{3}{5}=2\frac{12}{20}$$
$$+\,7\frac{1}{4}=7\frac{5}{20}$$
$$\overline{\qquad\quad 9\frac{17}{20}}$$

So the answer is $9\frac{17}{20}$.

Adding mixed numbers with carrying

In some cases when adding mixed numbers, you may find that the fraction addition results in an improper fraction. When this happens, change the improper fraction to a mixed number and carry the 1, just as you would do in regular addition.

For example:

$$8\frac{3}{5}$$
$$+\,6\frac{4}{5}$$

Begin by adding the fractional parts, and convert the resulting improper fraction to a mixed number (see Chapter 10 for more on converting improper fractions to mixed numbers):

$$\frac{3}{5} + \frac{4}{5} = \frac{7}{5} = 1\frac{2}{5}$$

Now, place the fractional part of this mixed number into the fractions column, and carry the 1 into the whole-numbers column:

$$
\begin{array}{r}
^{1} \\
8\frac{3}{5} \\
+\ 6\frac{4}{5} \\
\hline
\frac{2}{5}
\end{array}
$$

To complete the problem, add the whole numbers, $1 + 8 + 6 = 15$, and place this value into the whole-numbers column.

$$
\begin{array}{r}
^{1} \\
8\frac{3}{5} \\
+\ 6\frac{4}{5} \\
\hline
15\frac{2}{5}
\end{array}
$$

Therefore, the answer is $15\frac{2}{5}$.

Q. What is $4\frac{1}{8} + 2\frac{3}{8}$?

A. $6\frac{1}{2}$. To start out, set up the problem in column form:

$$
\begin{array}{r}
4\frac{1}{8} \\
+\ 2\frac{3}{8} \\
\hline
\end{array}
$$

Add the fractions and reduce the result: $\frac{1}{8} + \frac{3}{8} = \frac{4}{8} = \frac{1}{2}$.

Next, add the whole-number parts: $4 + 2 = 6$. Here's how the problem looks in column form:

$$
\begin{array}{r}
4\frac{1}{8} \\
+\ 2\frac{3}{8} \\
\hline
6\frac{1}{2}
\end{array}
$$

Q. Add $5\frac{7}{9}+6\frac{8}{9}$.

A. $12\frac{2}{3}$. To start out, set up the problem in column form:

$$5\frac{7}{9}$$
$$+6\frac{8}{9}$$

Add the fractions, change the result to a mixed number, and reduce:

$$\frac{7}{9}+\frac{8}{9}=\frac{15}{9}=1\frac{6}{9}=1\frac{2}{3}$$

Place the fraction in the fractions column, and carry the 1:

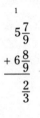

$$\overset{1}{}$$
$$5\frac{7}{9}$$
$$+6\frac{8}{9}$$
$$\frac{2}{3}$$

Next, add the whole-number parts and place this result in the whole-numbers column:

$$\overset{1}{}$$
$$5\frac{7}{9}$$
$$+6\frac{8}{9}$$
$$12\frac{2}{3}$$

Q. Add $5\frac{2}{3}+4\frac{5}{9}$.

A. $10\frac{2}{9}$. To start out, set up the problem in column form:

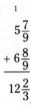

$$5\frac{2}{3}$$
$$+4\frac{5}{9}$$

The fractional parts have different denominators, but you can increase the terms of $\frac{2}{3}$ by a factor of 3 to $\frac{6}{9}$:

$$5\frac{2}{3}=5\frac{6}{9}$$
$$+4\frac{5}{9}=4\frac{5}{9}$$

Now you can add the fractional terms:

$$\frac{6}{9} + \frac{5}{9} = \frac{11}{9} = 1\frac{2}{9}$$

Place the fractional part in the fractions column, carry the 1, and add the whole-numbers column:

$$\overset{1}{}$$
$$5\frac{2}{3} = 5\frac{6}{9}$$
$$+ 4\frac{5}{9} = 4\frac{5}{9}$$
$$\overline{10\frac{2}{9}}$$

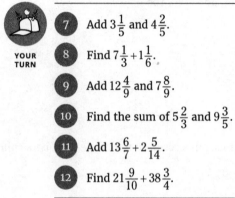

YOUR TURN

7 Add $3\frac{1}{5}$ and $4\frac{2}{5}$.

8 Find $7\frac{1}{3} + 1\frac{1}{6}$.

9 Add $12\frac{4}{9}$ and $7\frac{8}{9}$.

10 Find the sum of $5\frac{2}{3}$ and $9\frac{3}{5}$.

11 Add $13\frac{6}{7} + 2\frac{5}{14}$.

12 Find $21\frac{9}{10} + 38\frac{3}{4}$.

Subtracting Mixed Numbers

As with adding, subtracting mixed numbers uses columns just like subtraction with whole numbers: Stack them one on top of the other, draw a line, and subtract. In this section, you hone your mixed number subtraction skills.

Subtracting mixed numbers that have the same denominator

As with addition, fraction subtraction is much easier when the denominators are the same. For example, suppose you want to subtract $7\frac{5}{8} - 3\frac{3}{8}$. Here's what the problem looks like in column form:

$$7\frac{5}{8}$$
$$- 3\frac{3}{8}$$
$$\overline{}$$

To begin, subtract the fractional parts and reduce the result:

$$\frac{5}{8} - \frac{3}{8} = \frac{2}{8} = \frac{1}{4}$$

Place this value in the fractions column:

$$7\frac{5}{8}$$
$$-3\frac{3}{8}$$
$$\overline{\quad\frac{1}{4}}$$

Next, subtract $7 - 3 = 4$, placing this value in the whole-numbers column:

$$7\frac{5}{8}$$
$$-3\frac{3}{8}$$
$$\overline{4\frac{1}{4}}$$

So the answer is $4\frac{1}{4}$.

Subtracting mixed numbers that have different denominators

When a pair of mixed numbers have fractional parts with different denominators, your first step is to find a common denominator, as I show you in Chapter 11. For example:

$$7\frac{13}{16}$$
$$-4\frac{3}{4}$$

To begin, increase the terms of the fraction $\frac{3}{4}$ to $\frac{12}{16}$, as I show you in Chapter 11:

$$7\frac{13}{16} = 7\frac{13}{16}$$
$$-4\frac{3}{4} = 4\frac{12}{16}$$

Now, subtract in both the fractions column and the whole-numbers column:

$$7\frac{13}{16} = 7\frac{13}{16}$$
$$-4\frac{3}{4} = 4\frac{12}{16}$$
$$\overline{\qquad\quad 3\frac{1}{16}}$$

Therefore, the answer is $3\frac{1}{16}$.

Subtracting mixed numbers with borrowing

A complication arises when you try to subtract mixed number with a smaller fractional part minus one with a larger fractional part. Most students agree that this is the most difficult case when subtracting mixed numbers.

For example:

$$6\frac{1}{7}$$
$$-\ 2\frac{5}{7}$$

In this case, you can't subtract $\frac{1}{7}-\frac{5}{7}$, because the result would be a negative fraction.

To solve this problem, you need to make the fractional portion of $6\frac{1}{7}$ larger by *borrowing* from the column to the left. This idea is similar to the borrowing that you use in regular subtraction.

When borrowing in mixed-number subtraction, follow these steps:

1. **Borrow 1 from the whole-number portion and add it to the fractional portion, turning the fraction into a separate mixed number.**

2. **Change this new mixed number into an improper fraction.**

Here's how borrowing works with $6\frac{1}{7}$:

1. **Borrow 1 from the whole-number portion and add it to the fractional portion, turning the fraction into a separate mixed number:**

$$6\frac{1}{7}=5+1\frac{1}{7}$$

2. **Change this new mixed number into an improper fraction:**

$$=5\frac{8}{7}$$

This result may look odd, but it allows you to subtract both the fractional parts and the whole-number parts of the mixed numbers:

$$6\frac{1}{7}=5\frac{8}{7}$$
$$-\ 2\frac{5}{7}=2\frac{5}{7}$$
$$\overline{\qquad 3\frac{3}{7}}$$

Therefore, the answer is $3\frac{3}{7}$.

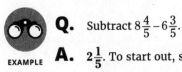

Q. Subtract $8\frac{4}{5} - 6\frac{3}{5}$.

A. $2\frac{1}{5}$. To start out, set up the problem in column form:

$$8\frac{4}{5}$$
$$-6\frac{3}{5}$$

Subtract both the fractional and whole-number parts:

$$8\frac{4}{5}$$
$$-6\frac{3}{5}$$
$$\overline{2\frac{1}{5}}$$

Q. Subtract $9\frac{1}{6} - 3\frac{5}{6}$.

A. $5\frac{1}{3}$. To start out, set up the problem in column form:

$$9\frac{1}{6}$$
$$-3\frac{5}{6}$$

Notice that $\frac{1}{6}$ is less than $\frac{5}{6}$, so you need to borrow 1 from 9:

$$9\frac{1}{6} = 8 + 1\frac{1}{6} = 8\frac{7}{6}$$

Substitute this equivalent value:

$$9\frac{1}{6} = 8\frac{7}{6}$$
$$-3\frac{5}{6} = 3\frac{5}{6}$$

Now, subtract in both the fractions and whole-numbers columns:

$$9\frac{1}{6} = 8\frac{7}{6}$$
$$-3\frac{5}{6} = 3\frac{5}{6}$$
$$\overline{5\frac{2}{6}}$$

To complete the problem, reduce the fractional part of this mixed number to lowest terms:

$$5\frac{2}{6} = 5\frac{1}{3}$$

Q. Subtract $19\frac{4}{11} - 6\frac{3}{8}$.

A. $12\frac{87}{88}$. To start out, set up the problem in column form:

$$19\frac{4}{11}$$
$$-\,6\,\frac{3}{8}$$

The fractional parts of the two mixed numbers have different denominators, so you need to rewrite them with a common denominator:

$$19\frac{4}{11} = 19\frac{32}{88}$$
$$-\,6\frac{3}{8}\; =\; 6\frac{33}{88}$$

Now, notice that $\frac{32}{88}$ is less than $\frac{33}{88}$, so you need to borrow before you can subtract:

$$19\frac{32}{88} = 18 + 1\frac{32}{88} = 1\frac{120}{88}$$

Substitute this value for the top mixed number:

$$19\frac{4}{11} = 19\frac{32}{88} = 18\frac{120}{88}$$
$$-6\frac{3}{8}\; =\; 6\frac{33}{88} = 6\frac{33}{88}$$

Subtract the fractional parts and the whole-number parts:

$$19\frac{4}{11} = 19\frac{32}{88} = 18\frac{120}{88}$$
$$-6\frac{3}{8}\; =\; 6\frac{33}{88} = \;6\frac{33}{88}$$
$$\rule{3cm}{0.4pt}$$
$$12\frac{87}{88}$$

YOUR TURN

13 Subtract $5\frac{7}{9} - 2\frac{4}{9}$.

14 Find $9\frac{1}{8} - 7\frac{5}{8}$.

15 Subtract $11\frac{3}{4} - 4\frac{2}{3}$.

16 What is $16\frac{2}{5} - 8\frac{4}{9}$?

Practice Questions Answers and Explanations

(1) $3\frac{1}{3}$. Change both mixed numbers to improper fractions:

$$2\frac{1}{3} = \frac{2\times 3+1}{3} = \frac{7}{3} \qquad 1\frac{3}{7} = \frac{1\times 7+3}{7} = \frac{10}{7}$$

Set up the multiplication, cancel out factors of 7 in the numerator and denominator, and then multiply:

$$\frac{7}{3} \times \frac{10}{7} = \frac{\cancel{7}^{1}}{3} \times \frac{10}{\cancel{7}_{1}} = \frac{10}{3}$$

Because the answer is an improper fraction, change it to a mixed number:

$$= 3\frac{1}{3}$$

(2) $4\frac{2}{5}$. Change both mixed numbers to improper fractions:

$$2\frac{2}{5} = \frac{2\times 5+2}{5} = \frac{12}{5} \qquad 1\frac{5}{6} = \frac{1\times 6+5}{6} = \frac{11}{6}$$

Set up the multiplication, cancel out 6s in the numerator and denominator, and then multiply:

$$\frac{12}{5} \times \frac{11}{6} = \frac{\cancel{12}^{2}}{5} \times \frac{11}{\cancel{6}_{1}} = \frac{22}{5}$$

Because the answer is an improper fraction, change it to a mixed number:

$$= 4\frac{2}{5}$$

(3) **15.** Change both mixed numbers to improper fractions:

$$4\frac{4}{5} = \frac{4\times 5+4}{5} = \frac{24}{5} \qquad 3\frac{1}{8} = \frac{3\times 8+1}{8} = \frac{25}{8}$$

Set up the multiplication, then cancel out factors of both 5 and 8 in the numerator and denominator, and multiply:

$$\frac{24}{5} \times \frac{25}{8} = \frac{\cancel{24}^{3}}{\cancel{5}_{1}} \times \frac{\cancel{25}^{5}}{\cancel{8}_{1}} = \frac{15}{1} = 15$$

(4) $2\frac{10}{13}$. Change both mixed numbers to improper fractions:

$$4\frac{1}{2} = \frac{4\times 2\times 1}{2} = \frac{9}{2} \qquad 1\frac{5}{8} = \frac{1\times 8+5}{8} = \frac{13}{8}$$

Set up the division, then use Keep–Change–Flip to change it to multiplication:

$$\frac{9}{2} \div \frac{13}{8} = \frac{9}{2} \times \frac{8}{13}$$

Cancel out a factor of 2 and multiply:

$$= \frac{9}{\cancel{2}_1} \times \frac{\cancel{8}_4}{13} = \frac{36}{13}$$

Because the answer is an improper fraction, change it to a mixed number:

$$= 2\frac{10}{13}$$

 $\frac{14}{15}$. Change both mixed numbers to improper fractions:

$$2\frac{1}{10} = \frac{2 \times 10 + 1}{10} = \frac{21}{10} \qquad 2\frac{1}{4} = \frac{2 \times 4 + 1}{4} = \frac{9}{4}$$

Set up the division, then use Keep–Change–Flip to change it to multiplication:

$$\frac{21}{10} \div \frac{9}{4} = \frac{21}{10} \times \frac{4}{9}$$

Before you multiply, cancel factors of 2 and 3 from the numerator and denominator:

$$= \frac{\cancel{21}_7}{5} \times \frac{2}{\cancel{9}_3} = \frac{14}{15}$$

⑥ $\frac{10}{49}$. Change both mixed numbers to improper fractions:

$$1\frac{2}{7} = \frac{1 \times 7 + 2}{7} = \frac{9}{7} \qquad 6\frac{3}{10} = \frac{6 \times 10 + 3}{10} = \frac{63}{10}$$

Set up the division, then use Keep–Change–Flip to change it to multiplication:

$$\frac{9}{7} \div \frac{63}{10} = \frac{9}{7} \times \frac{10}{63}$$

Before you multiply, cancel factors of 9 from the numerator and denominator:

$$= \frac{\cancel{9}_1}{7} \times \frac{10}{\cancel{63}_7} = \frac{10}{49}$$

⑦ $7\frac{3}{5}$. Set up the problem in column form:

$$3\frac{1}{5}$$
$$+\ 4\frac{2}{5}$$
$$\overline{}$$

Add the fractional and whole-number parts:

$$3\frac{1}{5}$$
$$+\,4\frac{2}{5}$$
$$7\frac{3}{5}$$

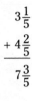 $8\frac{1}{2}$. To start out, set up the problem in column form:

$$7\frac{1}{3}$$
$$+1\frac{1}{6}$$

Increase the terms of $\frac{1}{3}$ so that the denominator is 6 by multiplying the numerator and denominator by 2:

$$7\frac{1}{3} = 7\frac{2}{6}$$
$$+1\frac{1}{6} = 1\frac{1}{6}$$

Now, add the fractional parts and the whole-number parts:

$$7\frac{1}{3} = 7\frac{2}{6}$$
$$+1\frac{1}{6} = 1\frac{1}{6}$$
$$8\frac{3}{6}$$

To complete the problem, reduce this mixed number to lowest terms:

$$8\frac{3}{6} = 8\frac{1}{2}$$

 $20\frac{1}{3}$. Set up the problem in column form:

$$12\frac{4}{9}$$
$$+\,7\frac{8}{9}$$

Add the fractional parts, convert the result to a mixed number, and reduce:

$$\frac{4}{9} + \frac{8}{9} = \frac{12}{9} = 1\frac{3}{9} = 1\frac{1}{3}$$

Place $\frac{1}{3}$ into the fractions column and carry the 1 into the whole-numbers column:

$$
\begin{array}{r}
^1 \\
12\frac{4}{9} \\
+\ 7\frac{8}{9} \\
\hline
\frac{1}{3}
\end{array}
$$

Add the values $1 + 12 + 7 = 20$, and place this result in the whole-numbers column:

$$
\begin{array}{r}
^1 \\
12\frac{4}{9} \\
+\ 7\frac{8}{9} \\
\hline
20\frac{1}{3}
\end{array}
$$

(10) $15\frac{4}{15}$. Set up the problem in column form:

$$
\begin{array}{r}
5\frac{2}{3} \\
+\ 9\frac{3}{5} \\
\hline
\end{array}
$$

Use cross-multiplication to express the two mixed numbers using a common denominator:

$$
\begin{array}{r}
5\frac{2}{3} = 5\frac{10}{15} \\
+\ 9\frac{3}{5} = 9\frac{9}{15} \\
\hline
\end{array}
$$

Add in the fractions column and convert the result to a mixed number:

$$
\frac{10}{15} + \frac{9}{15} = \frac{19}{15} = 1\frac{4}{15}
$$

Place the fractional part of this result into the fractions column and carry the 1:

$$
\begin{array}{r}
^1 \\
5\frac{2}{3} = 5\frac{10}{15} \\
+\ 9\frac{3}{5} = 9\frac{9}{15} \\
\hline
\frac{4}{15}
\end{array}
$$

Add in the whole-numbers column: $1 + 5 + 9 = 15$

$$\overset{1}{}$$

$$5\frac{10}{15}$$

$$+\,9\frac{9}{15}$$

$$\overline{15\frac{4}{15}}$$

(11) $16\frac{3}{14}$. Set up the problem in column form:

$$13\ \frac{6}{7}$$

$$+\,2\frac{5}{14}$$

Increase the terms of $\frac{6}{7}$ by a factor of 2:

$$13\ \frac{6}{7} = 13\frac{12}{14}$$

$$+\,2\frac{5}{14} =\ 2\frac{5}{14}$$

Add in the fractions column, and convert the result to a mixed number:

$$\frac{12}{14} + \frac{5}{14} = \frac{17}{14} = 1\frac{3}{14}$$

Place the fraction $\frac{3}{14}$ into the fractions column and carry the 1:

$$\overset{1}{}$$

$$13\ \frac{6}{7} = 13\frac{12}{14}$$

$$+\,2\frac{5}{14} =\ 2\frac{5}{14}$$

$$\overline{\frac{3}{14}}$$

Add $1 + 13 + 2 = 16$ and place this result in the whole-numbers column:

$$\overset{1}{}$$

$$13\ \frac{6}{7} = 13\frac{12}{14}$$

$$+\,2\frac{5}{14} =\ 2\frac{5}{14}$$

$$\overline{16\frac{3}{14}}$$

(12) $60\frac{13}{20}$. Set up the problem in column form:

$$21\frac{9}{10}$$

$$+\,38\ \frac{3}{4}$$

Cross-multiply to change the fractional parts of the two mixed numbers to a common denominator:

$$21\frac{9}{10} = 21\frac{36}{40}$$
$$+\ 38\frac{3}{4} = 38\frac{30}{40}$$

Add the fractions column, change to a mixed number, and reduce:

$$\frac{36}{40} + \frac{30}{40} = \frac{66}{40} = 1\frac{26}{40} = 1\frac{13}{20}$$

Place the fraction $\frac{13}{20}$ into the fractions column and carry the 1:

$$\overset{1}{}$$
$$21\frac{9}{10} = 21\frac{36}{40}$$
$$+\ 38\frac{3}{4} = 38\frac{30}{40}$$
$$\rule{3cm}{0.4pt}$$
$$\frac{13}{20}$$

Add $1 + 21 + 38 = 60$ in the whole-numbers column:

$$\overset{1}{}$$
$$21\frac{9}{10} = 21\frac{36}{40}$$
$$+\ 38\frac{3}{4} = 38\frac{30}{40}$$
$$\rule{3cm}{0.4pt}$$
$$60\frac{13}{20}$$

(13) $3\frac{1}{3}$. Set up the problem in column form:

$$5\frac{7}{9}$$
$$-\ 2\frac{4}{9}$$

Subtract the fractional parts and reduce:

$$\frac{7}{9} - \frac{4}{9} = \frac{3}{9} = \frac{1}{3}$$

Place this value into the fractions column:

$$5\frac{7}{9}$$
$$-\ 2\frac{4}{9}$$
$$\rule{2cm}{0.4pt}$$
$$\frac{1}{3}$$

Subtract the whole–number parts:

$$5\frac{7}{9}$$
$$-\,2\frac{4}{9}$$
$$3\frac{1}{3}$$

 $1\frac{1}{2}$. Set up the problem in column form:

$$9\frac{1}{8}$$
$$-\,7\frac{5}{8}$$

The fraction $\frac{1}{8}$ is less than $\frac{5}{8}$, so you need to borrow 1 from 9 before you can subtract:

$$9\frac{1}{8}=8+1\frac{1}{8}=8\frac{9}{8}$$

Substitute this value into the problem:

$$9\frac{1}{8}=8\frac{9}{8}$$
$$-\,7\frac{5}{8}=7\frac{5}{8}$$

Now you can subtract the fractional parts and reduce:

$$\frac{9}{8}-\frac{5}{8}=\frac{4}{8}=\frac{1}{2}$$

Subtract in the fractions and whole–numbers columns:

$$9\frac{1}{8}=8\frac{9}{8}$$
$$-\,7\frac{5}{8}=7\frac{5}{8}$$
$$1\frac{1}{2}$$

(15) $7\frac{1}{12}$. Set up the problem in column form:

$$11\frac{3}{4}$$
$$-\,4\frac{2}{3}$$

Use cross-multiplication to get common denominators for the fractions $\frac{3}{4}$ and $\frac{2}{3}$:

$$11\frac{3}{4} = 11\frac{9}{12}$$
$$-\ 4\frac{2}{3} = \ 4\frac{8}{12}$$

Subtract in the fractions and whole-numbers columns:

$$11\frac{3}{4} = 11\frac{9}{12}$$
$$-\ 4\frac{2}{3} = \ 4\frac{8}{12}$$
$$\overline{\hspace{2cm}}$$
$$7\frac{1}{12}$$

(16) $7\frac{43}{45}$. To start out, set up the problem in column form:

$$16\frac{2}{5}$$
$$-\ 8\frac{4}{9}$$
$$\overline{\hspace{1.5cm}}$$

Cross-multiply to find a common denominator for the fractions $\frac{2}{5}$ and $\frac{4}{9}$:

$$16\frac{2}{5} = 16\frac{18}{45}$$
$$-\ 8\frac{4}{9} = \ 8\frac{20}{45}$$

Because $\frac{18}{45}$ is less than $\frac{20}{45}$, you need to borrow before you can subtract fractions:

$$16\frac{18}{45} = 15 + 1\frac{18}{45} = 15\frac{63}{45}$$

Substitute this mixed number into the problem:

$$16\frac{2}{5} = 16\frac{18}{45} = 15\frac{63}{45}$$
$$-\ 8\frac{4}{9} = \ 8\frac{20}{45} = \ 8\frac{20}{45}$$

Now subtract in the fractions and whole-numbers columns:

$$16\frac{2}{5} = 16\frac{18}{45} = 15\frac{63}{45}$$
$$-\ 8\frac{4}{9} = \ 8\frac{20}{45} = \ 8\frac{20}{45}$$
$$\overline{\hspace{3cm}}$$
$$7\frac{43}{45}$$

If you're ready to test your skills a bit more, take the following chapter quiz in which you apply your knowledge of the Big Four operations to mixed numbers.

Whaddya Know? Chapter 12 Quiz

Quiz time! The 16 questions that follow cover the essential skills found in this chapter. When you're done, flip to the next section for answers and explanations.

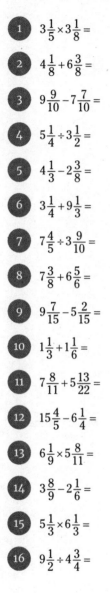

1 $3\frac{1}{5} \times 3\frac{1}{8} =$

2 $4\frac{1}{8} + 6\frac{3}{8} =$

3 $9\frac{9}{10} - 7\frac{7}{10} =$

4 $5\frac{1}{4} \div 3\frac{1}{2} =$

5 $4\frac{1}{3} - 2\frac{3}{8} =$

6 $3\frac{1}{4} + 9\frac{1}{3} =$

7 $7\frac{4}{5} \div 3\frac{9}{10} =$

8 $7\frac{3}{8} + 6\frac{5}{6} =$

9 $9\frac{7}{15} - 5\frac{2}{15} =$

10 $1\frac{1}{3} + 1\frac{1}{6} =$

11 $7\frac{8}{11} + 5\frac{13}{22} =$

12 $15\frac{4}{5} - 6\frac{1}{4} =$

13 $6\frac{1}{9} \times 5\frac{8}{11} =$

14 $3\frac{8}{9} - 2\frac{1}{6} =$

15 $5\frac{1}{3} \times 6\frac{1}{3} =$

16 $9\frac{1}{2} \div 4\frac{3}{4} =$

Answers to Chapter 12 Quiz

(1) **10.** First, change the mixed numbers to improper fractions.

$$3\frac{1}{5} \times 3\frac{1}{8} = \frac{16}{5} \times \frac{25}{8}$$

Then cancel factors and multiply.

$$= \frac{\cancel{16}^{2}}{\cancel{5}_{1}} \times \frac{\cancel{25}^{5}}{\cancel{8}_{1}} = \frac{10}{1} = 10$$

(2) $10\frac{1}{2}$. Set up the problem in column form.

$$\begin{array}{r} 4\frac{1}{8} \\ + 6\frac{3}{8} \\ \hline \end{array}$$

Add the numbers in the fractions column and reduce the result.

$$\frac{1}{8} + \frac{3}{8} = \frac{4}{8} = \frac{1}{2}$$

Place this result in the fractions column.

$$\begin{array}{r} 4\frac{1}{8} \\ + 6\frac{3}{8} \\ \hline \frac{1}{2} \end{array}$$

Now add the whole-number parts and place the result under the whole numbers.

$$\begin{array}{r} 4\frac{1}{8} \\ + 6\frac{3}{8} \\ \hline 10\frac{1}{2} \end{array}$$

(3) $2\frac{1}{5}$. Set up the problem in column form.

$$\begin{array}{r} 9\frac{9}{10} \\ - 7\frac{7}{10} \\ \hline \end{array}$$

Subtract the numbers in the fractions column and reduce.

$$\frac{9}{10} - \frac{7}{10} = \frac{2}{10} = \frac{1}{5}$$

Place this result in the fractions column.

$$9\frac{9}{10}$$
$$-7\frac{7}{10}$$
$$\overline{\frac{1}{5}}$$

Now subtract the whole numbers.

$$9\frac{9}{10}$$
$$-7\frac{7}{10}$$
$$\overline{2\frac{1}{5}}$$

(4) $1\frac{1}{2}$. First, change the mixed numbers to improper fractions.

$$5\frac{1}{4} \div 3\frac{1}{2} = \frac{21}{4} \div \frac{7}{2}$$

Next, change the division to multiplication using Keep-Change-Flip.

$$= \frac{21}{4} \times \frac{2}{7}$$

Now cancel factors and then multiply.

$$\frac{\cancel{21}^{3}}{\cancel{4}_{2}} \times \frac{\cancel{2}_{1}}{\cancel{7}_{1}} = \frac{3}{2}$$

Rewrite the resulting improper fraction as a mixed number.

$$\frac{3}{2} = 1\frac{1}{2}$$

(5) $1\frac{23}{24}$. First, set up the problem in column form.

$$4\frac{1}{3}$$
$$-2\frac{3}{8}$$
$$\overline{}$$

Use cross-multiplication to get a common denominator for the fractions.

$$4\frac{1}{3} = 4\frac{8}{24}$$
$$-2\frac{3}{8} = 2\frac{9}{24}$$
$$\overline{}$$

Borrow 1 from the whole-number portion of the top fraction and add it to the fraction.

$$4\frac{8}{24} = 3 + 1 + \frac{8}{24} = 3\frac{32}{24}$$

Put this result into the problem and subtract.

$$4\frac{1}{3} = 4\frac{8}{24} = 3\frac{32}{24}$$
$$-2\frac{3}{8} = 2\frac{9}{24} = 2\frac{9}{24}$$
$$\overline{\phantom{-2\frac{3}{8} = 2\frac{9}{24} = 2}1\frac{23}{24}}$$

(6) $12\frac{7}{12}$. First, set up the problem in column form.

$$3\frac{1}{4}$$
$$+\,9\frac{1}{3}$$

Use cross-multiplication to get a common denominator for the fractions.

$$\frac{1}{4} + \frac{1}{3} = \frac{1\times 3}{4\times 3} + \frac{1\times 4}{3\times 4} = \frac{3}{12} + \frac{4}{12}$$

Put these fractions in the fractions column.

$$3\frac{1}{4} = 3\frac{3}{12}$$
$$+\,9\frac{1}{3} = 9\frac{4}{12}$$

Add the fractional parts and then the whole-number parts.

$$3\frac{1}{4} = 3\frac{3}{12}$$
$$+\,9\frac{1}{3} = 9\frac{4}{12}$$
$$\overline{\phantom{+\,9\frac{1}{3} = 9}12\frac{7}{12}}$$

(7) **2.** First, change the mixed numbers to improper fractions.

$$7\frac{4}{5} \div 3\frac{9}{10} = \frac{39}{5} \div \frac{39}{10}$$

Next, change the division to multiplication using Keep-Change-Flip.

$$= \frac{39}{5} \times \frac{10}{39}$$

Reduce the fractions, and then multiply.

$$= \frac{\cancel{39}^{1}}{\cancel{5}_{1}} \times \frac{\cancel{10}^{2}}{\cancel{39}_{1}} = \frac{2}{1} = 2$$

8 $14\frac{5}{24}$. First, set up the problem in column form.

$$7\frac{3}{8}$$
$$+\ 6\frac{5}{6}$$

Use cross-multiplication to get a common denominator for the fractions.

$$\frac{3}{8}+\frac{5}{6}=\frac{3\times6}{8\times6}+\frac{5\times8}{6\times8}=\frac{18}{48}+\frac{40}{48}=\frac{9}{24}+\frac{20}{24}$$

Place the results into the fractional portion of the problem.

$$7\frac{3}{8}=7\frac{18}{48}$$
$$+\ 6\frac{5}{6}=6\frac{40}{48}$$

Begin by adding the fractional part. Convert the resulting improper fraction to a mixed number.

$$\frac{18}{48}+\frac{40}{48}=\frac{58}{48}=1\frac{10}{48}=1\frac{5}{24}$$

Place the fractional part in the fractions column and add the 1 to the whole-numbers column.

$$
\begin{array}{r}
^{1}\\
7\frac{3}{8}=\ 7\frac{18}{48}\\
+\ 6\frac{5}{6}=\ 6\frac{40}{48}\\
\hline
14\frac{10}{48}
\end{array}
$$

Now, simplify this result:

$$14\frac{10}{48}=14\frac{5}{24}$$

9 $4\frac{1}{3}$. First, set up the problem in column form.

$$9\frac{7}{15}$$
$$-\ 5\frac{2}{15}$$

Subtract the fractional parts and put the result in the fractions column. Then subtract the whole numbers.

$$\frac{7}{15}-\frac{2}{15}=\frac{5}{15}=\frac{1}{3}$$

$$9\frac{7}{15}$$
$$-5\frac{2}{15}$$
$$4\frac{1}{3}$$

 $2\frac{1}{2}$. First, set up the problem in column form.

$$1\frac{1}{3}$$
$$+1\frac{1}{6}$$

Increase the terms in the fraction $\frac{1}{3}$ by a factor of 2.

$$\frac{1}{3} = \frac{2}{6}$$

Place that adjusted fraction in the fractions column.

$$1\frac{2}{6}$$
$$+1\frac{1}{6}$$

Add the fractions and reduce the result. Add the whole numbers.

$$1\frac{1}{3} = 1\frac{2}{6}$$
$$+1\frac{1}{6} = 1\frac{1}{6}$$
$$2\frac{3}{6}$$

Simplify this result:

$$2\frac{3}{6} = 2\frac{1}{2}$$

11 $13\frac{7}{22}$. Set up the problem in column form:

$$7\frac{8}{11}$$
$$+5\frac{13}{22}$$

Increase the terms of $\frac{8}{11}$ by a factor of 2:

$$7\frac{8}{11} = 7\frac{16}{22}$$
$$+5\frac{13}{22} = 5\frac{13}{22}$$

Add in the fractions column, and convert the result to a mixed number:

$$\frac{16}{22} + \frac{13}{22} = \frac{29}{22} = 1\frac{7}{22}$$

Place the fraction $\frac{7}{22}$ into the fractions column and carry the 1:

$$
\begin{array}{r}
^1\\
7\frac{8}{11} = 7\frac{16}{22}\\
+\, 5\frac{13}{22} = 5\frac{13}{22}\\
\hline
\frac{7}{22}
\end{array}
$$

Add $1 + 7 + 5 = 13$ and place this result in the whole-numbers column:

$$
\begin{array}{r}
^1\\
7\frac{8}{11} = 7\frac{16}{22}\\
+\, 5\frac{13}{22} = 5\frac{13}{22}\\
\hline
13\frac{7}{22}
\end{array}
$$

(12) $9\frac{11}{20}$. Set up the problem in column form:

$$
\begin{array}{r}
15\frac{4}{5}\\
-\, 6\frac{1}{4}
\end{array}
$$

Use cross-multiplication to get common denominators for the fractions $\frac{4}{5}$ and $\frac{1}{4}$:

$$
\begin{array}{r}
15\frac{4}{5} = 15\frac{16}{20}\\
-\, 6\frac{1}{4} = \ 6\frac{5}{20}
\end{array}
$$

Subtract in the fractions and whole-numbers columns:

$$
\begin{array}{r}
15\frac{4}{5} = 15\frac{16}{20}\\
-\, 6\frac{1}{4} = \ 6\frac{5}{20}\\
\hline
9\frac{11}{20}
\end{array}
$$

(13) **35.** First, change the mixed numbers to improper fractions.

$$6\frac{1}{9} \times 5\frac{8}{11} = \frac{55}{9} \times \frac{63}{11}$$

Reduce the fractions and then multiply.

$$= \frac{\cancel{55}^5}{\cancel{9}_1} \times \frac{\cancel{63}^7}{\cancel{11}_1} = \frac{35}{1} = 35$$

(14) $1\frac{13}{18}$. First, set up the problem in column form.

$$3\frac{8}{9}$$
$$-2\frac{1}{6}$$

Use cross-multiplication to get a common denominator for the fractions.

$$\frac{8}{9}-\frac{1}{6}=\frac{8\times6}{9\times6}-\frac{1\times9}{6\times9}=\frac{48}{54}-\frac{9}{54}$$

Put the result into the problem in the fractions column.

$$3\frac{8}{9}=3\frac{48}{54}$$
$$-2\frac{1}{6}=2\frac{9}{54}$$

Subtract the fractions and reduce.

$$\frac{48}{54}-\frac{9}{54}=\frac{39}{54}=\frac{13}{18}$$

Place this result in the fractions column, then subtract in the whole-numbers column.

$$3\frac{8}{9}=3\frac{48}{54}$$
$$-2\frac{1}{6}=2\frac{9}{54}$$
$$\overline{\qquad 1\frac{13}{18}}$$

(15) $33\frac{7}{9}$. First, change the mixed numbers to improper fractions.

$$5\frac{1}{3}\times6\frac{1}{3}=\frac{16}{3}\times\frac{19}{3}$$

Now multiply and change the resulting improper fraction to a mixed number.

$$=\frac{304}{9}=33\frac{7}{9}$$

(16) **2.** First, change the mixed numbers to improper fractions.

$$9\frac{1}{2}\div4\frac{3}{4}=\frac{19}{2}\div\frac{19}{4}$$

Next, change the division to multiplication using Keep-Change-Flip.

$$=\frac{19}{2}\times\frac{4}{19}$$

Reduce the fractions and multiply.

$$=\frac{\cancel{19}^{1}}{\cancel{2}_{1}}\times\frac{\cancel{4}^{2}}{\cancel{19}_{1}}=\frac{2}{1}=2$$

5

Decimals and
Percents

In This Unit . . .

Chapter **13**

Getting to the Point with Decimals

Because early humans used their fingers for counting, the number system is based on the number 10. So numbers come in ones, tens, hundreds, thousands, and so on. A *decimal* — with its handy decimal point — allows people to work with numbers smaller than 1: tenths, hundredths, thousandths, and the like.

Here's some lovely news: Decimals are much easier to work with than fractions (which I discuss in Chapters 13, 14, and 15). Decimals look and feel more like whole numbers than fractions do, so when you're working with decimals, you don't have to worry about reducing and increasing terms, improper fractions, mixed numbers, and a lot of other stuff.

Performing the Big Four operations — addition, subtraction, multiplication, and division — on decimals is very close to performing them on whole numbers (which I cover in Unit 2 of the book). The numerals 0 through 9 work just like they usually do. As long as you get the decimal point in the right place, you're home free.

In this chapter, I show you all about working with decimals. I also show you how to convert fractions to decimals and decimals to fractions. Finally, I give you a peek into the strange world of repeating decimals.

Understanding Basic Decimal Stuff

The good news about decimals is that they look a lot more like whole numbers than fractions do. So a lot of what you find out about whole numbers in Chapter 3 applies to decimals as well. In this section, I introduce you to decimals, starting with place value.

When you understand place value of decimals, a lot falls into place. Then I discuss trailing zeros and what happens when you move the decimal point either to the left or to the right.

Counting dollars and decimals

You use decimals all the time when you count money. And a great way to begin thinking about decimals is as dollars and cents. For example, you know that $0.50 is half of a dollar (see Figure 13-1), so this information tells you:

$$0.5 = \frac{1}{2}$$

FIGURE 13-1:
One-half (0.5)
of a dollar bill.

Notice that, in the decimal 0.5, I drop the zero at the end. This practice is common with decimals.

You also know that $0.25 is a quarter — that is, one-fourth of a dollar (see Figure 13-2) — so:

$$0.25 = \frac{1}{4}$$

FIGURE 13-2:
One-fourth
(0.25) of a
dollar bill.

Similarly, you know that $0.75 is three quarters, or three-fourths, of a dollar (see Figure 13-3), so:

$$0.75 = \frac{3}{4}$$

FIGURE 13-3:
Three-fourths
(0.75) of a
dollar bill.

Taking this idea even further, you can use the remaining denominations of coins — dimes, nickels, and pennies — to make further connections between decimals and fractions.

$$\text{A dime} = \$0.10 = \frac{1}{10} \text{ of a dollar, so } \frac{1}{10} = 0.1$$

$$\text{A nickel} = \$0.05 = \frac{1}{20} \text{ of a dollar, so } \frac{1}{20} = 0.05$$

$$\text{A penny} = \$0.01 = \frac{1}{100} \text{ of a dollar, so } \frac{1}{100} = 0.01$$

REMEMBER

Notice that I again drop the final zero in the decimal 0.1, but I keep the zeros in the decimals 0.05 and 0.01. You can drop zeros from the right end of a decimal, but you can't drop zeros that fall between the decimal point and another digit.

Decimals are just as good for cutting up cake as for cutting up money. Figure 13-4 gives you a look at the four cut-up cakes that I show you in Chapter 10. This time, I give you the decimals that tell you how much cake you have. Fractions and decimals accomplish the same task: allowing you to cut a whole object into pieces and talk about how much you have.

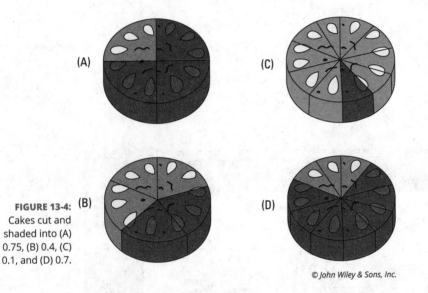

(A)

(C)

FIGURE 13-4: (B)
Cakes cut and
shaded into (A)
0.75, (B) 0.4, (C)
0.1, and (D) 0.7.

(D)

© John Wiley & Sons, Inc.

Identifying the place value of decimals

In Chapter 3, you find out about the place value of whole numbers. Table 13-1 shows how the whole number 4,672 breaks down in terms of place value.

Table 13-1   Breaking Down 4,672 in Terms of Place Value

Thousands	Hundreds	Tens	Ones
4	6	7	2

This number means $4,000 + 600 + 70 + 2$.

With decimals, this idea is extended. First, a decimal point is placed to the right of the ones place in a whole number. Then more numbers are appended to the right of the decimal point.

For example, the decimal 4,672.389 breaks down as shown in Table 13-2.

Table 13-2   Breaking Down the Decimal 4,672.389

Thousands	Hundreds	Tens	Ones	Decimal Point	Tenths	Hundredths	Thousandths
4	6	7	2	.	3	8	9

This decimal means $4,000 + 600 + 70 + 2 + \frac{3}{10} + \frac{8}{100} + \frac{9}{1000}$.

The connection between fractions and decimals becomes obvious when you look at place value. Decimals really are a shorthand notation for fractions. You can represent any fraction as a decimal.

Knowing the decimal facts of life

When you understand how place value works in decimals (as I explain in the preceding section), a whole lot of facts about decimals begin to make sense. Two key ideas are trailing zeros and what happens when you move a decimal point left or right.

Understanding trailing zeros

You probably know that you can attach zeros to the beginning of a whole number without changing its value. For example, these three numbers are all equal in value:

27   027   0,000,027

The reason becomes clear when you know about place value of whole numbers. See Table 13-3.

Table 13-3   Example of Attaching Leading Zeros

Millions	Hundred Thousands	Ten Thousands	Thousands	Hundreds	Tens	Ones
0	0	0	0	0	2	7

As you can see, 0,000,027 simply means $0 + 0 + 0 + 0 + 0 + 20 + 7$. No matter how many zeros you add to the beginning of a number, the number 27 doesn't change.

Zeros attached to the beginning of a number in this way are called *leading zeros*.

In decimals, this idea of zeros that don't add value to a number can be extended to trailing zeros.

REMEMBER

A *trailing zero* is any zero that appears to the right of both the decimal point and every digit other than zero.

For example:

34.8   34.80   34.8000

All three of these numbers are the same. The reason becomes clear when you understand how place value works in decimals. See Table 13-4.

Table 13-4   Example of Attaching Trailing Zeros

Tens	Ones	Decimal Point	Tenths	Hundredths	Thousandths	Ten Thousandths
3	4	.	8	0	0	0

In this example, 34.8000 means $30 + 4 + \frac{8}{10} + \frac{0}{100} + \frac{0}{1000} + \frac{0}{10000}$.

REMEMBER

You can attach or remove as many trailing zeros as you want without changing the value of a number.

When you understand trailing zeros, you can see that every whole number can easily be changed to a decimal. Just attach a decimal point and a 0 to the end of it. For example:

$$4 = 4.0$$
$$20 = 20.0$$
$$971 = 971.0$$

WARNING

Make sure that you don't attach or remove any non-leading or non-trailing zeros — it changes the value of the decimal.

For example, look at this number:

0450.0070

In this number, you can remove the leading and trailing zeros without changing the value, as follows:

450.007

The remaining zeros, however, need to stay where they are as *placeholders* between the decimal point and digits other than zero. See Table 13-5.

I continue to discuss zeros as placeholders in the next section.

Table 13-5   Example of Zeros as Placeholders

Thousands	Hundreds	Tens	Ones	Decimal Point	Tenths	Hundredths	Thousandths	Ten Thousandths
0	4	5	0	.	0	0	7	0

Moving the decimal point

When you're working with whole numbers, you can multiply any number by 10 just by adding a zero to the end of it. For example:

$$45,971 \times 10 = 459,710$$

To see why this answer is so, again think about the place value of digits and look at Table 13-6.

Table 13-6   Example of Decimal Points and Place Value of Digits

Millions	Hundred Thousands	Ten Thousands	Thousands	Hundreds	Tens	Ones
		4	5	9	7	1
4	5	9	7	1	0	

Here's what these two numbers really mean:

$$45,971 = 40,000 + 5,000 + 900 + 70 + 1$$
$$459,710 = 400,000 + 50,000 + 9,000 + 700 + 10 + 0$$

As you can see, that little zero makes a big difference: It causes the rest of the numbers to shift one place.

This concept makes even more sense when you think about the decimal point. See Table 13-7.

Table 13-7   Example of Numbers Shifting One Place

Hundred Thousands	Ten Thousands	Thousands	Hundreds	Tens	Ones	Decimal Point	Tenths	Hundredths
	4	5	9	7	1	.	0	0
4	5	9	7	1	0	.	0	0

In effect, adding a 0 to the end of a whole number moves the decimal point one place to the right. So for any decimal, when you move the decimal point one place to the right, you multiply that number by 10. This fact becomes clear when you start with a simple number like 7:

$$7.0$$
$$70.0$$
$$700.0$$
$$7,000.0$$

In this case, the net effect is that you moved the decimal point three places to the right, which is the same as multiplying 7 by 1,000.

Similarly, to divide any number by 10, move the decimal point one place to the left. For example:

7.0
0.7
0.07
0.007

This time, the net effect is that you moved the decimal point three places to the left, which is the same as dividing 7 by 1,000.

TIP

You can express any whole number as a decimal simply by attaching a decimal point and a trailing zero to the end of it. For example,

$$7 = 7.0 \quad 12 = 12.0 \quad 1,568 = 1,568.0$$

To sum up, moving the decimal point to the *right* is the same as multiplying that decimal by a power of 10. For example,

>> Moving the decimal point *one* place to the right is the same as multiplying by 10.

>> Moving the decimal point *two* places to the right is the same as multiplying by 100.

>> Moving the decimal point *three* places to the right is the same as multiplying by 1,000.

Similarly, moving the decimal point to the *left* is the same as dividing that decimal by a power of 10. For example,

>> Moving the decimal point *one* place to the left is the same as dividing by 10.

>> Moving the decimal point *two* places to the left is the same as dividing by 100.

>> Moving the decimal point *three* places to the left is the same as dividing by 1,000.

To multiply a decimal by any power of 10, count the number of zeros and move the decimal point that many places to the right. To divide a decimal by any power of 10, count the number of zeros and move the decimal point that many places to the left.

Rounding decimals is similar to rounding whole numbers (if you need a refresher, see Chapter 3). Generally speaking, to round a number to a given decimal place, focus on that decimal place and the place to its immediate right; then round as you would with whole numbers.

>> **Rounding down:** If the digit on the right is 0, 1, 2, 3, or 4, drop this digit and every digit to its right.

>> **Rounding up:** If the digit on the right is 5, 6, 7, 8, or 9, add 1 to the digit you're rounding to and then drop every digit to its right.

When rounding, people often refer to the first three decimal places in two different ways — by the number of the decimal place and by the name:

>> Rounding to *one* decimal place is the same as rounding to the nearest tenth.

>> Rounding to *two* decimal places is the same as rounding to the nearest hundredth.

>> Rounding to *three* decimal places is the same as rounding to the nearest thousandth.

When rounding to four or more decimal places, the names get longer, so they're usually not used.

Q. Expand the decimal 7,358.293.

A. $7,358.293 = 7,000 + 300 + 50 + 8 + \frac{2}{10} + \frac{9}{100} + \frac{3}{1,000}$

Q. Simplify the decimal 0400.0600 by removing all leading and trailing zeros, without removing placeholding zeros.

A. **400.06.** The first zero is a leading zero because it appears to the left of all non-zero digits. The last two zeros are trailing zeros because they appear to the right of all non-zero digits. The remaining three zeros are placeholding zeros.

Q. Multiply 3.458×100.

A. **345.8.** The number 100 has two zeros, so to multiply by 100, move the decimal point two places to the right.

Q. Divide $29.81 \div 10,000$.

A. **0.002981.** The number 10,000 has four zeros, so to divide by 10,000, move the decimal point four places to the left.

1   Expand the following decimals:

    (a)   2.7

    (b)   31.4

    (c)   86.52

    (d)   103.759

    (e)   1,040.0005

    (f)   16,821.1384

2   Simplify each of the following decimals by removing all leading and trailing zeros whenever possible, without removing placeholding zeros:

    (a)   5.80

    (b)   7.030

    (c)   90.0400

    (d)   9,000.005

    (e)   0108.0060

    (f)   00100.0102000

3   Do the following decimal multiplication and division problems by moving the decimal point the correct number of places:

    (a) 7.32×10

    (b) 9.04×100

    (c) $51.6 \times 100{,}000$

    (d) $2.786 \div 1{,}000$

    (e) $943.812 \div 1{,}000{,}000$

4   Round each of the following decimals to the number of places indicated:

    (a)   Round 4.777 to one decimal place.

    (b)   Round 52.305 to the nearest tenth.

    (c)   Round 191.2839 to two decimal places.

    (d)   Round 99.995 to the nearest hundredth.

    (e)   Round 0.00791 to three decimal places.

    (f)   Round 909.9996 to the nearest thousandth.

Performing the Big Four Operations with Decimals

Everything you already know about adding, subtracting, multiplying, and dividing whole numbers (see Chapter 2) carries over when you work with decimals. In fact, in each case, there's really only one key difference: how to handle that pesky little decimal point. In this section, I show you how to perform the Big Four math operations with decimals.

The most common use of adding and subtracting decimals is working with money — for example, balancing your checkbook. Later in this book, you find that multiplying and dividing by decimals is useful for calculating percentages (see Chapter 14), using scientific notation (see Chapter 17), and measuring with the metric system (see Chapter 18).

Adding decimals

Adding decimals is almost as easy as adding whole numbers. As long as you set up the problem correctly, you're in good shape. To add decimals, follow these steps:

1. **Arrange the numbers in a column and line up the decimal points vertically.**

2. **Add as usual, column by column, from right to left.**

3. **Place the decimal point in the answer in line with the other decimal points in the problem.**

For example, suppose you want to add the numbers 14.5 and 1.89. Line up the decimal points neatly, as follows:

$$\begin{array}{r} 14.5 \\ + 1.89 \\ \hline \end{array}$$

Begin adding from the right-most column. Treat the blank space after 14.5 as a 0 — you can write this in as a trailing 0 (see earlier in this chapter to see why adding zeros to the end of a decimal doesn't change its value). Adding this column gives you $0 + 9 = 9$:

$$\begin{array}{r} 14.50 \\ + 1.89 \\ \hline 9 \end{array}$$

Continuing to the left, $5 + 8 = 13$, so put down the 3 and carry the 1:

$$\begin{array}{r} 1\\[-6pt] 14.50 \\ + 1.89 \\ \hline 39 \end{array}$$

Complete the problem column by column, and at the end, put the decimal point directly below the others in the problem:

$$\begin{array}{r} 14.50 \\ + 1.89 \\ \hline 16.39 \end{array}$$

When adding more than one decimal, the same rules apply. For example, suppose you want to add $15.1 + 0.005 + 800 + 1.2345$. The most important idea is lining up the decimal points correctly:

$$\begin{array}{r} 15.1 \\ 0.005 \\ 800.0 \\ + 1.2345 \\ \hline \end{array}$$

TIP

To avoid mistakes, be especially neat when adding a lot of decimals.

Because the number 800 isn't a decimal, I place a decimal point and a 0 at the end of it, to be clear about how to line it up. If you like, you can make sure all numbers have the same number of decimal places (in this case, four) by adding trailing zeros. When you properly set up the problem, the addition is no more difficult than in any other addition problem:

$$
\begin{array}{r}
15.1000 \\
0.0050 \\
800.0000 \\
+\,1.2345 \\
\hline
816.3395
\end{array}
$$

Subtracting decimals

Subtracting decimals uses the same trick as adding them (which I talk about in the preceding section). Here's how you subtract decimals:

1. **Arrange the numbers in a column and line up the decimal points.**

2. **Subtract as usual, column by column from right to left.**

3. **When you're done, place the decimal point in the answer in line with the other decimal points in the problem.**

For example, suppose you want to figure out $144.87 - 0.321$. First, line up the decimal points:

$$
\begin{array}{r}
144.870 \\
-\,0.321 \\
\hline
\end{array}
$$

In this case, I add a zero at the end of the first decimal. This placeholder reminds you that, in the right-most column, you need to borrow to get the answer to $0 - 1$:

$$
\begin{array}{r}
144.8\overset{6}{\cancel{7}}{}^{1}0 \\
-\,0.32\,1 \\
\hline
49
\end{array}
$$

The rest of the problem is straightforward. Just finish the subtraction and drop the decimal point straight down:

$$
\begin{array}{r}
144.8\overset{6}{\cancel{7}}{}^{1}0 \\
-\ 0.321 \\
\hline
144.549
\end{array}
$$

So the answer is 144.549.

Q. Add the following decimals: $321.81 + 24.5 + 0.006 = ?$

EXAMPLE **A.** **346.316.** Place the decimal numbers in a column (as you would for column addition) with the decimal points lined up:

```
  321.810
   24.500
+  0.006
```

Notice that the decimal point in the answer lines up with the others. As you can see, I've also filled out the columns with trailing zeros. This is optional, but do it if it helps you to see how the columns line up.

Now add as you would when adding whole numbers, carrying when necessary (see Chapter 2 for more on carrying in addition):

```
   ¹
  321.810
   24.500
+  0.006
  346.316
```

Q. Subtract the following decimals: $978.245 - 29.03 = ?$

A. **949.215.** Place the decimals one on top of the other with the decimal points lined up, dropping the decimal point straight down in the answer:

```
  978.245
-  29.030
```

Now subtract as you would when subtracting whole numbers, borrowing when necessary (see Chapter 2 for more on borrowing in subtraction):

```
  9⁷̸¹8.245
-  29.030
  949.215
```

YOUR
TURN

⑤ Add these decimals: $17.4 + 2.18 = ?$

⑥ Compute the following decimal addition: $0.0098 + 10.101 + 0.07 + 33 = ?$

⑦ Add the following decimals: $1,000.001 + 75 + 0.03 + 800.2 = ?$

⑧ Subtract these decimals: $0.748 - 0.23 = ?$

⑨ Compute the following: $674.9 - 5.0001.$

⑩ Find the solution to this decimal subtraction problem: $100.009 - 0.68 = ?$

Multiplying decimals

Multiplying decimals is different from adding and subtracting them, in that you don't have to worry about lining up the decimal points (see the preceding sections). In fact, the only difference between multiplying whole numbers and decimals comes at the end.

Here's how to multiply decimals:

1. **Perform the multiplication as you do for whole numbers.**

2. **When you're done, count the number of digits to the right of the decimal point in each factor, and add the result.**

3. **Place the decimal point in your answer so that your answer has the same number of digits after the decimal point.**

This process sounds tricky, but multiplying decimals can actually be simpler than adding or subtracting them. Suppose, for instance, that you want to multiply 23.5 by 0.16. The first step is to pretend that you're multiplying numbers without decimal points:

$$
\begin{array}{r}
23.5 \\
\times\, 0.16 \\
\hline
1410 \\
2350 \\
\hline
3760
\end{array}
$$

This answer isn't complete, though, because you still need to find out where the decimal point goes in the answer. To do this, notice that 23.5 has one digit after the decimal point and that 0.16 has two digits after the decimal point. Because $1 + 2 = 3$, place the decimal point in the answer so that it has three digits after the decimal point. (You can put your pencil at the 0 at the end of 3760 and move the decimal point three places to the left.)

$$
\begin{array}{rl}
23.5 & \text{1 digit after the decimal point} \\
\times\, 0.16 & \text{2 digits after the decimal point} \\
\hline
1410 & \\
2350 & \\
\hline
3.760 & 1 + 2 = 3 \text{ digits after the decimal point}
\end{array}
$$

 WARNING Even though the last digit in the answer is a 0, you still need to count this as a digit when placing the decimal point. When the decimal point is in place, you can drop trailing zeros (flip to the section, "Understanding Basic Decimal Stuff," earlier in this chapter, to see why the zeros at the end of a decimal don't change the value of the number).

So the answer is 3.760, which is equal to 3.76.

Q. Multiply the following decimals: $74.2 \times 0.35 = ?$

EXAMPLE **A.** **25.97.** Ignoring the decimal points, perform the multiplication just as you would for whole numbers:

$$
\begin{array}{r}
74.2 \\
\times\ 0.35 \\
\hline
3710 \\
+\ 22260 \\
\hline
25970
\end{array}
$$

At this point, you're ready to find out where the decimal point goes in the answer. Count the number of decimal places in the two factors (74.2 and 0.35), add these two numbers together $(1 + 2 = 3)$, and place the decimal point in the answer so that it has three digits to the right of the decimal point:

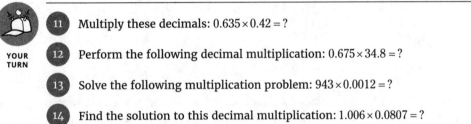

74.2 1 digit after the decimal point

× 0.35 2 digits after the decimal point
3710

22260 $1 + 2 = 3$ digits after the decimal point
25.970

YOUR TURN

11 Multiply these decimals: $0.635 \times 0.42 = ?$

12 Perform the following decimal multiplication: $0.675 \times 34.8 = ?$

13 Solve the following multiplication problem: $943 \times 0.0012 = ?$

14 Find the solution to this decimal multiplication: $1.006 \times 0.0807 = ?$

Dividing decimals

Long division has never been a crowd pleaser. Dividing decimals is almost the same as dividing whole numbers, which is why a lot of people don't particularly like dividing decimals, either.

But at least you can take comfort in the fact that, when you know how to do long division, figuring out how to divide decimals is easy. The main difference comes at the beginning, before you start dividing.

Here's how to divide decimals:

1. **Turn the *divisor* (the number you're dividing by) into a whole number by moving the decimal point all the way to the right; at the same time, move the decimal point in the *dividend* (the number you're dividing) the same number of places to the right.**

For example, suppose you want to divide 10.274 by 0.11. Write the problem as usual:

$$0.11\overline{)10.274}$$

Turn 0.11 into a whole number by moving the decimal point in 0.11 two places to the right, giving you 11. At the same time, move the decimal point in 10.274 two places to the right, giving you 1,027.4:

$$11.\overline{)1027.4}$$

2. **Place a decimal point in the *quotient* (the answer) directly above where the decimal point now appears in the dividend.**

Here's what this step looks like:

$$11.\overline{)1027.4}\,^{.}$$

3. **Divide as usual, being careful to line up the quotient properly so that the decimal point falls into place.**

To start out, notice that 11 is too large to go into either 1 or 10. However, 11 does go into 102 (nine times). So write the first digit of the quotient just above the 2 and continue:

$$
\begin{array}{r}
9\,. \\
11.\overline{)1027.4} \\
\underline{99} \\
37
\end{array}
$$

I paused after bringing down the next number, 7. This time, 11 goes into 37 three times. The important point is to place the next digit in the answer just above the 7:

$$
\begin{array}{r}
93. \\
11.\overline{)1027.4} \\
\underline{99} \\
37 \\
\underline{33} \\
44
\end{array}
$$

I paused after bringing down the next number, 4. Now, 11 goes into 44 four times. Again, be careful to place the next digit in the quotient just above the 4, and complete the division:

$$
\begin{array}{r}
93.4 \\
11.\overline{)1027.4} \\
\underline{99} \\
37 \\
\underline{33} \\
44 \\
\underline{44} \\
0
\end{array}
$$

So the answer is 93.4. As you can see, as long as you're careful when placing the decimal point and the digits, the correct answer appears with the decimal point in the right position.

Dealing with more zeros in the dividend

Sometimes you have to add one or more trailing zeros to the dividend. As I discuss earlier in this chapter, you can add as many trailing zeros as you like to a decimal without changing its value. For example, suppose you want to divide 67.8 by 0.333:

$$0.333 \overline{)67.8}$$

Follow these steps:

1. **Change 0.333 into a whole number by moving the decimal point three places to the right; at the same time, move the decimal point in 67.8 three places to the right:**

    $$333. \overline{)67800.}$$

    In this case, when you move the decimal point in 67.8, you run out of room, so you have to add a couple zeros to the dividend. This step is perfectly valid, and you need to do this whenever the divisor has more decimal places than the dividend.

2. **Place the decimal point in the quotient directly above where it appears in the dividend:**

    $$333. \overline{)67800.}$$

3. **Divide as usual, being careful to correctly line up the numbers in the quotient. This time, 333 doesn't go into 6 or 67, but it does go into 678 (two times). So place the first digit of the quotient directly above the 8:**

    ```
          2    .
    333.)67800.
         666
         120
    ```

    I've jumped forward in the division to the place where I bring down the first 0. At this point, 333 doesn't go into 120, so you need to put a 0 above the first 0 in 67,800 and bring down the second 0. Now, 333 does go into 1,200, so place the next digit in the answer (3) over the second 0:

    ```
         203.
    333.)67800.
         666
         1200
          999
          201
    ```

This time, the division doesn't work out evenly. If this were a problem with whole numbers, you'd finish by writing down a remainder of 201. (For more on remainders in division, see

Chapter 2.) But decimals are a different story. The next section explains why, with decimals, the show must go on.

Completing decimal division

When you're dividing whole numbers, you can complete the problem simply by writing down the remainder. But remainders are *never* allowed in decimal division.

A common way to complete a problem in decimal division is to round off the answer. In most cases, you're instructed to round your answer to the nearest whole number or to one or two decimal places (see earlier in this chapter to find out how to round off decimals).

To complete a decimal division problem by rounding it off, you need to add at least one trailing zero to the dividend:

>> To round a decimal to a whole number, add one trailing zero.

>> To round a decimal to one decimal place, add two trailing zeros.

>> To round a decimal to two decimal places, add three trailing zeros.

Here's what the problem looks like with a trailing zero attached:

$$
\begin{array}{r}
203.6 \\
333.\overline{)67800.0} \\
\underline{666} \\
1200 \\
\underline{999} \\
2010
\end{array}
$$

Attaching a trailing zero doesn't change a decimal, but it does allow you to bring down one more number, changing 201 into 2,010. Now you can divide 333 into 2,010:

$$
\begin{array}{r}
203.6 \\
333.\overline{)67800.0} \\
\underline{666} \\
1200 \\
\underline{999} \\
2010 \\
\underline{1998} \\
12
\end{array}
$$

At this point, you can round the answer to the nearest whole number, 204. I give you more practice dividing decimals later in this chapter.

Q. Divide the following: 9.152 ÷ 0.8 = ?

A. **11.44.** To start out, write the problem as usual:

$$0.8\overline{)9.152}$$

Turn 0.8 into the whole number 8 by moving the decimal point one place to the right. At the same time, move the decimal point in 9.152 one place to the right. Put your decimal point in the quotient directly above where it falls in 91.52:

$$8.\overline{)91.52}$$

Now you're ready to divide. Just be careful to line up the quotient properly so that the decimal point falls into place.

```
       11.44
   8.)91.52
      8
      11
       8
      35
      32
       32
       32
```

Q. Divide the following: 21.9 ÷ 0.015 = ?

A. **1,460.** Set up the problem as usual:

$$0.015\overline{)21.900}$$

Notice that I attach two trailing zeros to the dividend. I do this because I need to move the decimal points in each number three places to the right. Again, place the decimal point in the quotient directly above where it now appears in the dividend, 21900:

$$15.\overline{)21900.}$$

Now you're ready to divide. Line up the quotient carefully so the decimal point falls into place:

```
       1460.
  15.)21900.
      15
      69
      60
       90
       90
        0
```

Even though the division comes out even after you write the digit 6 in the quotient, you still need to add a placeholding zero so that the decimal point appears in the correct place.

YOUR TURN

15 Divide these two decimals: $9.345 \div 0.05 = ?$

16 Solve the following division: $3.15 \div 0.021 = ?$

17 Perform the following decimal division, rounding to one decimal place: $6.7 \div 10.1$.

18 Find the solution, rounding to the nearest hundredth: $9.13 \div 4.25$.

Converting between Decimals and Fractions

Fractions (see Chapters 10 and 11) and decimals are similar, in that they both allow you to represent parts of the whole — that is, these numbers fall on the number line *between* whole numbers.

In practice, though, sometimes one of these options is more desirable than the other. For example, calculators love decimals but aren't so crazy about fractions. To use your calculator, you may have to change fractions into decimals.

As another example, some units of measurement (such as inches) use fractions, whereas others (such as meters) use decimals. To change units, you may need to convert between fractions and decimals.

In this section, I show you how to convert back and forth between fractions and decimals. (If you need a refresher on fractions, review Chapters 10 and 11 before proceeding.)

Simple Decimal-Fraction Conversions

Some decimals are so common that you can memorize how to represent them as fractions. Here's how to convert all the one-place decimals to fractions:

0.1	0.2	0.3	0.4	0.5	0.6	0.7	0.8	0.9
$\frac{1}{10}$	$\frac{1}{5}$	$\frac{3}{10}$	$\frac{2}{5}$	$\frac{1}{2}$	$\frac{3}{5}$	$\frac{7}{10}$	$\frac{4}{5}$	$\frac{9}{10}$

And, here are a few more common decimals that translate easily to fractions:

0.125	0.25	0.375	0.625	0.75	0.875
$\frac{1}{8}$	$\frac{1}{4}$	$\frac{3}{8}$	$\frac{5}{8}$	$\frac{3}{4}$	$\frac{7}{8}$

Q. Convert 13.7 to a mixed number.

EXAMPLE **A.** $13\frac{7}{10}$. The whole-number part of the decimal (13) becomes the whole-number part of the mixed number. Use the conversion chart to change the rest of the decimal (0.7) to a fraction.

Q. Change $9\frac{4}{5}$ to a decimal.

A. **9.8.** The whole-number part of the mixed number (9) becomes the whole-number part of the decimal. Use the conversion chart to change the fractional part of the mixed number $\left(\frac{4}{5}\right)$ to a decimal.

YOUR
TURN

19 Convert the following decimals into fractions:

(a) 0.7

(b) 0.4

(c) 0.25

(d) 0.125

(e) 0.1

(f) 0.75

20 Change these fractions to decimals:

(a) $\frac{9}{10}$

(b) $\frac{2}{5}$

(c) $\frac{3}{4}$

(d) $\frac{3}{8}$

(e) $\frac{7}{8}$

(f) $\frac{1}{2}$

21 Change these decimals to mixed numbers:

(a) 1.6

(b) 3.3

(c) 14.5

(d) 20.75

(e) 100.625

(f) 375.375

22 Change these mixed numbers to decimals:

(a) $1\frac{1}{5}$

(b) $2\frac{1}{10}$

(c) $3\frac{1}{2}$

(d) $5\frac{1}{4}$

(e) $7\frac{1}{8}$

(f) $12\frac{5}{8}$

Changing decimals to fractions

Converting a decimal to a fraction is pretty simple. The only tricky part comes in when you have to reduce the fraction or change it to a mixed number.

In this section, I first show you the easy case, when no further work is necessary. Then I show you the harder case, when you need to tweak the fraction. I also show you a great time-saving trick.

Doing a basic decimal-to-fraction conversion

Here's how to convert a decimal to a fraction:

1. **Draw a line (fraction bar) under the decimal and place a 1 underneath it.**

    Suppose you want to turn the decimal 0.3763 into a fraction. Draw a line under 0.3763 and place a 1 underneath it:

    $$\frac{0.3763}{1}$$

    This number looks like a fraction, but technically it isn't one because the top number (the numerator) is a decimal.

2. **Move the decimal point one place to the right and add a 0 after the 1.**

    $$= \frac{3.763}{10}$$

3. **Repeat Step 2 until the decimal point moves all the way to the right so you can drop the decimal point entirely.**

    In this case, this is a three-step process:

    $$\frac{37.63}{100} = \frac{376.3}{1000} = \frac{3763}{10000}$$

As you can see in the last step, the decimal point in the numerator moves all the way to the end of the number, so dropping the decimal point is okay.

Note: Moving a decimal point one place to the right is the same thing as multiplying a number by 10. When you move the decimal point four places in this problem, you're essentially multiplying the 0.3763 and the 1 by 10,000. Notice that the number of digits after the decimal point in the original decimal is equal to the number of 0s that end up following the 1.

A quick way to make a fraction out of a decimal is to use the name of the smallest decimal place in that decimal. For example,

>> In the decimal 0.3, the smallest decimal place is the tenths place, so the equivalent fraction is $\frac{3}{10}$.

>> In the decimal 0.29, the smallest decimal place is the hundredths place, so the equivalent fraction is $\frac{29}{100}$.

>> In the decimal 0.817, the smallest decimal place is the thousandths place, so the equivalent fraction is $\frac{817}{1,000}$.

In the following sections, I show you how to convert decimals to fractions when you have to work with mixed numbers and reduce the terms.

Getting mixed results

When you convert a decimal greater than 1 to a fraction, the result is a mixed number. Fortunately, this process is easy because the whole number part is unaffected by the conversion. So focusing only on the decimal part, follow the same steps I outline in the preceding section.

For example, suppose you want to change 4.51 to a fraction. The result will be a mixed number with a whole number part of 4. To find the fractional part, follow these steps:

1. **Draw a line (fraction bar) under the decimal and place a 1 underneath it.**

   $$\frac{0.51}{1}$$

2. **Move the decimal point one place to the right and add a 0 after the 1.**

   $$= \frac{5.1}{10}$$

3. **Repeat Step 2 until the decimal point moves all the way to the right so you can drop the decimal point entirely.**

   In this case, you have only one additional step:

   $$= \frac{51}{100}$$

So the mixed-number equivalent of 4.51 is $4\frac{51}{100}$.

Q. Change the decimal 0.83 to a fraction.

EXAMPLE **A.** $\frac{83}{100}$. Create a "fraction" with 0.83 in the numerator and 1.0 in the denominator:

$$\frac{0.83}{1.0}$$

Move the decimal point in 0.83 two places to the right to turn it into a whole number; move the decimal point in the denominator the same number of places. I do this one decimal place at a time:

$$\frac{0.83}{1.0} = \frac{8.3}{10.0} = \frac{83.0}{100.0}$$

At this point, you can drop the decimal points and trailing zeros in both the numerator and denominator.

Q. Change the decimal 0.0205 to a fraction.

A. $\frac{41}{2,000}$. Create a "fraction" with 0.0205 in the numerator and 1.0 in the denominator:

$$\frac{0.0205}{1.0}$$

Move the decimal point in the 0.0205 four places to the right to turn the numerator into a whole number; move the decimal point in the denominator the same number of places:

$$\frac{0.0205}{1.0} = \frac{0.205}{10.0} = \frac{02.05}{100.0} = \frac{020.5}{1,000.0} = \frac{0205.0}{10,000.0}$$

Drop the decimal points, plus any leading or trailing zeros in both the numerator and denominator.

$$= \frac{205}{10,000}$$

Both the numerator and denominator are divisible by 5, so reduce this fraction:

$$= \frac{41}{2,000}$$

YOUR TURN

23 Change the decimal 0.27 to a fraction.

24 Convert the decimal 0.0315 to a fraction.

25 Write 45.12 as a mixed number.

26 Change 100.001 to a mixed number.

Changing fractions to decimals

Converting fractions to decimals isn't difficult, but to do it, you need to know about decimal division. If you need to get up to speed on this, check out the section, "Dividing decimals," earlier in this chapter.

To convert a fraction to a decimal, follow these steps:

1. **Set up the fraction as a decimal division, dividing the numerator (top number) by the denominator (bottom number).**

2. **Attach enough trailing zeros to the numerator so that you can continue dividing until you find that the answer is either a *terminating decimal* or a *repeating decimal*.**

Don't worry, I explain terminating and repeating decimals next.

The last stop: Terminating decimals

Sometimes when you divide the numerator of a fraction by the denominator, the division eventually works out evenly. The result is a terminating decimal.

For example, suppose you want to change the fraction $\frac{2}{5}$ to a decimal. Here's your first step:

$$5\overline{)2}$$

One glance at this problem, and it looks like you're doomed from the start because 5 doesn't go into 2. But watch what happens when I add a few trailing zeros. Notice that I also place another decimal point in the answer just above the first decimal point. This step is important — you can read more about it in the section "Dividing decimals":

$$5\overline{)2.000}$$

Now you can divide because, although 5 doesn't go into 2, 5 does go into 20 four times:

$$
\begin{array}{r}
0.4 \\
5\overline{)2.000} \\
\underline{20} \\
0
\end{array}
$$

You're done! As it turns out, you needed only one trailing zero, so you can ignore the rest:

$$\frac{2}{5} = 0.4$$

Because the division worked out evenly, the answer is an example of a *terminating decimal*.

As another example, suppose you want to find out how to represent $\frac{7}{16}$ as a decimal. As earlier, I attach three trailing zeros:

$$
\begin{array}{r}
0.437 \\
16\overline{)7.000} \\
\underline{64} \\
60 \\
\underline{48} \\
120 \\
\underline{112} \\
8
\end{array}
$$

This time, three trailing zeros aren't enough to get my answer, so I attach a few more and continue:

$$
\begin{array}{r}
0.4375 \\
16\overline{)7.000000} \\
\underline{64} \\
60 \\
\underline{48} \\
120 \\
\underline{112} \\
80 \\
\underline{80} \\
0
\end{array}
$$

At last, the division works out evenly, so again the answer is a terminating decimal. Therefore, $\frac{7}{16} = 0.4375$.

The endless ride: Repeating decimals

Sometimes when you try to convert a fraction to a decimal, the division *never* works out evenly. The result is a *repeating decimal* — a decimal that cycles through the same number pattern forever.

You may recognize these pesky little critters from your calculator, when an apparently simple division problem produces a long string of numbers.

For example, to change $\frac{2}{3}$ to a decimal, begin by dividing 2 by 3. As in the last section, start by adding three trailing zeros, and see where it leads:

$$
\begin{array}{r}
0.666 \\
3\overline{)2.000} \\
\underline{18} \\
20 \\
\underline{18} \\
20 \\
\underline{18} \\
2
\end{array}
$$

At this point, you still haven't found an exact answer. But you may notice that a repeating pattern has developed in the division. No matter how many trailing zeros you attach to the number 2, the same pattern continues forever. This answer, 0.666, is an example of a repeating decimal. You can write $\frac{2}{3}$ as

$$\frac{2}{3} = 0.\overline{6}$$

The bar over the 6 means that, in this decimal, the number 6 repeats forever. You can represent many simple fractions as repeating decimals. In fact, *every* fraction can be represented either as a repeating decimal or as a terminating decimal — that is, as an ordinary decimal that ends.

Now suppose you want to find the decimal representation of $\frac{5}{11}$. Here's how this problem plays out:

```
      0.4545
11)5.0000
   44
   ──
    60
    55
    ──
     50
     44
     ──
      60
      55
      ──
       5
```

This time, the pattern repeats every other number — 4, then 5, then 4 again, and then 5 again, forever. Attaching more trailing zeros to the original decimal only strings out this pattern indefinitely. So you can write

$$\frac{5}{11} = 0.\overline{45}$$

This time, the bar is over both the 4 and the 5, telling you that these two numbers alternate forever.

Repeating decimals are an oddity, but they aren't hard to work with. In fact, as soon as you can show that a decimal division is repeating, you've found your answer. Just remember to place the bar only over the numbers that keep on repeating.

Some decimals never end and never repeat. You can't write them as fractions, so mathematicians have agreed on some shorter ways of naming them so that writing them out doesn't take, well, forever.

When you're asked to find the exact decimal value of a fraction, feel free to attach trailing zeros to the *dividend* (the number you're dividing) as you go along. Keep dividing until either the division works out evenly (so the quotient is a terminating decimal) or a repeating pattern develops (so it's a repeating decimal).

EXAMPLE

Q. Convert the fraction $\frac{9}{16}$ to an exact decimal value.

$$16\overline{)9.000}$$

A. **0.5625.** Divide $9 \div 16$. Because 16 is too big to go into 9, I attached a decimal point and some trailing zeros to the 9. Now you can divide as I show you earlier in this chapter:

$$
\begin{array}{r}
0.5625 \\
16\overline{)9.0000} \\
-80 \\
\hline
100 \\
-96 \\
\hline
40 \\
-32 \\
\hline
80 \\
-80 \\
\hline
0
\end{array}
$$

Q. What is the exact decimal value of the fraction $\frac{5}{6}$?

A. **.08$\overline{3}$.** Divide $5 \div 6$. Because 6 is too big to go into 5, I attached a decimal point and some trailing zeros to the 5. Now divide:

$$
\begin{array}{r}
.8333 \\
6\overline{)5.0000} \\
-48 \\
\hline
20 \\
-18 \\
\hline
20 \\
-18 \\
\hline
20 \\
-18 \\
\hline
2
\end{array}
$$

As you can see, a pattern has developed. No matter how many trailing zeros you attach, the quotient will never come out evenly. Instead, the quotient is the repeating decimal .08$\overline{3}$. The bar over the 3 indicates that the number 3 repeats forever: 0.83333333.

YOUR TURN

27 Change $\frac{13}{16}$ to an exact decimal value.

28 Express $\frac{7}{9}$ exactly as a decimal.

Practice Questions Answers and Explanations

1

(a) $2.7 = 2 + \dfrac{7}{10}$

(b) $31.4 = 30 + 1 + \dfrac{4}{10}$

(c) $86.52 = 80 + 6 + \dfrac{5}{10} + \dfrac{2}{100}$

(d) $103.759 = 100 + 3 + \dfrac{7}{10} + \dfrac{5}{100} + \dfrac{9}{1,000}$

(e) $1,040.0005 = 1,000 + 40 + \dfrac{5}{10,000}$

(f) $16,821.1384 = 10,000 + 6,000 + 800 + 20 + 1 + \dfrac{1}{10} + \dfrac{3}{100} + \dfrac{8}{1,000} + \dfrac{4}{10,000}$

2

(a) **5.8**

(b) **7.03**

(c) **90.04**

(d) **9,000.005**

(e) **108.006**

(f) **100.0102**

3

(a) **73.2**

(b) **904**

(c) **5,160,000**

(d) **0.002786**

(e) **0.000943812**

4

(a) **4.8**

(b) **52.3**

(c) **191.28**

(d) $99.9\underline{95} \rightarrow \mathbf{100.00}$

(e) $0.00\underline{791} \rightarrow \mathbf{0.008}$

(f) $909.9\underline{996} \rightarrow \mathbf{910.000}$

5 **19.58.** Place the numbers in a column as you would for addition with whole numbers, but with the decimal points lined up. I've filled out the columns with trailing zeros to help show how the columns line up:

$$
\begin{array}{r}
17.40 \\
+\ 2.18 \\
\hline
19.58
\end{array}
$$

(6) **43.1808.** Notice that the decimal point in the answer lines up with the others.

$0.0098 + 10.101 + 0.07 + 33 = 43.1808.$ Line up the decimal points and do column addition:

$$
\begin{array}{r}
\overset{1}{.0098} \\
10.1010 \\
.0700 \\
+\ 33.0000 \\
\hline
43.1808
\end{array}
$$

(7) **1,875.231.** Place the decimal numbers in a column, lining up the decimal points:

$$
\begin{array}{r}
1000.001 \\
75.000 \\
0.030 \\
+\ 800.200 \\
\hline
1875.231
\end{array}
$$

(8) **0.518.** Place the first number on top of the second number, with the decimal points lined up. I've also added a trailing zero to the second number to fill out the right-hand column and emphasize how the columns line up:

$$
\begin{array}{r}
0.748 \\
-\ 0.230 \\
\hline
0.518
\end{array}
$$

(9) **669.8999.** Place the first number on top of the second number, with the decimal points lined up. I've filled out the right-hand column with trailing zeros so I can complete the math:

$$
\begin{array}{r}
6\ \overset{6}{\cancel{7}}\ 4.\ \overset{1}{8}\ \overset{8}{\cancel{9}}\ \overset{9}{\cancel{9}}\ \overset{9}{\cancel{9}}\ \overset{1}{0} \\
-\,5.\ 0\ 0\ 0\ 1 \\
\hline
6\ 6\ 9.\ 8\ 9\ 9\ 9
\end{array}
$$

(10) **99.329.** Place the first number on top of the second number, with the decimal points lined up:

$$
\begin{array}{r}
\overset{0}{\cancel{1}}\ \overset{9}{\cancel{0}}\ \overset{9}{\cancel{0}}.\ \overset{9}{\cancel{0}}\ \overset{1}{0}\ 9 \\
-\,0.\ 6\ 8\ 0 \\
\hline
9\ 9.\ 3\ 2\ 9
\end{array}
$$

(11) **0.2667.** Place the first number on top of the second number, ignoring the decimal points. Complete the multiplication as you would for whole numbers:

$$
\begin{array}{r}
0.635 \\
\times\ 0.42 \\
\hline
1270 \\
+\ 25400 \\
\hline
0.26670
\end{array}
$$

At this point, you're ready to find out where the decimal point goes in the answer. Count the number of decimal places in the two factors, add these two numbers together ($3 + 2 = 5$), and place the decimal point in the answer so that it has five digits after the decimal point. After you place the decimal point (but not before!), you can drop the trailing zero.

(12) **23.49.** Ignore the decimal points and simply place the first number on top of the second. Complete the multiplication as you would for whole numbers:

$$
\begin{array}{r}
0.675 \\
\times\ 34.8 \\
\hline
5400 \\
27000 \\
+\ 202500 \\
\hline
23.4900
\end{array}
$$

Count the number of decimal places in the two factors, add these two numbers together ($3 + 1 = 4$), and place the decimal point in the answer so that it has four digits after the decimal point (as shown). Last, you can drop the trailing zeros.

(13) **1.1316.** Complete the multiplication as you would for whole numbers:

$$
\begin{array}{r}
943 \\
\times\ 0.0012 \\
\hline
1886 \\
+\ 9430 \\
\hline
1.1316
\end{array}
$$

Zero digits come after the decimal point in the first factor, and you have four after-decimal digits in the second factor, for a total of $4(0 + 4 = 4)$; place the decimal point in the answer so that it has four digits after the decimal point.

(14) **0.0811842.** Complete the multiplication as you would for whole numbers:

$$
\begin{array}{r}
1.006 \\
\times\ 0.0807 \\
\hline
7042 \\
+\ 804800 \\
\hline
0.0811842
\end{array}
$$

You have a total of seven digits after the decimal points in the two factors — three in the first factor and four in the second ($3 + 4 = 7$) — so place the decimal point in the answer so that it has seven digits after the decimal point. Notice that I need to create an extra decimal place in this case by attaching an additional non-trailing zero.

(15) **186.9.** To start out, write the problem as usual:

$$
0.05\overline{)9.345}
$$

Turn the divisor (0.05) into a whole number by moving the decimal point two places to the right. At the same time, move the decimal point in the dividend (9.345) two places to the right. Place the decimal point in the quotient directly above where it now appears in the dividend:

$$5.\overline{)934.5}$$

Now you're ready to divide. Be careful to line up the quotient properly so that the decimal point falls into place.

$$
\begin{array}{r}
186.9 \\
5.\overline{)934.5} \\
\underline{-5} \\
43 \\
\underline{-40} \\
34 \\
\underline{-30} \\
45 \\
\underline{-45} \\
0
\end{array}
$$

(16) **150.** Write the problem as usual:

$$0.021\overline{)3.15}$$

You need to move the decimal point in the divisor (0.021) three places to the right, so attach an additional trailing zero to the dividend (3.15) to extend it to three decimal places:

$$0.021\overline{)3.150}$$

Now you can move both decimal points three places to the right. Place the decimal point in the quotient above the decimal point in the dividend:

$$21.\overline{)3150.}$$

Divide, being careful to line up the quotient properly:

$$
\begin{array}{r}
150. \\
21.\overline{)3150.} \\
\underline{-21} \\
105 \\
\underline{-105} \\
0
\end{array}
$$

Remember to insert a placeholding zero in the quotient so that the decimal point ends up in the correct place.

(17) **0.7.** To start out, write the problem as usual:

$$10.1\overline{)6.7}$$

Turn the divisor (10.1) into a whole number by moving the decimal point one place to the right. At the same time, move the decimal point in the dividend (6.7) one place to the right:

$$101.\overline{)67.}$$

The problem asks you to round the quotient to one decimal place, so fill out the dividend with trailing zeros to two decimal places:

$$101.\overline{)67.00}$$

Now you're ready to divide:

$$
\begin{array}{r}
0.66 \\
101.\overline{)67.00} \\
\underline{-606} \\
640 \\
\underline{-606} \\
34
\end{array}
$$

Round the quotient to one decimal place:

$$0.\underline{66} \rightarrow 0.7$$

(18) **2.15.** First, write the problem as usual:

$$4.25\overline{)9.13}$$

Turn the divisor (4.25) into a whole number by moving the decimal point two places to the right. At the same time, move the decimal point in the dividend (9.13) two places to the right:

$$425\overline{)913.}$$

The problem asks you to round the quotient to the nearest hundredth, so fill out the dividend with trailing zeros to three decimal places:

$$425.\overline{)913.000}$$

Now divide, carefully lining up the quotient:

$$\begin{array}{r} 2.148 \\ 425.\overline{)913.000} \\ -850 \\ \hline 630 \\ -425 \\ \hline 2050 \\ -1700 \\ \hline 3500 \\ -3400 \\ \hline 100 \end{array}$$

Round the quotient to the nearest hundredth:

$$2.14\underline{8} \rightarrow 2.15$$

(19)

(a) $0.7 = \dfrac{7}{10}$

(b) $0.4 = \dfrac{2}{5}$

(c) $0.25 = \dfrac{1}{4}$

(d) $0.125 = \dfrac{1}{8}$

(e) $0.1 = \dfrac{1}{10}$

(f) $0.75 = \dfrac{3}{4}$

(20)

(a) $\dfrac{9}{10} = \mathbf{0.9}$

(b) $\dfrac{2}{5} = \mathbf{0.4}$

(c) $\dfrac{3}{4} = \mathbf{0.75}$

(d) $\dfrac{3}{8} = \mathbf{0.375}$

(e) $\dfrac{7}{8} = \mathbf{0.875}$

(f) $\dfrac{1}{2} = \mathbf{0.5}$

(21)

(a) $1.6 = 1\dfrac{3}{5}$

(b) $3.3 = 3\dfrac{3}{10}$

(c) $14.5 = 14\dfrac{1}{2}$

(d) $20.75 = 20\dfrac{3}{4}$

(e) $100.625 = \mathbf{100\dfrac{5}{8}}$

(f) $375.375 = \mathbf{375\dfrac{3}{8}}$

(22)

(a) $1\frac{1}{5} = 1.2$

(b) $2\frac{1}{10} = 2.1$

(c) $3\frac{1}{2} = 3.5$

(d) $5\frac{1}{4} = 5.25$

(e) $7\frac{1}{8} = 7.125$

(f) $12\frac{5}{8} = 12.625$

(23) $\frac{27}{100}$. Create a "fraction" with 0.27 in the numerator and 1.0 in the denominator. Then move the decimal points to the right until both the numerator and denominator are whole numbers:

$$\frac{0.27}{1.0} = \frac{2.7}{10.0} = \frac{27.0}{100.0}$$

At this point, you can drop the decimal points and trailing zeros.

(24) $\frac{63}{2,000}$. Create a "fraction" with 0.0315 in the numerator and 1.0 in the denominator. Then move the decimal points in both the numerator and denominator to the right one place at a time. Continue until both the numerator and denominator are whole numbers:

$$\frac{0.0315}{1.0} = \frac{0.315}{10.0} = \frac{3.15}{100.0} = \frac{31.5}{1,000.0} = \frac{315.0}{10,000.0}$$

Drop the decimal points and trailing zeros. The numerator and denominator are both divisible by 5, so reduce the fraction:

$$\frac{315}{10,000} = \frac{63}{2,000}$$

(25) $45\frac{3}{25}$. Before you begin, separate out the whole-number portion of the decimal (45). Create a "fraction" with 0.12 in the numerator and 1.0 in the denominator. Move the decimal points in both the numerator and denominator to the right until both are whole numbers:

$$\frac{0.12}{1.0} = \frac{1.2}{10.0} = \frac{12.0}{100.0}$$

Drop the decimal points and trailing zeros. As long as the numerator and denominator are both divisible by 2 (that is, even numbers), you can reduce this fraction:

$$\frac{12}{100} = \frac{6}{50} = \frac{3}{25}$$

To finish up, reattach the whole-number portion that you separated at the beginning.

(26) $100\frac{1}{1,000}$. Separate out the whole-number portion of the decimal (100) and create a "fraction" with 0.001 in the numerator and 1.0 in the denominator. Move the decimal points in both the numerator and denominator to the right one place at a time until both are whole numbers:

$$\frac{0.001}{1.0} = \frac{0.01}{10.0} = \frac{0.1}{100.0} = \frac{1.0}{1,000.0}$$

Drop the decimal points and trailing zeros and reattach the whole-number portion of the number you started with:

$$100\frac{1}{1,000}$$

(27) **0.8125.** Divide $13 \div 16$, attaching plenty of trailing zeros to the 13:

```
        0.8125
   16)13.00000
      −128
        20
       −16
        40
       −32
        80
       −80
         0
```

This division eventually ends, so the quotient is a terminating decimal.

(28) $\frac{7}{9} = 0.\overline{7}$.

Divide $7 \div 9$, attaching plenty of trailing zeros to the 7:

```
        0.77
    9)7.000
     −63
       70
      −63
       70
```

A pattern has developed in the subtraction: $70 - 63 = 7$, so when you bring down the next 0, you'll get 70 again. Therefore, the quotient is a repeating decimal.

If you're ready to test your skills a bit more, take the following chapter quiz that incorporates all the chapter topics.

Whaddya Know? Chapter 13 Quiz

Quiz time! Complete each problem to test your knowledge on the various topics covered in this chapter. You can then find the solutions and explanations in the next section.

1. Subtract: $4.213 - 0.06$

2. Expand the number 6.41 using the appropriate fractions.

3. Multiply by moving the decimal point: 43.7×100

4. Round 83.4172 to the nearest tenth.

5. Add: $137.27 + 19.5$

6. Divide: $36.162 \div 6.3$

7. Convert to a fraction in lowest terms: 0.45

8. Subtract: $167.5 - 41.317$

9. Change to a decimal number: $\dfrac{9}{1000}$

10. Multiply: 18.6×2.45

11. Simplify the number 04080.06070 by removing all leading and trailing zeros.

12. Divide: $0.5026 \div 0.00007$

13. Change to a decimal number: $\dfrac{7}{22}$

14. Add: $16.1 + 610 + 4.036$

15. Multiply: 140×0.000005

16. Expand the number 201.0045 using the appropriate fractions.

17. Change to a decimal number: $\dfrac{27}{32}$

18. Divide by moving the decimal point: $67.4 \div 10,000$

19. Convert to a mixed number in lowest terms: 3.0005

20. Round 6.713895 to the nearest thousandth.

Answers to Chapter 13 Quiz

(1) **4.153.** First, arrange the two numbers in a column, lining up the decimal points. Insert any missing zeros if needed. Then subtract.

$$
\begin{array}{r}
4\;.\;2\;\;1\;\;3 \\
-\;\;0\;.\;0\;\;6\;\;0 \\
\hline
4\;.\;1\;\;5\;\;3
\end{array}
$$

(2) $6 + \dfrac{4}{10} + \dfrac{1}{100}$. Write the digits to the right of the decimal point over powers of 10.

$$6.41 = 6 + \frac{4}{10} + \frac{1}{100}$$

(3) **4,370.** Move the decimal point two places to the right after adding a 0.

$$43.7 \times 100 = 4370_{\wedge}$$

(4) **83.4.** The tenths place is the first digit to the right of the decimal point. Since the digit following the 4 is less than 5, just drop off the remaining digits.

$$83.\underline{4}172 \rightarrow 83.\underline{4}\cancel{172} \rightarrow 83.4$$

(5) **156.77.** First arrange the two numbers in a column, lining up the decimal points. Insert any missing zeros if needed. Then add.

$$
\begin{array}{r}
1\;\;3\;\;7\;.\;2\;\;7 \\
+\;\;\;\;\;1\;\;9\;.\;5\;\;0 \\
\hline
1\;\;5\;\;6\;.\;7\;\;7
\end{array}
$$

(6) **5.74.** Write the division problem. Then turn the divisor into a whole number and move the decimal point in the dividend the same number of places.

$$6.3_{\wedge}\overline{)36.1_{\wedge}62}$$

Place the decimal point in the quotient and perform the division.

$$
\begin{array}{r}
5.74 \\
63\overline{)361.62} \\
-315 \\
\hline
466 \\
-441 \\
\hline
252 \\
-252 \\
\hline
\end{array}
$$

(7) $\frac{9}{20}$. Write the two digits over the number 100 and reduce.

$$\frac{45}{100} = \frac{9}{20}$$

(8) **126.183.** First, arrange the two numbers in a column, lining up the decimal points. Insert any missing zeros if needed. Then subtract.

```
    1 6 7 . 5 0 0
 −      4 1 . 3 1 7
    1 2 6 . 1 8 3
```

(9) **0.009.** Place a decimal point and then two zeros to put the 9 in the thousandths place.

$$\frac{9}{1000} = 0.009$$

(10) **45.57.** Place the numbers on top of one another. Perform the multiplication on the digits.

```
        1 8 6
 ×      2 4 5
        9 3 0
      7 4 4
    3 7 2
    4 5 5 7 0
```

Counting the number of digits to the right of the decimal points in the two multipliers, 18.6×2.45, you see a total of 3 digits. Place the decimal point in the answer so there are 3 digits to the right of the decimal point. You can simplify by omitting the trailing zero.

```
        1 8 6
 ×      2 4 5
        9 3 0
      7 4 4
    3 7 2
    4 5 . 5 7 0
```

(11) **4,080.0607.** Removing the leading and trailing zeros:

$$04080.06070 = 4{,}080.0607$$

(12) **7,180.** Write the division problem. Then turn the divisor into a whole number and move the decimal point in the dividend the same number of places; you'll need to add a zero.

$$0.00007_\wedge \overline{)0.50260_\wedge}$$

Place the decimal point in the quotient and perform the division.

$$
\begin{array}{r}
7180. \\
7\overline{)50260.} \\
-49 \\
\hline
12 \\
-7 \\
\hline
56 \\
-56 \\
\hline
00 \\
-0 \\
\hline
\end{array}
$$

(13) **0.3$\overline{18}$.** Divide the numerator by the denominator.

$$
\begin{array}{r}
0.31818 \\
22\overline{)7.00000} \\
-66 \\
\hline
40 \\
-22 \\
\hline
180 \\
-176 \\
\hline
40 \\
-22 \\
\hline
180 \\
-176 \\
\hline
\end{array}
$$

The digits 18 keep repeating. Indicate this with a bar across the top.

(14) **630.136.** First, arrange the numbers in a column, lining up the decimal points. Insert any missing zeros if needed.

$$
\begin{array}{r}
1\ 6\ .\ 1\ 0\ 0 \\
6\ 1\ 0\ .\ 0\ 0\ 0 \\
+4\ .\ 0\ 3\ 6 \\
\hline
6\ 3\ 0\ .\ 1\ 3\ 6 \\
\end{array}
$$

(15) **0.0007.** Place the numbers on top of one another. Perform the multiplication on the digits.

$$
\begin{array}{r}
1\ 4\ 0 \\
\times5 \\
\hline
7\ 0\ 0 \\
\end{array}
$$

Counting the number of digits to the right of the decimal points in the two multipliers, 140×0.000005, you see a total of 6 digits — all in the second number. Place the decimal point in the answer so there are 6 to the right of the decimal point. You will have to add three zeros in front of the 7. And, to finish, you can eliminate the trailing zeros.

$$
\begin{array}{r}
1\ \ 4\ \ 0 \\
\times \qquad 5 \\
\hline
.0\ \ 0\ \ 0\ \ 7\ \ 0\ \ 0
\end{array}
$$

(16) $201 + \dfrac{4}{1000} + \dfrac{5}{10,000}$. Write the non-zero digits to the right of the decimal point over powers of 10.

$$201.0045 = 201 + \frac{4}{1000} + \frac{5}{10,000}$$

(17) **0.84375.** Divide the numerator by the denominator.

$$
\begin{array}{r}
0.84375 \\
32\overline{)27.00000} \\
\underline{-256} \\
140 \\
\underline{-128} \\
120 \\
\underline{-96} \\
240 \\
\underline{-224} \\
160 \\
\underline{-160}
\end{array}
$$

(18) **0.00674.** The number 10,000 has 4 zeros. You move the decimal point in 67.4 four places to the left. You need to add 2 zeros in front of the 6.

$$67.4 \div 10,000 \rightarrow_{\wedge} 0067.4$$
$$67.4 \div 10,000 = 0.00674$$

(19) $3\dfrac{1}{2000}$. Write the number as 3 plus the fraction 5 over 10,000 and reduce the fraction.

$$3.0005 = 3 + \frac{5}{10,000} = 3 + \frac{1}{2,000} = 3\frac{1}{2000}$$

(20) **6.714.** The thousandths place is three digits to the right of the decimal point. Since the digit following the 3 is greater than 5, round up the 3 to a 4 and drop the rest of the digits.

$$6.713895 \rightarrow 6.714895 \rightarrow 6.714$$

Chapter **14**

Playing the Percentages

L ike fractions and decimals, percentages are a way to talk about parts of a whole. The word *percent* means "out of 100." So if you have 50% of something, you have 50 out of 100. If you have 25% of it, you have 25 out of 100. Of course, if you have 100% of anything, you have all of it.

In this chapter, I show you how to work with percentages. Because percentages resemble decimals, I first show you how to convert numbers back and forth between percentages and decimals. No worries — this switch is easy to do. Next, I show you how to convert back and forth between percentages and fractions — also not too bad. When you understand how conversions work, I show you the three basic types of percent problems, plus a method that makes the problems simple.

Making Sense of Percentages

The word *percent* literally means "for 100," but in practice, it means closer to "out of 100." For example, suppose that a school has exactly 100 children — 50 girls and 50 boys. You can say that "50 out of 100" children are girls — or you can shorten it to simply "50 percent." Even shorter than that, you can use the symbol %, which means *percent*.

Saying that 50% of the students are girls is the same as saying that $\frac{1}{2}$ of them are girls.

Or if you prefer decimals, it's the same thing as saying that 0.5 of all the students are girls. This example shows you that percentages, like fractions and decimals, are just another way of talking about parts of the whole. In this case, the whole is the total number of children in the school.

You don't literally have to have 100 of something to use a percent. You probably won't ever really cut a cake into 100 pieces, but that doesn't matter. The values are the same. Whether you're talking about cake, a dollar, or a group of children, 50% is still half, 25% is still one-quarter, 75% is still three-quarters, and so on.

Any percentage smaller than 100% means less than the whole — the smaller the percentage, the less you have. You probably know this fact well from the school grading system. If you get 100%, you get a perfect score. And 90% is usually A work, 80% is a B, 70% is a C, and, well, you know the rest.

Of course, 0% means "0 out of 100" — any way you slice it, you have nothing.

Dealing with Percentages Greater than 100%

The term 100% means "100 out of 100" — in other words, everything. So when I say I have 100% confidence in you, I mean that I have complete confidence in you.

What about percentages more than 100%? Well, sometimes percentages like these don't make sense. For example, you can't spend more than 100% of your time playing basketball, no matter how much you love the sport; 100% is all the time you have, and there ain't no more.

But a lot of times, percentages larger than 100% are perfectly reasonable. For example, suppose I own a hot dog wagon and sell the following:

10 hot dogs in the morning

30 hot dogs in the afternoon

The number of hot dogs I sell in the afternoon is 300% of the number I sold in the morning. It's three times as many.

Here's another way of looking at this: I sell 20 more hot dogs in the afternoon than in the morning, so this is a *200% increase* in the afternoon — 20 is twice as many as 10.

Spend a little time thinking about this example until it makes sense. You visit some of these ideas again in Chapter 15, when I show you how to do word problems involving percentages.

Converting to and from Percentages, Decimals, and Fractions

To solve many percentage problems, you need to change the percent to either a decimal or a fraction. Then you can apply what you know about solving decimal and fraction problems. For this reason, I show you how to convert to and from percentages before I show you how to solve percent problems.

Percentages and decimals are similar ways of expressing parts of a whole. This similarity makes converting percentages to decimals, and vice versa, mostly a matter of moving the decimal point. It's so simple that you can probably do it in your sleep (but you should probably stay awake when you first read about the concept).

Percentages and fractions both express the same idea — parts of a whole — in different ways. So converting back and forth between percentages and fractions isn't quite as simple as just moving the decimal point back and forth. In this section, I cover the ways to convert to and from percentages, decimals, and fractions, starting with percentages to decimals.

Converting Percentages to Decimals

Percentages and decimals are very similar forms, so everything you know about decimals (see Chapter 13) carries over when you're working with percentages. All you need to do is convert your percent to a decimal, and you're good to go.

To change a whole-number percent to a decimal, simply replace the percent sign with a decimal point and then move this decimal point two places to the left; after this, you can drop any trailing zeros. Here are a few common conversions between percentages and decimals:

$$100\% = 1 \qquad 75\% = 0.75 \qquad 50\% = 0.5 \qquad 25\% = 0.25 \qquad 20\% = 0.2 \qquad 10\% = 0.1$$

Sometimes a percent already has a decimal point. In this case, just drop the percent sign and move the decimal point two places to the left. For instance, $12.5\% = 0.125$.

EXAMPLE

Q. Change 80% to a decimal.

A. **0.8.** Replace the percent sign with a decimal point — changing 80% to 80.0 — and then move the decimal point two places to the left:

$$80\% = 0.80$$

At the end, you can drop the trailing zero to get 0.8.

Q. Change 37.5% to a decimal.

A. **0.375.** Drop the percent sign and move the decimal point two places to the left:

$$37.5\% = 0.375$$

1. Change 90% to a decimal.

2. A common interest rate on an investment such as a bond is 4%. Convert 4% to a decimal.

3. Find the decimal equivalent of 99.44%.

4. What is 243.1% expressed as a decimal?

Changing Decimals to Percentages

Calculating with percentages is often easier when you convert to decimals first. When you're done calculating, however, you frequently need to change your answer from a decimal back to a percent. This is especially true when you're working with interest rates, taxes, or the likelihood of a big snowfall the night before a big test. All these numbers are most commonly expressed as percentages.

To change a decimal to a percent, move the decimal point two places to the right and attach a percent sign. If the result is a whole number, you can drop the decimal point.

Q. Change 0.6 to a percent.

EXAMPLE **A.** **60%.** Move the decimal point two places to the right and attach a percent sign:

$0.6 = 60\%$

YOUR
TURN

5. Convert 0.57 to a percent.

6. What is 0.3 expressed as a percent?

7. Change 0.015 to a percent.

8. Express 2.222 as a percent.

Switching from Percentages to Fractions

Some percentages are easy to convert to fractions. Here are a few quick conversions that are worth knowing:

$$1\% = \frac{1}{100} \qquad 5\% = \frac{1}{20} \qquad 10\% = \frac{1}{10} \qquad 20\% = \frac{1}{5}$$

$$25\% = \frac{1}{4} \qquad 50\% = \frac{1}{2} \qquad 75\% = \frac{3}{4} \qquad 100\% = 1$$

Beyond these simple conversions, changing a percent to a fraction isn't a skill you're likely to use much outside of a math class. Decimals are much easier to work with.

However, teachers often test you on this skill to make sure you understand the ins and outs of percentages, so here's the scoop on converting percentages to fractions: To change a percent to a fraction, use the percent without the percent sign as the *numerator* (top number) of the fraction and use 100 as the *denominator* (bottom number). When necessary, reduce this fraction to lowest terms or change it to a mixed number. (For a refresher on reducing fractions, see Chapter 10.)

Q. Change 35% to a fraction.

A. $\frac{7}{20}$. Place 35 in the numerator and 100 in the denominator:

$$35\% = \frac{35}{100}$$

You can reduce this fraction because the numerator and denominator are both divisible by 5:

$$\frac{7}{20}$$

YOUR TURN

9   Change 19% to a fraction.

10   A common interest rate on credit cards and other types of loans is 8%. What is 8% expressed as a fraction?

11   Switch 123% to a fraction.

12   Convert 375% to a fraction.

Converting Fractions to Percentages

Knowing how to make a few simple conversions from fractions to percentages is a useful real-world skill. Here are some of the most common conversions:

$$\frac{1}{100} = 1\% \qquad \frac{1}{20} = 5\% \qquad \frac{1}{10} = 10\% \qquad \frac{1}{5} = 20\%$$

$$\frac{1}{4} = 25\% \qquad \frac{1}{2} = 50\% \qquad \frac{3}{4} = 75\% \qquad 1 = 100\%$$

Beyond these, you're not all that likely to need to convert a fraction to a percent outside of a math class. But then, passing your math class is important, so in this section I show you how to make this type of conversion.

Converting a fraction to a percent is a two-step process:

1. **Convert the fraction to a decimal, as I show you in Chapter 13.**

    In some problems, the result of this step may be a repeating decimal. This is fine — in this case, the percent will also contain a repeating decimal.

2. **Convert this decimal to a percent.**

    Move the decimal point two places to the right and add a percent sign.

Q. Change the fraction $\frac{1}{9}$ to a percent.

A. $11.\overline{1}\%$. First, change $\frac{1}{9}$ to a decimal:

$$
\begin{array}{r}
0.111 \\
9\overline{)1.000} \\
\underline{9} \\
10 \\
\underline{9} \\
10 \\
\underline{9} \\
1
\end{array}
$$

The result is the repeating decimal $0.\overline{1}$. Now change this repeating decimal to a percent:

$$0.\overline{1} = 11.\overline{1}\%$$

13 Express $\frac{2}{5}$ as a percent.

14 Change $\frac{3}{20}$ to a percent.

15 Convert $\frac{7}{8}$ to a percent.

16 Change $\frac{2}{11}$ to a percent.

Solving Percentage Problems

When you know the connection between percentages and fractions, which I discuss earlier in the section "Converting to and from Percentages, Decimals, and Fractions," you can solve a lot of percent problems with a few simple tricks. Other problems, however, require a bit more work. In this section, I show you how to tell an easy percent problem from a tough one, and I give you the tools to solve all of them.

Figuring out simple percent problems

TIP

A lot of percent problems turn out to be easy when you give them a little thought. In many cases, just remember the connection between percentages and fractions, and you're halfway home.

>> **Finding 100% of a number:** Remember that 100% means the whole thing, so 100% of any number is simply the number itself:

> 100% of 5 is 5.
> 100% of 91 is 91.
> 100% of 732 is 732.

- **Finding 50% of a number:** Remember that 50% means half, so to find 50% of a number, just divide it by 2:

> 50% of 20 is 10.
> 50% of 88 is 44.
> 50% of 7 is $3\frac{1}{2}$ or 3.5.

- **Finding 25% of a number:** Remember that 25% equals $\frac{1}{4}$, so to find 25% of a number, divide it by 4:

> $25\% \text{ of } 40 = 10$
> $25\% \text{ of } 88 = 22$
> $25\% \text{ of } 15 = \frac{15}{4} = 3\frac{3}{4} = 3.75$

>> **Finding 20% of a number:** Finding 20% of a number is handy if you like the service you've received in a restaurant, because a good tip is 20% of the check. Because 20% equals $\frac{1}{5}$, you can find 20% of a number by dividing it by 5. But I can show you an easier way: Remember that 20% is 2 times 10%, so to find 20% of a number, move the decimal point one place to the left and double the result:

> $20\% \text{ of } 80 = 8 \times 2 = 16$
> $20\% \text{ of } 300 = 30 \times 2 = 60$
> $20\% \text{ of } 41 = 4.1 \times 2 = 8.2$

>> **Finding 10% of a number:** Finding 10% of any number is the same as finding $\frac{1}{10}$ of that number. To do this, just move the decimal point one place to the left:

> $10\% \text{ of } 30 = 3$
> $10\% \text{ of } 41 = 4.1$
> $10\% \text{ of } 7 = 0.7$

>> **Finding 200%, 300%, and so on of a number:** Working with percentages that are multiples of 100 is easy. Just drop the two 0s and multiply by the number that's left:

$$200\% \text{ of } 7 = 2 \times 7 = 14$$
$$300\% \text{ of } 10 = 3 \times 10 = 30$$
$$1{,}000\% \text{ of } 45 = 10 \times 45 = 450$$

(See the earlier section, "Dealing with Percentages Greater than 100%," for details on what having more than 100% really means.)

EXAMPLE

Q. What is 10% of 64?

A. **6.4.** Move the decimal point one place to the left: 10% of 64 = 6.4.

Q. What is 20% of 52?

A. **10.4.** Move the decimal point one place to the left and double the result: 20% of 52 = $5.2 \times 2 = 10.4$.

Q. What is 400% of 11?

A. **44.** Multiply the number by 4: 400% of 11 = $11 \times 4 = 44$.

YOUR TURN

17 What is 10% of 83?

18 What is 20% of 75?

19 What is 25% of 84?

20 What is 50% of 172?

21 What is 100% of 9.9?

22 What is 500% of 36?

Turning the problem around

Here's a trick that makes certain tough-looking percent problems so easy that you can do them in your head. Simply move the percent sign from one number to the other and flip the order of the numbers.

Suppose someone (like a teacher) wants you to figure out the following:

88% of 50

Finding 88% of anything isn't an activity anybody looks forward to. But an easy way of solving the problem is to switch it around:

88% of 50 = 50% of 88

This move is perfectly valid, and it makes the problem a lot easier. It works because the word *of* really means multiplication, and you can multiply either backward or forward and get the same answer. As I discuss in the preceding section, "Figuring out simple percent problems," 50% of 88 is simply half of 88:

88% of 50 = 50% of 88 = 44

As another example, suppose you want to find

7% of 200

Again, finding 7% is tricky, but finding 200% is simple, so switch the problem around:

7% of 200 = 200% of 7

In the preceding section, I tell you that to find 200% of any number, you just multiply that number by 2:

$7\% \text{ of } 200 = 200\% \text{ of } 7 = 2 \times 7 = 14$

Q. What is 18% of 10?

A. 1.8. Reverse the two numbers in the problem and solve: 18% of 10 = 10% of 18 = 1.8.

23 What is 48% of 25?

24 What is 61% of 300?

25 What is 98% of 50?

26 What is 132% of 20?

Deciphering more-difficult percent problems

You can solve a lot of percent problems using the tricks I show you earlier in this chapter. For more difficult problems, you may want to switch to a calculator. If you don't have a calculator at hand, solve percent problems by turning them into decimal multiplication, as follows:

1. **Change the word of to a multiplication sign and the percent to a decimal (as I show you earlier in this chapter).**

    Suppose you want to find 35% of 80. Here's how you start:

    $35\% \text{ of } 80 = 0.35 \times 80$

2. **Solve the problem using decimal multiplication (see Chapter 13).**

Here's what the example looks like:

$$
\begin{array}{r}
0.35 \\
\times\ \ 80 \\
\hline
28.00
\end{array}
$$

So 35% of 80 is 28.

Q. What is 16% of 30?

EXAMPLE **A.** 4.8. Change the word *of* to a multiplication sign and the percent to a decimal, then solve using decimal multiplication: 16% of $30 = 0.16 \times 30 = 4.8$.

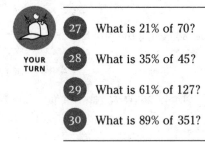

YOUR
TURN

27 What is 21% of 70?

28 What is 35% of 45?

29 What is 61% of 127?

30 What is 89% of 351?

Putting All the Percent Problems Together

In the preceding section, "Solving Percentage Problems," I give you a few ways to find any percent of any number. This type of percent problem is the most common, which is why it gets top billing.

But percentages crop up in a wide range of business applications, such as banking, real estate, payroll, and taxes. (I show you some real-world applications when I discuss word problems in Chapter 15.) And depending on the situation, two other common types of percent problems may present themselves.

In this section, I show you these two additional types of percent problems and how they relate to the type you now know how to solve. I also give you a simple tool to make quick work of all three types.

Identifying the three types of percent problems

Earlier in this chapter, I show you how to solve problems that look like this:

50% of 2 is ?

The answer, of course, is 1. (See the section, "Solving Percentage Problems," for details on how to get this answer.) Given two pieces of information — the percent and the number to start with — you can figure out what number you end up with.

Now suppose instead that I leave out the percent but give you the starting and ending numbers:

? % of 2 is 1

You can still fill in the blank without too much trouble. Similarly, suppose that I leave out the starting number but give the percent and the ending number:

50% of ? is 1

Again, you can fill in the blank.

If you get this basic idea, you're ready to solve percent problems. When you boil them down, nearly all percent problems are like one of the three types I show in Table 14-1.

In each case, the problem gives you two of the three pieces of information, and your job is to figure out the remaining piece. In the next section, I give you a simple tool to help you solve all three of these types of percent problems.

TABLE 14-1 **The Three Main Types of Percent Problems**

Problem Type	What to Find	Example
Type #1	The ending number	50% of 2 is *what*?
Type #2	The percentage	*What* percent of 2 is 1?
Type #3	The starting number	50% of *what* is 1?

Solving Percent Problems with Equations

Here's how to solve any percent problem:

REMEMBER 1. **Change the word *of* to a multiplication sign and the percent to a decimal (as I show you earlier in this chapter).**

This step is the same as for more straightforward percent problems. For example, consider this problem:

60% of what is 75?

Begin by changing the problem as follows:

60%	of	what	is	75
0.6	×			75

2. **Turn the word *is* to an equals sign and the word *what* into the letter n.**

   Here's what this step looks like:

60%	of	what	is	75
0.6	×	n	=	75

   This equation looks more normal, as follows:

   $$0.6 \times n = 75$$

3. **Find the value of n.**

   Technically, the last step involves a little bit of algebra, but I know you can handle it. (For a complete explanation of algebra, see Unit 7 of this book.) In the equation, n is being multiplied by 0.6. You want to "undo" this operation by *dividing* by 0.6 on both sides of the equation:

   $$0.6 \times n \div 0.6 = 75 \div 0.6$$

   Almost magically, the left side of the equation becomes a lot easier to work with because multiplication and division by the same number cancel each other out:

   $$n = 75 \div 0.6$$

   Remember that n is the answer to the problem. If your teacher lets you use a calculator, this last step is easy; if not, you can calculate it using some decimal division, as I show you in Chapter 13:

   $$n = 125$$

   Either way, the answer is 125 — so 60% of 125 is 75.

As another example, suppose you're faced with this percent problem:

   What percent of 250 is 375?

To begin, change the *of* into a multiplication sign and the percent into a decimal.

What	percent	of	250	is	375
	× 0.01	×	250		375

Notice here that, because I don't know the percent, I change the word *percent* to × 0.01. Next, change *is* to an equals sign and *what* to the letter n:

What	percent	of	250	is	375
n	× 0.01	×	250	=	375

Consolidate the equation and then multiply:

$$n \times 2.5 = 375$$

Now divide both sides by 2.5:

$$n = 375 \div 2.5 = 150$$

Therefore, the answer is 150 — so 150% of 250 is 375.

Here's one more problem: 49 is what percent of 140? Begin, as always, by translating the problem into words:

49	is	what	percent	of	140
49	=	n	$\times 0.01$	$\times$	140

Simplify the equation:

$$49 = n \times 1.4$$

Now divide both sides by 1.4:

$$49 \div 1.4 = n \times 1.4 \div 1.4$$

Again, multiplication and division by the same number allows you to cancel on the right side of the equation and complete the problem:

$$49 \div 1.4 = n$$
$$35 = n$$

Therefore, the answer is 35, so 49 is 35% of 140.

In summary, here are the three pieces — and the kinds of questions that ask for each piece.

>> **The percent:** The problem may give you the starting and ending numbers and ask you to find the percent. Here are some ways this problem can be asked:

?% of 4 is 1.
What percent of 4 is 1?
1 is what percent of 4?

The answer is 25%, because $25\% \times 4 = 1$.

>> **The starting number:** The problem may give you the percent and the ending number and ask you to find the starting number:

10% of ? is 40.
10% of what number is 40?
40 is 10% of what number?

This time, the answer is 400, because $10\% \times 400 = 40$.

>> **The ending number:** The most common type of percent problem gives you a percentage and a starting number and asks you to figure out the ending number:

> 50% of 6 is ?
>
> 50% of 6 equals what number?
>
> Can you find 50% of 6?

No matter how I phrase it, notice that the problem always includes 50% of 6. The answer is 3, because $50\% \times 6 = 3$.

Each type of percent problem gives you *two* pieces of information and asks you to find the third. Place the information into an equation using the following translations from words into symbols:

$$\text{What (number)} \to n$$
$$\text{is} \to =$$
$$\text{percent} \to \times 0.01$$
$$\text{of} \to \times$$

Q. Place the statement *25% of 12 is 3* into an equation.

A. $25 \times 0.01 \times 12 = 3.$

This is a direct translation, as follows:

25	%	of	12	is	3
25	× 0.01	×	12	=	3

Q. What is 18% of 90?

A. 16.2.

Translate the problem into an equation:

What	is	18	%	of	90
n	=	18	× 0.01	×	90

Solve this equation:

$$n = 18 \times 0.01 \times 90 = 16.2$$

31 Place the statement *20% of 350 is 70* into an equation. Check your work by simplifying the equation.

32 What percent of 150 is 25.5?

33 What is 79% of 11?

34 30% of what number is 10?

Practice Questions Answers and Explanations

 1 **0.9.** Replace the percent sign with a decimal point and then move this decimal point two places to the left:

$$90\% = 0.90$$

At the end, drop the trailing zero to get 0.9.

2 **0.04.** Replace the percent sign with a decimal point and then move this decimal point two places to the left:

$$4\% = 0.04$$

3 **0.9944.** Drop the percent sign and move the decimal point two places to the left:

$$99.44\% = 0.9944$$

4 **2.431.** Drop the percent sign and move the decimal point two places to the left:

$$243.1\% = 2.431$$

5 **57%.** Move the decimal point two places to the right and attach a percent sign:

$$0.57 = 057\%$$

At the end, drop the leading zero to get 57%.

6 **30%.** Move the decimal point two places to the right and attach a percent sign:

$$0.3 = 030\%$$

At the end, drop the leading zero to get 30%.

7 **1.5%.** Move the decimal point two places to the right and attach a percent sign:

$$0.015 = 01.5\%$$

At the end, drop the leading zero to get 1.5%.

8 **222.2%.** Move the decimal point two places to the right and attach a percent sign:

$$2.222 = 222.2\%$$

9 $\frac{19}{100}$. Place 19 in the numerator and 100 in the denominator.

10 $\frac{2}{25}$. Place 8 in the numerator and 100 in the denominator:

$$\frac{8}{100}$$

You can reduce this fraction by 2 twice:

$$= \frac{4}{50} = \frac{2}{25}$$

(11) $1\frac{23}{100}$. Place 123 in the numerator and 100 in the denominator:

$$\frac{123}{100}$$

You can change this improper fraction to a mixed number:

$$= 1\frac{23}{100}$$

(12) $3\frac{3}{4}$. Place 375 in the numerator and 100 in the denominator:

$$\frac{375}{100}$$

Change the improper fraction to a mixed number:

$$= 3\frac{75}{100}$$

Reduce the fractional part of this mixed number, first by 5 and then by another 5:

$$= 3\frac{15}{20} = 3\frac{3}{4}$$

(13) **40%.** First, change $\frac{2}{5}$ to a decimal:

$$2.0 \div 5 = 0.4$$

Now change 0.4 to a percent by moving the decimal point two places to the right and adding a percent sign:

$$0.4 = 40\%$$

(14) **15%.** First, change $\frac{3}{20}$ to a decimal:

$$3.00 \div 20 = 0.15$$

Then change 0.15 to a percent:

$$0.15 = 15\%$$

(15) **87.5%.** First, change $\frac{7}{8}$ to a decimal:

$$7.000 \div 8 = 0.875$$

Now change 0.875 to a percent:

$$0.875 = 87.5\%$$

(16) $18.\overline{18}\%$. First, change $\frac{2}{11}$ to a decimal:

$$
\begin{array}{r}
0.181 \\
11\overline{)2.000} \\
\underline{11} \\
90 \\
\underline{88} \\
20 \\
\underline{11} \\
9
\end{array}
$$

The result is the repeating decimal $0.\overline{18}$. Now change this repeating decimal to a percent:

$$0.\overline{18} = 18.\overline{18}\%$$

(17) **8.3.** Move the decimal point one place to the left: 10% of $83 = 8.3$.

(18) **15.** Move the decimal one place to the left and multiply by 2: 20% of $75 = 7.5 \times 2 = 15$.

(19) **21.** Divide the number by 4: 25% of $84 = 84 \div 4 = 21$.

(20) **86.** Divide the number by 2: 50% of $172 = 172 \div 2 = 86$.

(21) **9.9.** 100% of any number is the number itself: 100% of $9.9 = 9.9$.

(22) **180.** Multiply the number by 5: 500% of $36 = 36 \times 5 = 180$.

(23) **12.** Reverse the two numbers in the problem and solve: 48% of $25 = 25\%$ of $48 = 48 \div 4 = 12$.

(24) **183.** Reverse the two numbers in the problem and solve: 61% of $300 = 300\%$ of $61 = 61 \times 3 = 183$.

(25) **49.** Reverse the two numbers in the problem and solve: 98% of $50 = 50\%$ of $98 = 98 \div 2 = 49$.

(26) **26.4.** Reverse the two numbers in the problem and solve: 132% of $20 = 20\%$ of $132 = 13.2 \times 2 = 26.4$.

(27) **14.7.** Change the word *of* to a multiplication sign and the percent to a decimal, then solve using decimal multiplication: 21% of $70 = 0.21 \times 70 = 14.7$.

(28) **15.75.** Change the word *of* to a multiplication sign and the percent to a decimal, then solve using decimal multiplication: 35% of $45 = 0.35 \times 45 = 15.75$.

(29) **77.47.** Change the word *of* to a multiplication sign and the percent to a decimal, then solve using decimal multiplication: 61% of $127 = 0.61 \times 127 = 77.47$.

(30) **312.39.** Change the word *of* to a multiplication sign and the percent to a decimal, then solve using decimal multiplication: 89% of $351 = 0.89 \times 351 = 312.39$.

(31) **70.**

Turn the problem into an equation as follows:

20	percent	of	350	is	70
20	$\times 0.01$	$\times$	350	$=$	70

Check this equation:

$$20 \times 0.01 \times 350 = 70$$
$$0.2 \times 350 = 70$$
$$70 = 70$$

(32) 17%.

Turn the problem into an equation:

What	percent	of	150	is	25.5
n	$\times 0.01$	$\times$	150	=	25.5

Solve the equation for n:

$$n \times 0.01 \times 150 = 25.5$$
$$1.5n = 25.5$$
$$\frac{1.5n}{1.5} = \frac{25.5}{1.5}$$
$$n = 17$$

(33) 8.69.

Turn the problem into an equation:

What	is	79	%	of	11
n	=	79	$\times 0.01$	$\times$	11

To find the answer, solve this equation for n: $n = 79 \times 0.01 \times 11 = 8.69$.

(34) 33.$\overline{3}$.

Turn the problem into an equation:

30	%	of	what number	is	10
30	$\times 0.01$	$\times$	n	=	10

Solve for n:

$$30 \times 0.01 \times n = 10$$
$$0.3n = 10$$
$$\frac{0.3n}{0.3} = \frac{10}{0.3}$$
$$n = 33.\overline{3}$$

The answer is the repeating decimal $33.\overline{3}$.

When you're ready to test your skills, try out the quiz in the following section.

Whaddya Know? Chapter 14 Quiz

Here are 18 questions covering the range of percentage problems covered in this chapter. When you're done, turn to the next section for answers and explanations.

1. Change 75% to a decimal.

2. Change 0.4 to %.

3. Change 45% to a fraction.

4. Change $\frac{1}{8}$ to %.

5. What is 15% of 60?

6. Change 2% to a fraction.

7. Change $\frac{5}{6}$ to %.

8. Change 0.0003 to %.

9. What percent of 200 is 40?

10. What is 33% of 5?

11. Change 0.75% to a fraction.

12. What is 95% of 12?

13. Change 10 to %.

14. Change $\frac{3}{4}$ to %.

15. What is 16% of 25?

16. Change 15.6% to a decimal.

17. What is 150% of 40?

18. 80% of what number is 200?

Answers to Chapter 14 Quiz

(1) **0.75.** Move the decimal point two places to the left.

$_\wedge$75.

(2) **40%.** Move the decimal 2 places to the right; you'll have to add a 0. Then put the % sign after the result.

0.40$_\wedge$

(3) $\frac{9}{20}$. Place the 45 in the numerator of a fraction with 100 in the denominator. Then reduce the fraction.

$$\frac{45}{100} = \frac{9}{20}$$

(4) **12.5%.** Divide 1 by 8 to determine the decimal equivalent of the fraction.

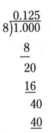

Now move the decimal point 2 places to the right and add the % sign.

0.12$_\wedge$5

(5) **9.** Using "percent of starting number = ending number," write:

$$15\% \text{ of } 60 = 0.15 \times 60 = 9$$

(6) $\frac{1}{50}$. Place the 2 in the numerator of a fraction with 100 in the denominator. Then reduce the fraction.

$$\frac{2}{100} = \frac{1}{50}$$

(7) $83\frac{1}{3}$%. Divide 5 by 6 to determine the decimal equivalent of the fraction.

$$\begin{array}{r} 0.833 \\ 6\overline{)5.000} \\ \underline{48} \\ 20 \\ \underline{18} \\ 20 \\ \underline{18} \\ 2 \end{array}$$

The quotient is a repeating decimal and can be written as $0.83\overline{3}$ or $0.83\frac{1}{3}$.

The percentage version, written by moving the decimal point two places to the right, can be written as $83\frac{1}{3}\%$. Note that the repeating remainder is 2, and $\frac{2}{6} = \frac{1}{3}$.

(8) **0.03%.** Move the decimal 2 places to the right and put the % sign at the end.

$0.00_{\wedge}03$

(9) **20%.** Using "percent of starting number = ending number," write:

$n\%$ of 200 is 40

Rewrite this word equation as a math equation and solve for n:

$$n \times 0.01 \times 200 = 40$$
$$2n = 40$$
$$n = 20$$

(10) **1.65.** Using "percent of starting number = ending number," write:

33% of $5 = 0.33 \times 5 = 1.65$

(11) $\frac{3}{400}$. Place the 0.75 in the numerator of a fraction with 100 in the denominator. Then multiply the numerator and denominator by 100 to get rid of the decimal in the numerator. Finally, reduce the fraction.

$$\frac{0.75}{100} \times \frac{100}{100} = \frac{75}{10,000} = \frac{3}{400}$$

(12) **11.4.** Using "percent of starting number = ending number," write:

95% of $12 = 0.95 \times 12 = 11.4$

(13) **1,000%.** Move the decimal 2 places to the right; you'll have to add two 0's. Then put the % sign after the result.

$10.00_{\wedge}$

(14) **75%.** First, write the fraction as a decimal. Then move the decimal point 2 places to the right.

$$\frac{3}{4} = 0.75$$
$$0.75_{\wedge} = 75\%$$

(15) **4.** Using "percent of starting number = ending number," write:

16% of $25 = 0.16 \times 25 = 4$

(16) **0.156.** Move the decimal point two places to the left.

$_{\wedge}15.6$

(17) **60.** Using "percent of starting number = ending number," write:

$$150\% \text{ of } 40 = 1.50 \times 40 = 60$$

(18) **250.** Using "percent of starting number = ending number," write:

80% of n is 200

Translate this word equation into a math equation and solve for n:

$$80 \times 0.01 \times n = 200$$
$$0.8n = 200$$
$$n = \frac{200}{0.8}$$
$$n = 250$$

equations

» **Translating the word *of* as multiplication**

» **Changing percentages to decimals in word problems**

» **Tackling business problems involving percent increase and decrease**

Chapter **15**

Word Problems with Fractions, Decimals, and Percentages

I n Chapter 7, I show you how to solve word problems (also known as story problems) by setting up word equations that use the Big Four operations (adding, subtracting, multiplying, and dividing). In this chapter, I show you how to extend these skills to solve word problems with fractions, decimals, and percentages.

First, I show you how to solve relatively easy problems, in which all you need to do is add or subtract fractions, decimals, or percentages. Next, I show you how to solve problems that require you to multiply fractions. Such problems are easy to spot because they almost always contain the word *of*. After that, you discover how to solve percent problems by setting up a word equation and changing the percent to a decimal. Finally, I show you how to handle problems of percent increase and decrease. These problems are often practical money problems in which you figure out information about raises and salaries, costs and discounts, or amounts before and after taxes.

Adding and Subtracting Parts of the Whole in Word Problems

Certain word problems involving fractions, decimals, and percentages are really just problems in adding and subtracting. You may add fractions, decimals, or percentages in a variety of real-world settings that rely on weights and measures — such as cooking and carpentry. (In Chapter 18, I discuss these applications in depth.)

To solve these problems, you can use the skills that you pick up in Chapters 11 and 12 (for adding and subtracting fractions), Chapter 13 (for adding and subtracting decimals), and Chapter 14 (for adding and subtracting percentages).

Sharing a pizza: Fractions

You may have to add or subtract fractions in problems that involve splitting up part of a whole. For example, consider the following:

> Joan ate $\frac{1}{6}$ of a pizza, Tony ate $\frac{1}{4}$ and Sylvia ate $\frac{1}{3}$. What fraction of the pizza was left when they were finished?

In this problem, just jot down the information that's given as word equations:

$$\text{Joan} = \frac{1}{6} \qquad \text{Tony} = \frac{1}{4} \qquad \text{Sylvia} = \frac{1}{3}$$

These fractions are part of one total pizza. To solve the problem, you need to find out how much all three people ate, so form the following word equation:

$$\text{All three} = \text{Joan} + \text{Tony} + \text{Sylvia}$$

Now you can substitute as follows:

$$\text{All three} = \frac{1}{6} + \frac{1}{4} + \frac{1}{3}$$

Chapter 11 gives you several ways to add these fractions. Here's one way:

$$\text{All three} = \frac{2}{12} + \frac{3}{12} + \frac{4}{12} = \frac{9}{12} = \frac{3}{4}$$

However, the question asks what fraction of the pizza was left after they finished, so you have to subtract that amount from the whole:

$$1 - \frac{3}{4} = \frac{1}{4}$$

Thus, the three people left $\frac{1}{4}$ of a pizza.

Buying by the pound: Decimals

You frequently work with decimals when dealing with money, metric measurements (see Chapter 18), and food sold by the pound. The following problem requires you to add and subtract decimals, which I discuss in Chapter 13. Even though the decimals may look intimidating, this problem is fairly simple to set up:

> Antonia bought 4.53 pounds of beef and 3.1 pounds of lamb. Lance bought 5.24 pounds of chicken and 0.7 pound of pork. Which of them bought more meat, and how much more?

To solve this problem, you first find out how much each person bought:

$$\text{Antonia} = 4.53 + 3.1 = 7.63$$
$$\text{Lance} = 5.24 + 0.7 = 5.94$$

You can already see that Antonia bought more than Lance. To find how much more, subtract:

$$7.63 - 5.94 = 1.69$$

So Antonia bought 1.69 pounds more than Lance.

Splitting the vote: Percentages

When percentages represent answers in polls, votes in an election, or portions of a budget, the total often has to add up to 100%. In real life, you may see such info organized as a pie chart (which I discuss in Chapter 14). Solving problems about this kind of information often involves nothing more than adding and subtracting percentages. Here's an example:

> In a recent mayoral election, five candidates were on the ballot. Faber won 39% of the vote, Gustafson won 31%, Ivanovich won 18%, Dixon won 7%, Obermayer won 3%, and the remaining votes went to write-in candidates. What percentage of voters wrote in their selection?

The candidates were in a single election, so all the votes have to total 100%. The first step here is just to add up the five percentages. Then subtract that value from 100%:

$$39\% + 31\% + 18\% + 7\% + 3\% = 98\%$$
$$100\% - 98\% = 2\%$$

Because 98% of voters voted for one of the five candidates, the remaining 2% wrote in their selections.

EXAMPLE

Q. Jeff is 1.13 meters tall and his younger brother, Ryan, is 0.25 meter shorter than he is. Their older sister, Trisha, is 0.4 meter shorter than the combined height of the two boys. How tall is Trisha?

A. **1.61 meters.** Jeff is 1.13 meters tall and Ryan is 0.25 meter less $(1.13 - 0.25 = 0.88)$, so Ryan is 0.88 meter tall. Trisha is 0.4 meter shorter than their combined height $(1.13 + 0.88 - 0.4 = 1.61)$.

YOUR TURN

1. Anita spends 25% of her monthly income on rent, 20% on bills, and 14% paying off her student loans. What percentage of her income does this leave her for other things?

2. On a four-day car trip from Chicago to San Diego, Phyllis drove $\frac{3}{10}$ of the total distance the first day, $\frac{1}{3}$ the second day, and $\frac{1}{4}$ the third day. What fraction of the distance was left for her on the fourth day?

3. Kevin always changes the oil in his car every 2,000 miles. Last month, he changed his oil, and then filled up his tank four times, driving the car 297.8 miles, 317.4 miles, 304.8 miles, and 315.6 miles on these four tanks of gasoline. How many more miles will he have to drive before it's time to change his oil again?

4. A cake recipe calls for $3\frac{1}{2}$ cups of flour, $1\frac{1}{3}$ cups of white sugar, $\frac{3}{4}$ cup of brown sugar, and $\frac{1}{8}$ cup of shaved chocolate. How many cups do these four ingredients total?

Problems about Multiplying Fractions

REMEMBER

In word problems, the word *of* almost always means multiplication. So whenever you see the word *of* following a fraction, decimal, or percent, you can usually replace it with a times sign.

When you think about it, *of* means multiplication even when you're not talking about fractions. For example, when you point to an item in a store and say, "I'll take three of those," in a sense you're saying, "I'll take that one multiplied by three."

The following examples give you practice turning word problems that include the word *of* into multiplication problems that you can solve with fraction multiplication.

Renegade grocery shopping: Buying less than they tell you to

When you understand that the word *of* means multiplication, you have a powerful tool for solving word problems. For instance, you can figure out how much you'll spend if you don't buy food in the quantities listed on the signs. Here's an example:

If beef costs $4 a pound, how much does $\frac{5}{8}$ of a pound cost?

Here's what you get if you simply change the *of* to a multiplication sign:

$\frac{5}{8} \times 1$ pound of beef

So you know how much beef you're buying. However, you want to know the cost. Because the problem tells you that 1 pound = $4, you can replace 1 pound of beef with $4:

$= \frac{5}{8} \times \$4$

Now you have an expression you can evaluate. Use the rules of multiplying fractions from Chapter 11 and solve:

$$= \frac{5 \times \$4}{8} = \$\frac{20}{8}$$

This fraction reduces to $\$\frac{5}{2}$. However, the answer looks weird because dollars are usually expressed in decimals, not fractions. So convert this fraction to a decimal using the rules I show you in Chapter 13:

$$\$\frac{5}{2} = \$2.5 = \$2.50$$

At this point, you recognize that $2.5 is more commonly written as $2.50, and you have your answer.

Easy as pie: Working out what's left on your plate

Sometimes when you're sharing something such as a pie, not everyone gets to it at the same time. The eager pie-lovers snatch the first slice, not bothering to divide the pie into equal servings, and the people who were slower, more patient, or just not that hungry cut their own portions from what's left over. When someone takes a part of the leftovers, you can do a bit of multiplication to see how much of the whole pie that portion represents.

Consider the following example:

Jerry bought a pie and ate $\frac{1}{5}$ of it. Then his wife, Doreen, ate $\frac{1}{6}$ of what was left. How much of the total pie was left?

To solve this problem, begin by jotting down what the first sentence tells you:

$$\text{Jerry} = \frac{1}{5}$$

Doreen ate part of what was left, so write a word equation that tells you how much of the pie was left after Jerry was finished. He started with a whole pie, so subtract his portion from 1:

$$\text{Pie left after Jerry} = 1 - \frac{1}{5} = \frac{4}{5}$$

Next, Doreen ate $\frac{1}{6}$ of this amount. Rewrite the word *of* as multiplication and solve as follows. This answer tells you how much of the whole pie Doreen ate:

$$\text{Doreen} = \frac{1}{6} \times \frac{4}{5} = \frac{4}{30}$$

To make the numbers a little smaller before you go on, notice that you can reduce the fraction:

$$\text{Doreen} = \frac{2}{15}$$

Now you know how much Jerry and Doreen both ate, so you can add these amounts together:

$$\text{Jerry} + \text{Doreen} = \frac{1}{5} + \frac{2}{15}$$

Solve this problem as I show you in Chapter 11:

$$= \frac{3}{15} + \frac{2}{15} = \frac{5}{15}$$

This fraction reduces to $\frac{1}{3}$. Now you know that Jerry and Doreen ate $\frac{1}{3}$ of the pie, but the problem asks you how much is left. So finish up with some subtraction and write the answer:

$$1 - \frac{1}{3} = \frac{2}{3}$$

The amount of pie left over was $\frac{2}{3}$.

EXAMPLE

Q. A coffee canister holds 10 cups of coffee when filled in the morning. At the morning meeting, employees drink $\frac{2}{3}$ of the coffee in the canister. Then at lunch, another employee drinks $\frac{3}{4}$ of what's left. How many cups of coffee are in the canister after that?

A. $\frac{5}{6}$ **cup.** At the morning meeting, employees drink $\frac{2}{3}$ of the contents of the full canister, so $\frac{1}{3}$ of 10 cups is left:

$$10 \times \frac{1}{3} = \frac{10}{3}$$

At lunch, another employee drinks $\frac{3}{4}$ of what's left, leaving $\frac{1}{4}$ of $\frac{10}{3}$ of a cup, so calculate this amount and reduce:

$$\frac{1}{4} \times \frac{10}{3} = \frac{10}{12} = \frac{5}{6}$$

YOUR TURN

5 A recipe calls for $\frac{3}{4}$ of a cup of chocolate chips. How many cups do you need if you triple the recipe?

6 Brynn lives $\frac{5}{8}$ of a mile from school. If she walks to school and back every day for five days, how many miles does she walk?

7 A rectangular field is $1\frac{2}{5}$ of a mile long and $\frac{9}{10}$ of a mile wide. What is the total area of the field in square miles? (*Hint:* To find the area of a rectangle, multiply the length times the width.)

8 A king gives a parcel of land to his four daughters. He gives $\frac{1}{3}$ of the land to his eldest daughter, $\frac{2}{5}$ of what remains to his second daughter, and $\frac{3}{4}$ of what's left to his third daughter. What fraction of the land remains for the king's youngest daughter?

Multiplying Decimals and Percentages in Word Problems

In the preceding section, "Problems about Multiplying Fractions," I show you how the word *of* in a fraction word problem usually means multiplication. This idea is also true in word problems involving decimals and percentages. The method for solving these two types of problems is similar, so I lump them together in this section.

TIP

You can easily solve word problems involving percentages by changing the percentages into decimals (see Chapter 14 for details). Here are a few common percentages and their decimal equivalents:

$$25\% = 0.25 \qquad 50\% = 0.5 \qquad 75\% = 0.75 \qquad 99\% = 0.99$$

To the end: Figuring out how much money is left

One common type of problem gives you a starting amount — and a bunch of other information — and then asks you to figure out how much you end up with. Here's an example:

> Maria's grandparents gave her $125 for her birthday. She put 40% of the money in the bank, spent 35% of what was left on a pair of shoes, and then spent the rest on a dress. How much did the dress cost?

Start at the beginning, forming a word equation to find out how much money Maria put in the bank:

Money in bank = 40% of $125

To solve this word equation, change the percent to a decimal and the word *of* to a multiplication sign; then multiply:

Money in bank = $0.4 \times \$125 = \$50$

TIP

Pay special attention to whether you're calculating how much of something was used up or how much of something is left over. If you need to work with the portion that remains, you may have to subtract the amount used from the amount you started with.

Because Maria started with $125, she had $75 left to spend:

Money left to spend
= Money from grandparents – Money in bank
= $125 – $50
= $75

The problem then says that she spent 35% of this amount on a pair of shoes. Again, change the percent to a decimal and the word *of* to a multiplication sign:

Shoes = 35% of $75 = $0.35 \times \$75 = \$26.25$

She spent the rest of the money on a dress, so

Dress = $75 – $26.25 = $48.75

Therefore, Maria spent $48.75 on the dress.

Finding out how much you started with

Some problems give you the amount that you end up with and ask you to find out how much you started with. In general, these problems are harder because you're not used to thinking backward. Here's an example, and it's kind of a tough one, so fasten your seat belt:

> Maria received some birthday money from her aunt. She put her usual 40% in the bank and spent 75% of the rest on a purse. When she was done, she had $12 left to spend on dinner. How much did her aunt give her?

This problem is similar to the one in the preceding section, but you need to start at the end and work backward. Notice that the only dollar amount in the problem comes after the two percent amounts. The problem tells you that she ends up with $12 after two transactions — putting money in the bank and buying a purse — and asks you to find out how much she started with.

To solve this problem, set up two word equations to describe the two transactions:

Money from aunt – Money for bank = Money after bank
Money after bank – Money for purse = $12

Notice what these two word equations are saying. The first tells you that Maria took the money from her aunt, subtracted some money to put in the bank, and left the bank with a new amount of money, which I'm calling *Money after bank*. The second word equation starts where the first leaves off. It tells you that Maria took the money left over from the bank, subtracted some money for a purse, and ended up with $12.

This second equation already has an amount of money filled in, so start here. To solve this problem, realize that Maria spent 75% of her money *at that time* on the purse — that is, 75% of the money she still had after the bank:

Money after bank – 75% of money after bank = $12

I'm going to make one small change to this equation so you can see what it's really saying:

100% of money after bank – 75% of money after bank = $12

Adding *100% of* doesn't change the equation because it really just means you're multiplying by 1. In fact, you can slip these two words in anywhere without changing what you mean, though you may sound ridiculous saying "Last night, I drove 100% of my car home from work, walked 100% of my dog, then took 100% of my wife to see 100% of a movie."

In this particular case, however, these words help you to make a connection because 100% – 75% = 25%; here's an even better way to write this equation:

25% of money after bank = $12

Before moving on, make sure you understand the steps that have brought you here.

You know now that 25% of money after bank is $12, so the total amount of money after bank is 4 times this amount — that is, $48. Therefore, you can plug this number into the first equation:

Money from aunt – money for bank = $48

Now you can use the same type of thinking to solve this equation (and it goes a lot more quickly this time!). First, Maria placed 40% of the money from her aunt in the bank:

Money from aunt – 40% of money from aunt = $48

Again, rewrite this equation to make what it's saying clearer:

100% of money from aunt – 40% of money from aunt = $48

Now, because $100\% - 40\% = 60\%$, rewrite it again:

60% of money from aunt = $48

Thus:

$0.6 \times$ money from aunt = $48

Divide both sides of this equation by 0.6:

Money from aunt = $48 \div 0.6 = \$80$

So Maria's aunt gave her $80 for her birthday.

EXAMPLE

Q. Kylie brings home a monthly paycheck totaling $3,400. She has a savings plan that automatically places 8% of this amount in a retirement account. Then, her monthly mortgage and other regular bills consume 65% of what's left. How much does this leave Kylie to spend on other things?

A. **$1,094.80.** Begin by finding 8% of $3,400 $(\$3,400 \times 0.08 = \$272)$ and subtracting this from $3,400 $(\$3,400 - \$272 = \$3,128)$. Next, find 65% of $3,128 $(\$3,128 \times 0.65 = \$2,033.20)$ and subtracting this from $3,128 $(\$3,128 - \$2,033.20 = \$1,094.80)$.

YOUR TURN

9 If you bought a stock for $1,200 and sold it for $1,428, what percent profit did you make?

10 Jacob bought a television for $752.50, using a special coupon that gave him 14% off the retail price. What would he have paid without the coupon?

11 Antonietta received a $12,500 inheritance. She invested this in a mutual fund for one year and earned 7%, then rolled the money over into a CD and earned an additional 4%. How much money did she have after these two investments?

12 Jack won some money playing poker with his friends. He spent 70% of his winnings taking his girlfriend out for dinner, and then spent 40% of the remaining money on tickets to a basketball game. This left him with exactly $25.92. How much did he win originally at poker?

Handling Percent Increases and Decreases in Word Problems

Word problems that involve increasing or decreasing by a percentage add a final spin to percent problems. Typical percent-increase problems involve calculating the amount of a salary plus a raise, the cost of merchandise plus tax, or an amount of money plus interest or dividend. Typical percent decrease problems involve the amount of a salary minus taxes or the cost of merchandise minus a discount.

To tell you the truth, you may have already solved problems of this kind earlier in the section, "Multiplying Decimals and Percentages in Word Problems." But people often get thrown by the language of these problems — which, by the way, is the language of business — so I want to give you some practice in solving them.

Raking in the dough: Finding salary increases

A little street smarts should tell you that the words *salary increase* and *raise* mean more money, so get ready to do some addition. Here's an example:

> Alison's salary was $40,000 last year, and at the end of the year, she received a 5% raise. What will she earn this year?

To solve this problem, first realize that Alison got a raise. So whatever she makes this year, it will be more than she made last year. The key to setting up this type of problem is to think of percent increase as "100% of last year's salary plus 5% of last year's salary." Here's the word equation:

> This year's salary = 100% of last year's salary + 5% of last year's salary

Now you can just add the percentages:

> This year's salary = 105% of last year's salary

Change the percent to a decimal and the word *of* to a multiplication sign; then fill in the amount of last year's salary:

> This year's salary = $1.05 \times \$40{,}000$

Now you're ready to multiply:

> This year's salary = $\$42{,}000$

So Alison's new salary is $42,000.

Earning interest on top of interest

The word *interest* means more money. When you receive interest from the bank, you get more money. And when you pay interest on a loan, you pay more money. Sometimes people earn interest on the interest they earned earlier, which makes the dollar amounts grow even faster. Here's an example:

Bethany placed $9,500 in a one-year CD that paid 4% interest. The next year, she rolled this over into a bond that paid 6% per year. How much did Bethany earn on her investment in those two years?

This problem involves interest, so it's another problem in percent increase — only this time, you have to deal with two transactions. Take them one at a time.

The first transaction is a percent increase of 4% on $9,500. The following word equation makes sense:

$$\text{Money after first year} = 100\% \text{ of initial deposit} + 4\% \text{ of initial deposit}$$
$$= 104\% \text{ of initial deposit}$$

Now, substitute $9,500 for the initial deposit and calculate:

$$= 104\% \text{ of } \$9,500$$
$$= 1.04 \times \$9,500$$
$$= \$9,880$$

At this point, you're ready for the second transaction. This is a percent increase of 6% on $9,880:

$$\text{Final amount} = 106\% \text{ of } \$9,880$$
$$= 1.06 \times \$9,880$$
$$= \$10,472.80$$

Then subtract the initial deposit from the final amount:

$$\text{Earnings} = \text{Final amount} - \text{Initial deposit}$$
$$= \$10,472.80 - \$9,500$$
$$= \$972.80$$

So Bethany earned $972.80 on her investment.

Getting a deal: Calculating discounts

When you hear the words *discount* or *sale price*, think of subtraction. Here's an example:

Greg has his eye on a television with a listed price of $2,100. The salesman offers him a 30% discount if he buys it today. What will the television cost with the discount?

In this problem, you need to realize that the discount lowers the price of the television, so you have to subtract:

$$\text{Sale price} = 100\% \text{ of regular price} - 30\% \text{ of regular price}$$
$$= 70\% \text{ of regular price}$$
$$= 0.7 \times \$2,100 = \$1,470$$

Thus, the television costs $1,470 with the discount.

Q. After lunch at a restaurant, Beth wants to leave her server a tip of at least 18% on a meal that cost $16.50. What is the minimum amount that Beth should pay, including the tip?

A. **$19.47.** Calculate a percent increase of 18% on $16.50:

$16.50 \times 1.18 = 19.47

YOUR
TURN

13 The Johnsons bought their home for $220,000, and sold it ten years later for an increase of 37%. What was the selling price of the house?

14 Finding that a car he wanted to buy had unusually high mileage, Myron negotiated a 7% discount on the Bluebook price of $11,000. What price did he pay for the car?

15 Andrea invested a total of $12,000 this year. She invested $7,500 on a tech stock, which gained 17%. But she also invested $4,500 on a friend's business, losing 24% of her investment. How much money did Andrea have after these investments?

16 Fred bought a lawnmower that normally sold for $640, using a coupon that saved him 15%. But he still had to pay 7.5% sales tax on the discounted cost of his purchase. How much did he end up paying for the lawnmower?

Practice Questions Answers and Explanations

(1) **41%.** Add up the three percentages $(25\% + 20\% + 14\% = 59\%)$, and subtract the result from 100% $(100\% - 59\% = 41\%)$.

(2) $\frac{7}{60}$. Begin by adding the three fractions:

$$\frac{3}{10} + \frac{1}{3} + \frac{1}{4} = \frac{18}{60} + \frac{20}{60} + \frac{15}{60} = \frac{53}{60}$$

Now, subtract the result from 1:

$$1 - \frac{53}{60} = \frac{60}{60} - \frac{53}{60} = \frac{7}{60}$$

(3) **764.4 miles.** Kevin drove the car a total of 1,235.6 miles $(297.8 + 317.4 + 304.8 + 315.6 = 1,235.6)$. Subtract this amount from 2,000 to find the answer $(2,000 - 1,235.6 = 764.4)$.

(4) $5\frac{17}{24}$ **cups.** Add the two mixed numbers and the two fractions by changing to a common denominator:

$$3\frac{1}{2} + 1\frac{1}{3} + \frac{3}{4} + \frac{1}{8} = 3\frac{12}{24} + 1\frac{8}{24} + \frac{18}{24} + \frac{3}{24} = 5\frac{17}{24}$$

(5) $2\frac{1}{4}$ **cups.** Multiply $\frac{3}{4}$ by 3:

$$\frac{3}{4} \times 3 = \frac{9}{4}$$

Change this improper fraction to a mixed number:

$$= 2\frac{1}{4}$$

(6) $6\frac{1}{4}$ **miles.** Multiply $\frac{5}{8}$ by 10:

$$\frac{5}{8} \times 10 = \frac{50}{8}$$

Change this improper fraction to a mixed number, then reduce it to lowest terms:

$$= 6\frac{2}{8} = 6\frac{1}{4}$$

(7) $1\frac{13}{50}$ **square miles.** To find the area, multiply $1\frac{2}{5}$ by $\frac{9}{10}$. Begin by changing $1\frac{2}{5}$ to an improper fraction, then multiply:

$$1\frac{2}{5} \times \frac{9}{10} = \frac{7}{5} \times \frac{9}{10} = \frac{63}{50}$$

Change this improper fraction to a mixed number:

$$= 1\frac{13}{50}$$

(8) $\frac{1}{10}$. The king gives $\frac{1}{3}$ of his land to his eldest daughter, so $\frac{2}{3}$ of the land remains. Of this, the king gives $\frac{2}{5}$ to his second daughter, which leaves $\frac{3}{5}$ of $\frac{2}{3}$:

$$\frac{3}{5} \times \frac{2}{3} = \frac{6}{15} = \frac{2}{5}$$

Of this remaining $\frac{2}{5}$ of the original parcel of land, $\frac{3}{4}$ goes to the third daughter, so $\frac{1}{4}$ of $\frac{2}{5}$ goes to the youngest daughter:

$$\frac{1}{4} \times \frac{2}{5} = \frac{2}{20} = \frac{1}{10}$$

(9) **19%.** The profit on the investment is $228, so divide 228 by 1,200 and convert the answer from a decimal to a percent:

$$228 \div 1,200 = 0.19 = 19\%$$

(10) **$875.** The cost of the television with a 14% discount is $752.50, so 86% of the cost of the television equals $752.50:

86% of n is $752.50

Change this to an equation:

$$0.86n = 752.5$$
$$n = \frac{752.5}{0.86}$$
$$n = 875$$

(11) **$13,910.** Antonietta earned 7% on her first investment $\left(\$12,500 \times 0.07 = \$875 \right)$, so she had $13,375 after the first year $\left(\$12,500 + \$875 = \$13,375 \right)$. She earned 4% on this investment $\left(13,375 \times 0.04 = \$535 \right)$, so she ended up with $13,910 $\left(\$13,375 + \$535 = \$13,910 \right)$.

(12) **$144.** To begin, find out how much money Jack had before he bought the basketball tickets. He spent 40% of some of this amount on the tickets and ended up with $25.92, so 60% of this amount equals $25.92:

$$0.6n = 25.92$$
$$n = \frac{25.92}{0.6}$$
$$n = 43.20$$

Thus, Jack had $43.20 before buying the basketball tickets, but after buying dinner. He spent 70% of his poker winnings on this dinner, so $43.20 represents the remaining 30% of his poker winnings:

$$0.3p = 43.20$$
$$p = \frac{43.20}{0.3}$$
$$n = 144$$

Therefore, Jack won $144 at poker.

(13) **$301,400.** Find a percent increase of 37% on $220,000:

$$\$220,000 \times 1.37 = \$301,400$$

(14) **$10,230.** Calculate a percent decrease of 7% on $11,000:

$$\$11,000 \times 0.93 = \$10,230$$

(15) **$12,195.** First, find a 17% percent increase on $7,500:

$$\$7,500 \times 1.17 = \$8,775$$

Next, find a 24% decrease on $4,500:

$$\$4,500 \times 0.76 = \$3,420$$

To complete the problem, add the two numbers together:

$$\$8,775 - \$3,420 = \$5355$$

(16) **$584.80.** To begin, find the cost of the lawnmower with a percent decrease of 15%:

$$\$640 \times 0.85 = \$544$$

Next, find a percent increase of 7.5% on this amount:

$$\$544 \times 1.075 = \$584.80$$

In the next section, you can test your skills with the chapter quiz.

Whaddya Know? Chapter 15 Quiz

The 15 questions in this quiz test you on the problem-solving skills discussed in this chapter. When you're finished, check your answers and find explanations in the next section.

1 You spent four days panning for gold in the Colorado Mountains. Your totals each day were: $\frac{1}{3}$ oz., $\frac{1}{8}$ oz., $\frac{1}{4}$ oz., and $\frac{1}{6}$ oz. If gold is worth \$1,600 per ounce, how much money did you make?

2 George wants to give his waitress a 20% tip. If the total cost of his dinner is \$48.75 and he has \$55 in cash, will he have enough? If not, how much will he have to borrow?

3 A Weight Watcher has been going in for regular weigh-ins and has lost: 3.6 lbs., 4.2 lbs., 5 lbs., 6 lbs., and 0.3 lb. What is the total weight loss?

4 A recipe for a mixture of juices calls for $1\frac{1}{2}$ qt. grapefruit juice, $\frac{1}{6}$ qt. lemon juice, and $2\frac{1}{4}$ qt. orange juice. How large a container will be needed?

5 In a certain class, 23% of the students were born in January through March, 29% were born in April through June, and the rest were evenly split between the last six months. What percent of the class was born in December?

6 The cost of gas went from \$3.50 per gallon to \$3.85 per gallon. What was the percent increase?

7 Albert, Benjamin, and Craig finally caught up to their naughty dog Toby and had to carry him back home — a total of 5 miles. If Albert carried Toby the first 1.47 miles and Benjamin carried him the next 2.4 miles, then how far did Craig have to carry him?

8 A customer wants to purchase a television that costs \$800. The sales tax is 7%. But he will pay for the television over the next year and will have to pay 20% interest on the total cost (television plus tax). What will the total be? And if he pays monthly installments, what will they be?

9 The area of a triangle is found with $A = \frac{1}{2}bh$, where b is the base and h is the height. What is the area of a triangle whose base measures $4\frac{2}{3}$ in. and whose height is $12\frac{3}{7}$ in.?

10 Joe's commissions totaled \$50,000 two years ago. Last year the commissions increased by 2.4%, but this year there will be a 1.75% decrease over the previous year. What will this year's commissions be?

11 Over a four-day period, the snowfall totals measured: $\frac{1}{2}$ in., $1\frac{1}{8}$ in., $2\frac{3}{4}$ in., and 5 in. If 10 inches of snow is equivalent to 1 inch of rain, then how much rain fell during that time?

12 The five Miller children learned that they will share an inheritance from their great-great uncle. This uncle stipulated that the oldest get $\frac{2}{3}$ of the estate and that the other four equally share the rest. How much of the estate do these other four get?

13 The cost of a turkey is $2.25 per pound. What is the total cost of a 21.8-pound turkey?

14 A cookie recipe calls for $4\frac{1}{2}$ cups of flour. You only have $2\frac{5}{8}$ cups. How much more flour do you need?

15 When Stella was 40 years old, she was 63.2 inches tall. Now, at the age of 60 she's only 61.62 inches tall. What is the percent decrease in her height?

Answers to Chapter 15 Quiz

1. **$1,400.** First, find the total number of ounces of gold you've accumulated:

$$\frac{1}{3}+\frac{1}{8}+\frac{1}{4}+\frac{1}{6}=\frac{8}{24}+\frac{3}{24}+\frac{6}{24}+\frac{4}{24}=\frac{21}{24}=\frac{7}{8}$$

Now multiply that total by $1,600.

$$\frac{7}{8}\times\frac{1600}{1}=\frac{7}{\cancel{8}_1}\times\frac{\cancel{1600}^{200}}{1}=\frac{1400}{1}=1400$$

2. **$3.50.** Multiply $48.75 by 20%: $48.75\times0.20=9.75$.

   Add that to the cost of the meal: $9.75+48.75=58.50$.

   Subtract: $58.50-55.00=3.50$.

   He'll have to borrow $3.50.

3. **19.1 lbs.** Find the sum of the amounts lost: $3.6+4.2+5+6+0.3=19.1$.

4. **$3\frac{11}{12}$ qts.** Find the sum of the amounts of juice: $1\frac{1}{2}+\frac{1}{6}+2\frac{1}{4}=1\frac{6}{12}+\frac{2}{12}+2\frac{3}{12}=3\frac{11}{12}$.

   The container has to hold $3\frac{11}{12}$ quarts. A 4-quart container should do.

5. **8%.** First, add the two percentages: $23\%+29\%=52\%$. That leaves $100\%-52\%=48\%$ to split evenly between the last 6 months. Since $\frac{48\%}{6}=8\%$, 8% were born in December.

6. **10%.** First, find the increase in cost per gallon: $3.85-\$3.50=\$0.35$.

   Next, divide the increase by the original cost: $\frac{\$0.35}{\$3.50}=0.10$.

   To find the percentage, just move the decimal point two places to the right: $0.10=10\%$.

7. **1.13 mi.** First, add the number of miles Albert and Benjamin carried Toby: $1.47+2.4=3.87$.

   Now subtract that sum from 5: $5-3.87=1.13$.

8. **$1,027.20; $85.60 monthly.** First, find the total cost of the television plus sales tax:

$$\$800\times0.07=\$56$$
$$\$800+\$56=\$856$$

   Now determine what 20% interest will be, and add that to the total amount.

$$\$856\times0.20=\$171.20$$
$$\$856+\$171.20=\$1,027.20$$

   Dividing this total by 12, $\frac{\$1027.20}{12}=\$85.60$.

9. **29 sq. in.** Multiply $\frac{1}{2}$ times the base times the height.

$$A=\frac{1}{2}\times4\frac{2}{3}\times12\frac{3}{7}=\frac{1}{2}\times\frac{14}{3}\times\frac{87}{7}=\frac{1}{\cancel{2}_1}\times\frac{\cancel{14}^1}{3}\times\frac{87}{\cancel{7}_1}=\frac{1}{1}\times\frac{1}{\cancel{3}_1}\times\frac{\cancel{87}^{29}}{1}=\frac{29}{1}=29$$

(10) **$50,304.** First, determine the increased commission he received last year.

$$\$50,000 \times 0.024 = \$1,200$$
$$\$50,000 + \$1,200 = \$51,200$$

Now determine the decreased commission.

$$\$51,200 \times 0.0175 = \$896$$
$$\$51,200 - \$896 = \$50,304$$

(11) $\frac{15}{16}$ **in.** First, find the total number of inches of snow.

$$\frac{1}{2} + 1\frac{1}{8} + 2\frac{3}{4} + 5 = \frac{4}{8} + 1\frac{1}{8} + 2\frac{6}{8} + 5 = 8\frac{11}{8} = 8 + 1\frac{3}{8} = 9\frac{3}{8}$$

Write a proportion to determine the amount of rain.

$$\frac{10 \text{ in. snow}}{1 \text{ in. rain}} = \frac{9\frac{3}{8} \text{ in. snow}}{x \text{ in. rain}}$$

Now cross-multiply:

$$10x = 9\frac{3}{8}$$
$$x = 9\frac{3}{8} \div 10 = \frac{75}{8} \times \frac{1}{10} = \frac{\cancel{75}^{15}}{8} \times \frac{1}{\cancel{10}_2} = \frac{15}{16}$$

(12) $\frac{1}{12}$. Subtracting the $\frac{2}{3}$ share from 1, $1 - \frac{2}{3} = \frac{1}{3}$.
Now divide that $\frac{1}{3}$ share by 4: $\frac{1}{3} \div 4 = \frac{1}{3} \times \frac{1}{4} = \frac{1}{12}$.

(13) **$49.05.** Multiply the price per pound times the number of pounds: $\$2.25 \times 21.8 = \$49.05$.

(14) $1\frac{7}{8}$ **cups.** Subtract $2\frac{5}{8}$ cups from $4\frac{1}{2}$ cups: $4\frac{1}{2} - 2\frac{5}{8} = 4\frac{4}{8} - 2\frac{5}{8} = 3\frac{12}{8} - 2\frac{5}{8} = 1\frac{7}{8}$.

(15) **2.5%.** Find the difference in height measures. Then divide the difference by the beginning height, 63.2 inches.

$$63.2 - 61.62 = 1.58$$
$$\frac{1.58}{63.2} = 0.025 = 2.5\%$$

6

Reaching the Summit: Advanced Pre-Algebra Topics

In This Unit . . .

Chapter **16**

Powers and Roots

I n Chapter 5, I show you how powers and roots are a compact way to express repeated multiplication of positive integers — for example, $4^2 = 4 \times 4 = 16$ and $\sqrt{16} = 4$. In this chapter, you extend your understanding of powers and roots to include negative numbers and fractional values.

To begin, I provide a list of powers and roots that show up so frequently in higher math that they're worth memorizing. Then, I show you how to apply exponents to both negative numbers and fractions using the order of operations (PEMDAS), as explained in Chapter 6.

Next, I show you how to make sense of exponents of 0 and negative numbers, and I give you practice handling these computations. And finally, you discover how fractional exponents are an alternative way to express square roots and other higher-order roots.

Memorizing Powers and Roots

By the time you were finished with third or fourth grade, you'd probably memorized most of the multiplication table up to $10 \times 10 = 100$.

If you're up for a similar challenge that can really boost your confidence (and your grades!) in math, I recommend memorizing the powers and roots listed in this section. These numbers appear over and over again in the math that you study through high school and college. Knowing them by heart can be surprisingly helpful.

Remembering square numbers and square roots

Square numbers arise when you multiply a positive integer by itself. You probably know the first 10 positive square numbers from the multiplication table. Because these numbers are handy to know, I always advise my students to learn a few extras by heart.

In this table, I list the first 20 positive square numbers and their related square roots.

Squares	Square roots
$1^2 = 1$	$\sqrt{1} = 1$
$2^2 = 4$	$\sqrt{4} = 2$
$3^2 = 9$	$\sqrt{9} = 3$
$4^2 = 16$	$\sqrt{16} = 4$
$5^2 = 25$	$\sqrt{25} = 5$
$6^2 = 36$	$\sqrt{36} = 6$
$7^2 = 49$	$\sqrt{49} = 7$
$8^2 = 64$	$\sqrt{64} = 8$
$9^2 = 81$	$\sqrt{81} = 9$
$10^2 = 100$	$\sqrt{100} = 10$
$11^2 = 121$	$\sqrt{121} = 11$
$12^2 = 144$	$\sqrt{144} = 12$
$13^2 = 169$	$\sqrt{169} = 13$
$14^2 = 196$	$\sqrt{196} = 14$
$15^2 = 225$	$\sqrt{225} = 25$
$16^2 = 256$	$\sqrt{256} = 16$
$17^2 = 289$	$\sqrt{289} = 17$
$18^2 = 324$	$\sqrt{324} = 18$
$19^2 = 361$	$\sqrt{361} = 19$
$20^2 = 400$	$\sqrt{400} = 20$

Keeping track of cubic numbers and cube roots

Cubic numbers are the result of multiplying a positive by itself three times. For example, 64 is a cubic number because $4^3 = 4 \times 4 \times 4 = 64$.

This table lists the first ten cubic numbers, from 1 to 1,000, plus their related cube roots.

Cubes	Cube roots
$1^3 = 1$	$\sqrt[3]{1} = 1$
$2^3 = 8$	$\sqrt[3]{8} = 2$
$3^3 = 27$	$\sqrt[3]{27} = 3$
$4^3 = 64$	$\sqrt[3]{64} = 4$
$5^3 = 125$	$\sqrt[3]{125} = 5$
$6^3 = 216$	$\sqrt[3]{216} = 6$
$7^3 = 343$	$\sqrt[3]{343} = 7$
$8^3 = 512$	$\sqrt[3]{512} = 8$
$9^3 = 729$	$\sqrt[3]{729} = 9$
$10^3 = 1{,}000$	$\sqrt[3]{1{,}000} = 10$

Knowing a few powers of 2 and their related roots

If you start with the number 2 and keep doubling it, you get the following sequence:

2, 4, 8, 16, 32, 64, ...

These numbers are the powers of 2, which arise when you apply a positive exponent to the number 2. They're handy to know, so in the following table, I list the first ten of these numbers. I also give you their related higher-power roots, which undo the powers and return the output value back to 2.

Powers of 2	Related roots
$2^1 = 2$	$\sqrt[1]{2} = 2$
$2^2 = 4$	$\sqrt{4} = 2$
$2^3 = 8$	$\sqrt[3]{8} = 2$
$2^4 = 16$	$\sqrt[4]{16} = 2$
$2^5 = 32$	$\sqrt[5]{32} = 2$
$2^6 = 64$	$\sqrt[6]{64} = 2$
$2^7 = 128$	$\sqrt[7]{128} = 2$
$2^8 = 256$	$\sqrt[8]{256} = 2$
$2^9 = 512$	$\sqrt[9]{512} = 2$
$2^{10} = 1{,}024$	$\sqrt[10]{1{,}024} = 2$

Changing the Base

In Chapter 5, you discover how to raise a natural number (that is, a positive integer) to the power of another natural number.

But you also change the base number (the number you're applying an exponent to) to a negative number or a fraction. When you do this, use the order of operations (PEMDAS) to evaluate the result. In this section, I show you how.

Negating a number raised to an exponent

You can negate a number raised to an exponent simply by attaching a negative sign to the beginning of it. For example:

$$-10^2 = -(10 \times 10) = -100$$
$$-2^5 = -(2 \times 2 \times 2 \times 2 \times 2) = -32$$
$$-7^3 = -(7 \times 7 \times 7) = -343$$

As you can see, when negating a negative number in this way, you follow the order of operations (PEMDAS), evaluating the exponent first and then the negative sign. (For more on PEMDAS, flip to Chapter 6.)

EXAMPLE

Q. What is the most simplified form of -4^3?

A. $-4^3 = -(4 \times 4 \times 4) = -64$

Q. Evaluate -10^6.

A. $-10^6 = -(10 \times 10 \times 10 \times 10 \times 10 \times 10) = -1,000,000$

Finding powers of negative numbers

To raise a negative number to the power of another number, use parentheses to change the order of operations (PEMDAS). For example:

$$(-11)^2 = (-11)(-11) = 121$$
$$(-4)^3 = (-4)(-4)(-4) = -64$$
$$(-10)^4 = (-10)(-10)(-10)(-10) = 10,000$$
$$(-2)^7 = (-2)(-2)(-2)(-2)(-2)(-2)(-2) = -128$$

As you can see, when you raise a negative number to an even power, the result is a positive number. This happens because multiplying an even number of negative numbers together always gives you a positive number.

In contrast, when you raise a negative number to an odd power, the result is a negative number, because multiplying an odd number of negative numbers together gives you an odd number.

Q. Simplify the expression $(-5)^3$.

A. $(-5)^3 = (-5)(-5)(-5) = -125$

Q. What is the value of $(-2)^6$?

A. $(-2)^6 = (-2)(-2)(-2)(-2)(-2)(-2) = 64$

Finding powers of fractions

You can also apply an exponent to a fractional base. For example:

$$\left(\frac{1}{2}\right)^3 = \left(\frac{1}{2}\right)\left(\frac{1}{2}\right)\left(\frac{1}{2}\right) = \frac{1}{8}$$
$$\left(\frac{9}{10}\right)^2 = \left(\frac{9}{10}\right)\left(\frac{9}{10}\right) = \frac{81}{100}$$
$$\left(\frac{3}{4}\right)^5 = \left(\frac{3}{4}\right)\left(\frac{3}{4}\right)\left(\frac{3}{4}\right)\left(\frac{3}{4}\right)\left(\frac{3}{4}\right) = \frac{243}{1,024}$$

As you can see, to raise a fraction to a power, you use the rules for multiplying fractions, as described in Chapter 11.

Q. Simplify $\left(\frac{2}{5}\right)^3$.

A. $\left(\frac{2}{5}\right)^3 = \left(\frac{2}{5}\right)\left(\frac{2}{5}\right)\left(\frac{2}{5}\right) = \frac{8}{125}$

Q. Evaluate following expression: $\left(\frac{3}{10}\right)^4$.

A. $\left(\frac{3}{10}\right)^4 = \left(\frac{3}{10}\right)\left(\frac{3}{10}\right)\left(\frac{3}{10}\right)\left(\frac{3}{10}\right) = \frac{81}{10,000}$

Mixing negative numbers and fractions with exponents

To negate a fraction that's raised to an exponent, place the negative sign *outside* the parentheses. For example:

Q. Write $-\left(\frac{1}{3}\right)^4$ in its most simplified fractional form.

A. $-\left(\frac{1}{3}\right)^4 = -\left(\frac{1}{3} \times \frac{1}{3} \times \frac{1}{3} \times \frac{1}{3}\right) = -\frac{1}{81}$

Q. Evaluate $-\left(\frac{9}{10}\right)^3$.

A. $-\left(\frac{9}{10}\right)^3 = -\left(\frac{9}{10} \times \frac{9}{10} \times \frac{9}{10}\right) = -\frac{729}{1,000}$

In contrast, if you wish to raise a negative fraction to the power of a number, place the negative sign *inside* the parentheses. For example:

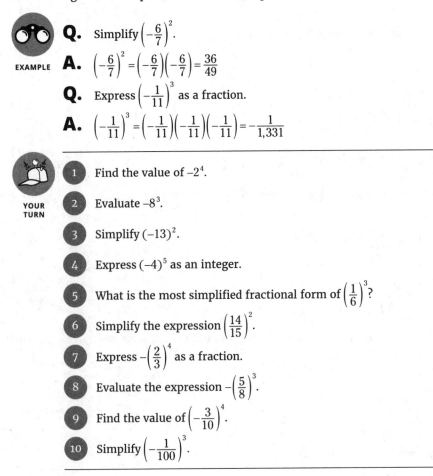

Q. Simplify $\left(-\frac{6}{7}\right)^2$.

A. $\left(-\frac{6}{7}\right)^2 = \left(-\frac{6}{7}\right)\left(-\frac{6}{7}\right) = \frac{36}{49}$

Q. Express $\left(-\frac{1}{11}\right)^3$ as a fraction.

A. $\left(-\frac{1}{11}\right)^3 = \left(-\frac{1}{11}\right)\left(-\frac{1}{11}\right)\left(-\frac{1}{11}\right) = -\frac{1}{1,331}$

YOUR TURN

1. Find the value of -2^4.

2. Evaluate -8^3.

3. Simplify $(-13)^2$.

4. Express $(-4)^5$ as an integer.

5. What is the most simplified fractional form of $\left(\frac{1}{6}\right)^3$?

6. Simplify the expression $\left(\frac{14}{15}\right)^2$.

7. Express $-\left(\frac{2}{3}\right)^4$ as a fraction.

8. Evaluate the expression $-\left(\frac{5}{8}\right)^3$.

9. Find the value of $\left(-\frac{3}{10}\right)^4$.

10. Simplify $\left(-\frac{1}{100}\right)^3$.

Exponents of 0 and Negative Numbers

In the last section, you discovered how to evaluate powers that have negative and fractional bases by expanding these expressions — that is, by changing them to multiplication.

In this section, you extend your understanding of powers further by exploring exponents that are numbers other than positive integers (natural numbers). First, you work with exponents of 0 and negative integers. Then, I show you how to interpret fractional exponents as radicals (roots).

Exponents of 0

When you raise a number to the power of 0, the result is 1. This is true for every number *except* 0, because 0^0 is undefined.

To get a sense of why raising a number to the power of 0 results in a value of 1, look at the following pattern of equations:

$$10^4 = 10,000$$
$$10^3 = 1,000$$
$$10^2 = 100$$
$$10^1 = 10$$
$$10^0 = 1$$

As you can see, when evaluating powers of 10, every time you subtract 1 from the exponent, you divide the previous value by 10. So when you continue this pattern, the result is that $10^0 = 1$.

Q. What is the value of $8,125^0$?

A. $8,125^0 = 1$

Q. Evaluate $\left(-\dfrac{21}{89}\right)^0$.

A. $\left(-\dfrac{21}{89}\right)^0 = 1$

Negative exponents

When you apply a negative exponent to a number, the result is the *reciprocal* of the result that you would have gotten if the exponent had been positive.

I know that this rule can be confusing for a lot of students. So to help you understand, recall that in the previous section, I include a chart giving you a way to think about why $10^0 = 1$. I extend that chart to exponents with negative numbers:

$$10^4 = 10,000$$
$$10^3 = 1,000$$
$$10^2 = 100$$
$$10^1 = 10$$
$$10^0 = 1$$
$$10^{-1} = \frac{1}{10} = 0.1$$
$$10^{-2} = \frac{1}{100} = 0.01$$
$$10^{-3} = \frac{1}{1,000} = 0.001$$
$$10^{-4} = \frac{1}{10,000} = 0.0001$$

Notice that in these equations, I continue the pattern of subtracting 1 from the exponent and dividing the result by 10. (By the way, this pattern comes in handy when you're working with scientific notation, which I introduce in Chapter 17.)

As you can see, when you apply a negative exponent to a base that's greater than 1, the result is a fractional or decimal value between 0 and 1.

Applying an exponent of –1 changes a number to its reciprocal. (For a refresher on reciprocals, see Chapter 10.) For example:

$$2^{-1} = \frac{1}{2}$$

$$14^{-1} = \frac{1}{14}$$

$$\left(\frac{3}{8}\right)^{-1} = \frac{8}{3}$$

$$\left(\frac{1}{5}\right)^{-1} = \frac{5}{1} = 5$$

TIP

When evaluating a negative exponent other than –1, change the base to its reciprocal and change the exponent from negative to positive. Getting rid of the negative number first usually makes the problem look a lot simpler to solve.

Applying an exponent of –2 changes the number to the reciprocal of its square. In the following examples, I do this calculation in two steps so you can see how it works:

$$3^{-2} = \frac{1}{3^2} = \frac{1}{9}$$

$$11^{-2} = \frac{1}{11^2} = \frac{1}{121}$$

$$\left(\frac{4}{7}\right)^{-2} = \left(\frac{7}{4}\right)^2 = \frac{49}{16}$$

$$\left(\frac{1}{6}\right)^{-2} = \left(\frac{6}{1}\right)^2 = 36$$

Generally speaking, when you apply a negative exponent to any number (*except* 0!), the result is the same as the reciprocal of that number raised to the positive value of that exponent. If that seems less than crystal clear, here are some examples:

$$8^{-3} = \frac{1}{8^3} = \frac{1}{512}$$

$$3^{-5} = \frac{1}{3^5} = \frac{1}{243}$$

$$\left(\frac{2}{5}\right)^{-4} = \left(\frac{5}{2}\right)^4 = \left(\frac{5}{2}\right)\left(\frac{5}{2}\right)\left(\frac{5}{2}\right)\left(\frac{5}{2}\right) = \frac{625}{16}$$

$$\left(\frac{1}{2}\right)^{-8} = (2)^8 = 256$$

EXAMPLE

Q. Simplify the expression 63^{-1}.

A. $63^{-1} = \frac{1}{63}$

Q. Find the value of $\left(\frac{35}{41}\right)^{-1}$.

A. $\left(\frac{35}{41}\right)^{-1} = \frac{41}{35}$

Q. What is the value of 4^{-3}?

A. $4^{-3} = \dfrac{1}{4^3} = \dfrac{1}{64}$

Q. Simplify $\left(\dfrac{9}{17}\right)^{-2}$.

A. $\left(\dfrac{9}{17}\right)^{-2} = \left(\dfrac{17}{9}\right)^{2} = \left(\dfrac{17}{9}\right)\left(\dfrac{17}{9}\right) = \dfrac{289}{81}$

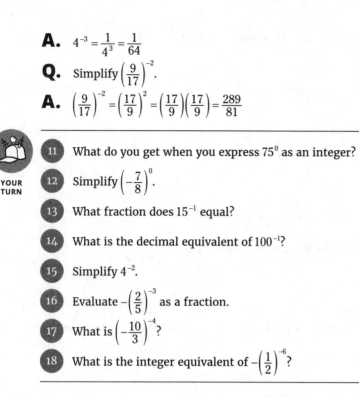

YOUR TURN

11 What do you get when you express 75^0 as an integer?

12 Simplify $\left(-\dfrac{7}{8}\right)^{0}$.

13 What fraction does 15^{-1} equal?

14 What is the decimal equivalent of 100^{-1}?

15 Simplify 4^{-2}.

16 Evaluate $-\left(\dfrac{2}{5}\right)^{-3}$ as a fraction.

17 What is $\left(-\dfrac{10}{3}\right)^{-4}$?

18 What is the integer equivalent of $-\left(\dfrac{1}{2}\right)^{-6}$?

Fractional Exponents

Many students are really confused by fractional exponents, such as $32^{\frac{3}{5}}$. Just what the heck does it mean to multiply the number 32 by itself $\frac{3}{5}$ times?

In this section, I show you that raising a number to the power of $\frac{1}{2}$ is the same as taking the square root of a number. Then, you extend this understanding to powers of other fractions.

Exponents of $\dfrac{1}{2}$

You can rewrite an exponent of $\frac{1}{2}$ as a square root. So, when you raise a square number to the power of $\frac{1}{2}$, the result is a positive integer. For example:

$$4^{\frac{1}{2}} = \sqrt{4} = 2$$
$$100^{\frac{1}{2}} = \sqrt{100} = 10$$
$$225^{\frac{1}{2}} = \sqrt{225} = 15$$

When you raise a non-square number to an exponent of $\frac{1}{2}$, the result is a radical expression — that is, a root that cannot be simplified further. For example:

$$2^{\frac{1}{2}} = \sqrt{2}$$
$$3^{\frac{1}{2}} = \sqrt{3}$$
$$5^{\frac{1}{2}} = \sqrt{5}$$

Radicals become more important in algebra and higher math, but for now you don't have to worry too much about them.

Q. Evaluate the expression $36^{\frac{1}{2}}$.

A. $36^{\frac{1}{2}} = \sqrt{36} = 6$

Q. Simplify $289^{\frac{1}{2}}$ as a positive integer.

A. $289^{\frac{1}{2}} = \sqrt{289} = 17$

Q. Evaluate $\left(\frac{25}{49}\right)^{\frac{1}{2}}$ as a fraction.

A. Begin by rewriting the problem as a square root:

$$\left(\frac{25}{49}\right)^{\frac{1}{2}} = \sqrt{\frac{25}{49}}$$

Now, evaluate the numerator and denominator as separate square roots and simplify:

$$= \frac{\sqrt{25}}{\sqrt{49}} = \frac{5}{7}$$

Exponents of $\frac{1}{3}$

The exponent of $\frac{1}{3}$ can be simplified as a cube root. So, when you raise a cubic number to the power of $\frac{1}{3}$, the result is a positive integer. For example:

$$8^{\frac{1}{3}} = \sqrt[3]{8} = 2$$
$$343^{\frac{1}{3}} = \sqrt[3]{343} = 7$$
$$1,000^{\frac{1}{3}} = \sqrt[3]{1,000} = 10$$

Raising a non-cubic number to the power of $\frac{1}{3}$ results in a radical expression that cannot be simplified further.

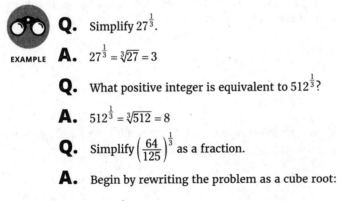

Q. Simplify $27^{\frac{1}{3}}$.

A. $27^{\frac{1}{3}} = \sqrt[3]{27} = 3$

Q. What positive integer is equivalent to $512^{\frac{1}{3}}$?

A. $512^{\frac{1}{3}} = \sqrt[3]{512} = 8$

Q. Simplify $\left(\frac{64}{125}\right)^{\frac{1}{3}}$ as a fraction.

A. Begin by rewriting the problem as a cube root:

$$\left(\frac{64}{125}\right)^{\frac{1}{3}} = \sqrt[3]{\frac{64}{125}}$$

Now split the numerator and denominator into separate cube roots and simplify:

$$= \frac{\sqrt[3]{64}}{\sqrt[3]{125}} = \frac{4}{5}$$

Exponents of $\frac{1}{4}$, $\frac{1}{5}$, $\frac{1}{6}$, and so forth

Exponents of other fractions that have 1 in the numerator are equivalent to higher-order roots, such as fourth roots, fifth roots, and so on. For example:

$$81^{\frac{1}{4}} = \sqrt[4]{81} = 3 \qquad \text{because } 3^4 = 81$$

$$32^{\frac{1}{5}} = \sqrt[5]{32} = 2 \qquad \text{because } 2^5 = 1$$

$$1{,}000{,}000^{\frac{1}{6}} = \sqrt[6]{1{,}000{,}000} = 10 \quad \text{because } 10^6 = 1{,}000{,}000$$

As with other fractional exponents, don't worry right now about radical results that don't work out nicely as whole numbers.

Q. What positive integer equals $10{,}000^{\frac{1}{4}}$?

EXAMPLE **A.** $10{,}000^{\frac{1}{4}} = \sqrt[4]{10{,}000} = 10$

Q. Simplify $128^{\frac{1}{7}}$.

A. $128^{\frac{1}{7}} = \sqrt[7]{128} = 2$

Other fractional exponents

When an exponent has a number other than 1 in the numerator, rewrite it using a combination of a power and a root, and then simplify. Here's a formula you can use to rewrite it:

$$b^{\frac{x}{y}} = (\sqrt[y]{b})^x$$

To make sense of this confusing formula, consider $1{,}000^{\frac{2}{3}}$. To use the formula, rewrite this expression as follows:

$$1{,}000^{\frac{2}{3}} = (\sqrt[3]{1{,}000})^2$$

Next, remember PEMDAS and simplify the root that's inside the parentheses:

$$= 10^2$$

Finally, simplify the exponent:

$$= 100$$

Here are a few more examples:

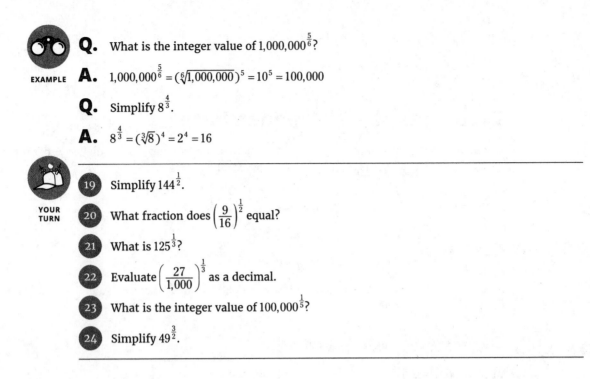

Q. What is the integer value of $1,000,000^{\frac{5}{6}}$?

EXAMPLE **A.** $1,000,000^{\frac{5}{6}} = (\sqrt[6]{1,000,000})^5 = 10^5 = 100,000$

Q. Simplify $8^{\frac{4}{3}}$.

A. $8^{\frac{4}{3}} = (\sqrt[3]{8})^4 = 2^4 = 16$

YOUR TURN

19 Simplify $144^{\frac{1}{2}}$.

20 What fraction does $\left(\dfrac{9}{16}\right)^{\frac{1}{2}}$ equal?

21 What is $125^{\frac{1}{3}}$?

22 Evaluate $\left(\dfrac{27}{1,000}\right)^{\frac{1}{3}}$ as a decimal.

23 What is the integer value of $100,000^{\frac{1}{5}}$?

24 Simplify $49^{\frac{3}{2}}$.

Practice Questions Answers and Explanations

(1) $-2^4 = -(2 \times 2 \times 2 \times 2) = \mathbf{-16}$

(2) $-8^3 = -(8 \times 8 \times 8) = \mathbf{-512}$

(3) $(-13)^2 = (-13)(-13) = \mathbf{169}$

(4) $(-4)^5 = (-4)(-4)(-4)(-4)(-4) = \mathbf{-1{,}024}$

(5) $\left(\frac{1}{6}\right)^3 = \left(\frac{1}{6}\right)\left(\frac{1}{6}\right)\left(\frac{1}{6}\right) = \mathbf{\frac{1}{216}}$

(6) $\left(\frac{14}{15}\right)^2 = \left(\frac{14}{15}\right)\left(\frac{14}{15}\right) = \mathbf{\frac{196}{225}}$

(7) $-\left(\frac{2}{3}\right)^4 = -\left(\frac{2}{3} \times \frac{2}{3} \times \frac{2}{3} \times \frac{2}{3}\right) = \mathbf{-\frac{16}{81}}$

(8) $-\left(\frac{5}{8}\right)^3 = -\left(\frac{5}{8} \times \frac{5}{8} \times \frac{5}{8}\right) = \mathbf{-\frac{125}{256}}$

(9) $\left(-\frac{3}{10}\right)^4 = \left(-\frac{3}{10}\right)\left(-\frac{3}{10}\right)\left(-\frac{3}{10}\right)\left(-\frac{3}{10}\right) = \mathbf{\frac{81}{10{,}000}}$

(10) $\left(-\frac{1}{100}\right)^3 = \left(-\frac{1}{100}\right)\left(-\frac{1}{100}\right)\left(-\frac{1}{100}\right) = \mathbf{-\frac{1}{1{,}000{,}000}}$

(11) $75^0 = \mathbf{1}$

(12) $\left(-\frac{7}{8}\right)^0 = \mathbf{1}$

(13) $15^{-1} = \mathbf{\frac{1}{15}}$

(14) $100^{-1} = \frac{1}{100} = \mathbf{0.01}$

(15) $4^{-2} = \frac{1}{4^2} = \mathbf{\frac{1}{16}}$

(16) $-\left(\frac{2}{5}\right)^{-3} = -\left(\frac{5}{2}\right)^3 = -\left(\frac{5}{2}\right)\left(\frac{5}{2}\right)\left(\frac{5}{2}\right) = \mathbf{-\frac{125}{8}}$

(17) $\left(-\frac{10}{3}\right)^{-4} = \left(-\frac{3}{10}\right)^4 = \left(-\frac{3}{10}\right)\left(-\frac{3}{10}\right)\left(-\frac{3}{10}\right)\left(-\frac{3}{10}\right) = \mathbf{\frac{81}{10{,}000}}$

(18) $-\left(\frac{1}{2}\right)^{-6} = -2^6 = -(2 \times 2 \times 2 \times 2 \times 2 \times 2) = \mathbf{-64}$

(19) $144^{\frac{1}{2}} = \sqrt{144} = \mathbf{12}$

(20) $\left(\dfrac{9}{16}\right)^{\frac{1}{2}} = \sqrt{\dfrac{9}{16}} = \dfrac{\sqrt{9}}{\sqrt{16}} = \dfrac{3}{4}$

(21) $125^{\frac{1}{3}} = \sqrt[3]{125} = \mathbf{5}$ because $5^3 = 125$.

(22) $\left(\dfrac{27}{1,000}\right)^{\frac{1}{3}} = \sqrt[3]{\dfrac{27}{1,000}} = \dfrac{\sqrt[3]{27}}{\sqrt[3]{1,000}} = \dfrac{3}{10} = \mathbf{0.3}$

(23) $100,000^{\frac{1}{5}} = \sqrt[5]{100,000} = \mathbf{10}$ because $10^5 = 100,000$.

(24) $49^{\frac{3}{2}} = \left(\sqrt{49}\right)^3 = 7^3 = \mathbf{343}$

When you feel ready, test your skills working with powers and roots with the chapter quiz in the next section.

Whaddya Know? Chapter 16 Quiz

This 20-question quiz tests you on the skills discussed in this chapter. When you've completed the problems, or if you get stuck, flip to the next section to find the answer to every question with a complete explanation.

Simplify the expressions.

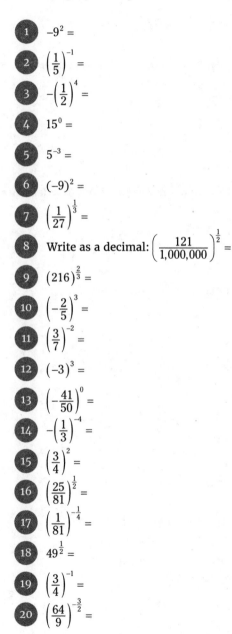

1. $-9^2 =$

2. $\left(\frac{1}{5}\right)^{-1} =$

3. $-\left(\frac{1}{2}\right)^4 =$

4. $15^0 =$

5. $5^{-3} =$

6. $(-9)^2 =$

7. $\left(\frac{1}{27}\right)^{\frac{1}{3}} =$

8. Write as a decimal: $\left(\frac{121}{1,000,000}\right)^{\frac{1}{2}} =$

9. $(216)^{\frac{2}{3}} =$

10. $\left(-\frac{2}{5}\right)^3 =$

11. $\left(\frac{3}{7}\right)^{-2} =$

12. $(-3)^3 =$

13. $\left(-\frac{41}{50}\right)^0 =$

14. $-\left(\frac{1}{3}\right)^{-4} =$

15. $\left(\frac{3}{4}\right)^2 =$

16. $\left(\frac{25}{81}\right)^{\frac{1}{2}} =$

17. $\left(\frac{1}{81}\right)^{-\frac{1}{4}} =$

18. $49^{\frac{1}{2}} =$

19. $\left(\frac{3}{4}\right)^{-1} =$

20. $\left(\frac{64}{9}\right)^{-\frac{3}{2}} =$

Answers to Chapter 16 Quiz

(1) **−81.** First square the 9, and then apply the negative sign to the result.

$$-9^2 = -81$$

(2) **5.** Write the reciprocal of the fraction.

$$\left(\frac{1}{5}\right)^{-1} = \frac{5}{1} = 5$$

(3) **$-\frac{1}{16}$.** First perform the power by raising both numerator and denominator to the fourth. Then apply the negative sign.

$$-\left(\frac{1}{2}\right)^4 = -\left(\frac{1^4}{2^4}\right) = -\left(\frac{1}{16}\right) = -\frac{1}{16}$$

(4) **1.** All non-zero numbers raised to the 0 power are equal to 1.

(5) **$\frac{1}{125}$.** First write the 5 as a fraction. Then write the reciprocal of that fraction raised to the third power.

$$5^{-3} = \left(\frac{5}{1}\right)^{-3} = \left(\frac{1}{5}\right)^3$$

Raise both the numerator and denominator to the third power.

$$= \frac{1^3}{5^3} = \frac{1}{125}$$

(6) **81.** Raising a negative integer to the second power results in a positive integer.

(7) **$\frac{1}{3}$.** Raise both the numerator and denominator to the $\frac{1}{3}$ power.

$$\left(\frac{1}{27}\right)^{\frac{1}{3}} = \frac{1^{\frac{1}{3}}}{27^{\frac{1}{3}}}$$

The one-third power is the same as the cube root.

$$= \frac{\sqrt[3]{1}}{\sqrt[3]{27}} = \frac{1}{3}$$

(8) **0.011.** Raise both the numerator and denominator to the $\frac{1}{2}$ power.

$$\left(\frac{121}{1,000,000}\right)^{\frac{1}{2}} = \frac{121^{\frac{1}{2}}}{1,000,000^{\frac{1}{2}}}$$

The one-half power is the same as the square root.

$$= \frac{\sqrt{121}}{\sqrt{1,000,000}} = \frac{11}{1,000}$$

The fraction written as a decimal has three decimal places.

(9) **36.** The two-thirds power is equivalent to finding the cube root and squaring the result.

$$\left(216\right)^{\frac{2}{3}} = \left(\sqrt[3]{216}\right)^2 = \left(6\right)^2 = 36$$

(10) $-\frac{8}{125}$. Raise both the numerator and denominator to the third power.

$$\left(-\frac{2}{5}\right)^3 = -\frac{2^3}{5^3} = -\frac{8}{125}$$

(11) $\frac{49}{9}$. First write the reciprocal of the fraction raised to the second power. Then square both the numerator and denominator.

$$\left(\frac{3}{7}\right)^{-2} = \left(\frac{7}{3}\right)^2 = \frac{7^2}{3^2} = \frac{49}{9}$$

(12) **−27.** Multiply −3 times itself three times.

$$\left(-3\right)^3 = \left(-3\right)\left(-3\right)\left(-3\right) = -27$$

(13) **1.** Any non-zero rational number raised to the zero power is equal to 1.

$$\left(-\frac{41}{50}\right)^0 =$$

(14) **−81.** First write the reciprocal of the fraction raised to the fourth power.

$$-\left(\frac{1}{3}\right)^{-4} = -\left(\frac{3}{1}\right)^4$$

Next, raise both the numerator and denominator to the fourth power.

$$-\left(\frac{3}{1}\right)^4 = -\frac{3^4}{1^4} = -\frac{81}{1} = -81$$

The negative sign applies to the result.

(15) $\frac{9}{16}$. Raise both the numerator and denominator to the second power.

$$\left(\frac{3}{4}\right)^2 = \frac{3^2}{4^2} = \frac{9}{16}$$

(16) $\frac{5}{9}$. Raise both the numerator and denominator to the one-half power.

$$\left(\frac{25}{81}\right)^{\frac{1}{2}} = \frac{25^{\frac{1}{2}}}{81^{\frac{1}{2}}}$$

The one–half power is equivalent to finding the square root.

$$= \frac{\sqrt{25}}{\sqrt{81}} = \frac{5}{9}$$

(17) **3.** Write the reciprocal of the fraction raised to the positive one-fourth power.

$$\left(\frac{1}{81}\right)^{-\frac{1}{4}} = \left(\frac{81}{1}\right)^{\frac{1}{4}}$$

Next, raise both the numerator and denominator to that one-fourth power.

$$= \frac{81^{\frac{1}{4}}}{1^{\frac{1}{4}}}$$

The one-fourth power is equivalent to finding the fourth root.

$$= \frac{\sqrt[4]{81}}{\sqrt[4]{1}} = \frac{3}{1} = 3$$

(18) **7.** The one–half power is equivalent to finding the square root.

$$49^{\frac{1}{2}} = \sqrt{49} = 7$$

(19) $\frac{4}{3}$. Write the reciprocal of the fraction.

$$\left(\frac{3}{4}\right)^{-1} = \frac{4}{3}$$

(20) $\frac{27}{512}$. Write the reciprocal of the fraction and change the exponent to its positive value.

$$\left(\frac{64}{9}\right)^{-\frac{3}{2}} = \left(\frac{9}{64}\right)^{\frac{3}{2}}$$

Next, raise both the numerator and denominator to that one-half power.

$$= \frac{9^{\frac{3}{2}}}{64^{\frac{3}{2}}}$$

The power of $\frac{3}{2}$ is equivalent to finding the square root and cubing the result.

$$= \frac{\left(\sqrt{9}\right)^3}{\left(\sqrt{64}\right)^3} = \frac{3^3}{8^3} = \frac{27}{512}$$

Chapter **17**

A Perfect Ten: Condensing Numbers with Scientific Notation

Scientists often work with very small or very large measurements — the distance to the next galaxy, the size of an atom, the mass of the Earth, or the number of bacteria cells growing in last week's leftover Chinese takeout. To save on time and space — and to make calculations easier — people developed a sort of shorthand called *scientific notation.*

Scientific notation uses a sequence of numbers known as the powers of ten, which I introduce in Chapter 16:

   1   10   100   1,000   10,000   100,000   1,000,000   10,000,000...

Each number in the sequence is 10 times more than the preceding number:

   10^0 10^1 10^2 10^3 10^4 10^5 10^6 10^7...

Powers of ten are easy to work with, especially when you're multiplying and dividing, because you can just add or drop zeros or move the decimal point.

Scientific notation is a handy system for writing very large and very small numbers without writing a bunch of 0s. It uses both decimals and exponents (so if you need a little brushing up on decimals, flip to Chapter 13; likewise, if you need to refresh on exponents, take a look at Chapter 16). In this chapter, I introduce you to this powerful method of writing numbers. I also explain the order of magnitude of a number. Finally, I show you how to multiply numbers written in scientific notation.

First Things First: Using Powers of Ten as Exponents

Scientific notation uses powers of ten expressed as exponents, so you need a little background before you can jump in. In this section, I round out your knowledge of exponents, which I first introduce in Chapter 5 and then expand upon in Chapter 16.

Counting zeros and writing exponents

Numbers starting with a 1 and followed by only 0s (such 10, 100, 1,000, 10,000, and so forth) are called powers of ten, and they're easy to represent as exponents. Powers of ten are the result of multiplying 10 times itself any number of times.

TIP

To represent a number that's a power of 10 as an exponential number, count the zeros and raise 10 to that exponent. For example, 1,000 has three zeros, so 1,000 = 10^3 (10^3 means to take 10 times itself three times, so it equals $10 \times 10 \times 10$). Table 17-1 shows a list of some powers of ten.

When you know this trick, representing a lot of large numbers as powers of ten is easy — just count the 0s! For example, the number 1 trillion — 1,000,000,000,000 — is a 1 with twelve 0s after it, so

$$1,000,000,000,000 = 10^{12}$$

Table 17-1   Powers of Ten Expressed as Exponents

Standard Notation	Exponential Notation
1	10^0
10	10^1
100	10^2
1,000	10^3
10,000	10^4
100,000	10^5
1,000,000	10^6

This trick may not seem like a big deal, but the higher the numbers get, the more space you save by using exponents. For example, a really big number is a googol, which is 1 followed by a hundred 0s. You can write it like this:

10,000,000,000,000,000,000,000,000,000,000,000,000,000,000,000,000,000,000,000,000,000,000,000,000,000,000,000,000,000,000,000,000,000

As you can see, a number of this size is practically unmanageable. You can save yourself some trouble and write 10^{100}.

A 10 raised to a negative number is also a power of ten.

REMEMBER You can also represent decimals using negative exponents. For example,

$$10^{-1} = 0.1 \qquad 10^{-2} = 0.01 \qquad 10^{-3} = 0.001 \qquad 10^{-4} = 0.0001$$

Although the idea of negative exponents may seem strange, it makes sense when you think about it alongside what you know about positive exponents. For example, to find the value of 10^7, start with 1 and make it larger by moving the decimal point seven spaces to the right:

$$10^7 = 10,000,000$$

Similarly, to find the value of 10^{-7}, start with 1 and make it smaller by moving the decimal point seven spaces to the left:

$$10^{-7} = 0.0000001$$

When you write the decimal form of a negative power of 10, always include a leading zero before the decimal point. When you do this, the decimal always has the same number of zeros that the power indicates. For example:

TIP

$$10^{-1} = 0.1$$
$$10^{-2} = 0.01$$
$$10^{-3} = 0.001$$

As with very large numbers, using exponents to represent very small decimals makes practical sense. For example,

$$10^{-23} = 0.00000000000000000000001$$

As you can see, this decimal is easy to work with in its exponential form but almost impossible to read otherwise.

Q. Write 10^6 in standard notation.

EXAMPLE **A.** **1,000,000.** The exponent is 6, so the standard notation is a 1 with six 0s after it.

Q. Write 100,000 in exponential notation.

A. **10^5.** The number 100,000 has five 0s, so the exponential notation has 5 in the exponent.

Q. Write 10^{-5} in standard notation.

A. **0.00001.** The exponent is –5, so the standard notation is a decimal with five 0s (including the leading 0) followed by a 1.

Q. Write 0.0000001 in exponential notation.

A. **10^{-7}.** The decimal has seven 0s (including the leading 0), so the exponential notation has –7 in the exponent.

YOUR
TURN

1   Write each of the following powers of ten in standard notation:

    (a) 10^4

    (b) 10^7

    (c) 10^{14}

    (d) 10^{22}

2   Write each of the following powers of ten in exponential notation:

    (a) 1,000,000,000

    (b) 1,000,000,000,000

    (c) 10,000,000,000,000,000

    (d) 100,000,000,000,000,000,000,000,000,000,000

3   Write each of the following powers of ten in standard notation:

    (a) 10^{-1}

    (b) 10^{-5}

    (c) 10^{-11}

    (d) 10^{-16}

4   Write each of the following powers of ten in exponential notation:

    (a) 0.01

    (b) 0.000001

    (c) 0.000000000001

    (d) 0.000000000000000001

Exponential Arithmetic: Multiplying and Dividing Powers of Ten

TIP

Multiplying and dividing powers of ten in exponential notation is a snap because you don't have to do any multiplying or dividing at all — it's nothing more than simple addition and subtraction.

>> **Multiplication:** To multiply two powers of ten in exponential notation, find the sum of the numbers' exponents; then write a power of ten using that sum as the exponent.

>> **Division:** To divide one power of ten by another, subtract the second exponent from the first; then write a power of ten using this resulting sum as the exponent.

This rule works equally well when one or both exponents are negative — just use the rules for adding negative numbers, which I discuss in Chapter 4.

EXAMPLE

Q. Multiply 10^7 by 10^4.

A. 10^{11}. Add the exponents $7 + 4 = 11$, and use this as the exponent of your answer:

$$10^7 \times 10^4 = 10^{7+4} = 10^{11}$$

Q. Find $10^9 \div 10^6$.

A. 10^3. For division, you subtract. Subtract the exponents $9 - 6 = 3$ and use this as the exponent of your answer:

$$10^9 \div 10^6 = 10^{(9-6)} = 10^3$$

YOUR TURN

⑤ Multiply each of the following powers of ten:

    (a) $10^9 \times 10^2$

    (b) $10^5 \times 10^5$

    (c) $10^{13} \times 10^{-16}$

    (d) $10^{100} \times 10^{21}$

    (e) $10^{-15} \times 10^0$

    (f) $10^{-10} \times 10^{-10}$

⑥ Divide each of the following powers of ten:

    (a) $10^6 \div 10^4$

    (b) $10^{12} \div 10^1$

    (c) $10^{-7} \div 10^{-7}$

    (d) $10^{18} \div 10^0$

    (e) $10^{100} \div 10^{-19}$

    (f) $10^{-50} \div 10^{50}$

Working with Scientific Notation

Scientific notation is a system for writing very large and very small numbers that makes them easier to work with. Every number can be written in scientific notation as the product of two numbers (two numbers multiplied together):

>> A decimal greater than or equal to 1 and less than 10 (see Chapter 13 for more on decimals)

>> A power of ten written as an exponent (see the preceding section)

Writing in scientific notation

Here's how to write any number in scientific notation:

REMEMBER **1. Write the number as a decimal (if it isn't one already).** Suppose you want to change the number $360,000,000$ to scientific notation. First, write it as a decimal:

$360,000,000.0$

2. Move the decimal point just enough places to change this number to a new number that's between 1 and 10.

Move the decimal point to the right or left so that only one non-zero digit comes before the decimal point. Drop any leading or trailing zeros as necessary.

Using $360,000,000.0$, only the 3 should come before the decimal point. So move the decimal point eight places to the left, drop the trailing zeros, and get 3.6:

$360,000,000.0$ becomes 3.6.

3. Multiply the new number by 10 raised to the number of places you moved the decimal point in Step 2.

You moved the decimal point eight places, so multiply the new number by 10^8:

3.6×10^8

4. If you moved the decimal point to the right in Step 2, put a negative sign on the exponent.

You moved the decimal point to the left, so you don't have to take any action here. Thus, $360,000,000$ in scientific notation is 3.6×10^8.

Changing a decimal to scientific notation basically follows the same process. For example, suppose you want to change the number 0.00006113 to scientific notation.

1. Write 0.00006113 as a decimal (this step's easy because it's already a decimal):

0.00006113

2. **To change 0.00006113 to a new number between 1 and 10, move the decimal point five places to the right and drop the leading zeros:**

   6.113

3. **Because you moved the decimal point five places to the right, multiply the new number by 10^{-5}:**

   6.113×10^{-5}

So 0.00006113 in scientific notation is 6.113×10^{-5}.

When you get used to writing numbers in scientific notation, you can do it all in one step. Here are a few examples:

$$17{,}400 = 1.74 \times 10^{4}$$
$$212.04 = 2.1204 \times 10^{2}$$
$$0.003002 = 3.002 \times 10^{-3}$$

Q. Change the number 70,000 to scientific notation.

EXAMPLE

A. 7.0×10^{4}. First, write the number as a decimal:

   70,000.0

Move the decimal point just enough places to change this decimal to a new decimal that's between 1 and 10. In this case, move the decimal four places to the left. You can drop all but one trailing zero:

   7.0

You moved the decimal point four places, so multiply the new number by 10^4:

   7.0×10^{4}

Because you moved the decimal point to the left (you started with a big number), the exponent is a positive number, so you're done.

Q. Change the decimal 0.000000439 to scientific notation.

A. 4.39×10^{-7}. You're starting with a decimal, so Step 1 — writing the number as a decimal — is already taken care of:

   0.000000439

To change 0.000000439 to a decimal that's between 1 and 10, move the decimal point seven places to the right and drop the leading zeros:

   4.39

Because you moved seven places to the right, multiply the new number by 10^{-7}:

$$4.39 \times 10^{-7}$$

YOUR TURN

7   Change 2,591 to scientific notation.

8   Write the decimal 0.087 in scientific notation.

9   Write 1.00000783 in scientific notation.

10 Convert 20,002.00002 to scientific notation.

Understanding order of magnitude

A good question to ask is why scientific notation always uses a decimal between 1 and 10. The answer has to do with order of magnitude. Order of magnitude is a simple way to keep track of roughly how large a number is so you can compare numbers more easily. The order of magnitude of a number is its exponent in scientific notation. For example,

$$703 = 7.03 \times 10^2 \,(\text{order of magnitude is } 2)$$
$$600,000 = 6 \times 10^5 \,(\text{order of magnitude is } 5)$$
$$0.00095 = 9.5 \times 10^{-4} \,(\text{order of magnitude is } -4)$$

Every number starting with 10 but less than 100 has an order of magnitude of 1. Every number starting with 100 but less than 1,000 has an order of magnitude of 2.

Q. What is the order of magnitude of 5.6×10^7?

EXAMPLE

A. 7. The number is expressed in scientific notation and the exponent is 7, so its magnitude is 7.

Q. What is the order of magnitude of 893,441?

A. 5. The number is expressed in standard notation, so change it to scientific notation:

$$893,441 = 8.93441 \times 10^5$$

Therefore, its order of magnitude is 5.

YOUR TURN

11 What is the order of magnitude of each of the following numbers?

(a) 8×10^4

(b) 6.02×10^{23}

(c) 7.77×10^{-8}

(d) 9×10^1

12 What is the order of magnitude of each of the following numbers?

(a) 0.8

(b) 538

(c) 6,000,000,000

(d) 0.000004321

Multiplying with scientific notation

Multiplying numbers that are in scientific notation is fairly simple because multiplying powers of ten is easy, as you see in the earlier section, "Exponential Arithmetic: Multiplying and Dividing Powers of Ten." Here's how to multiply two numbers that are in scientific notation:

1. **Multiply the two decimal parts of the numbers.**

    Suppose you want to multiply the following:

    $$\left(6.02\times10^{23}\right)\left(9\times10^{-28}\right)$$

    Multiplication is commutative (see Chapter 5), so you can change the order of the numbers without changing the result. And because of the associative property, you can also change how you group the numbers. Therefore, you can rewrite this problem as

    $$\left(6.02\times9\right)\left(10^{23}\times10^{-28}\right)$$

    Multiply what's in the first set of parentheses — 6.02×9 — to find the decimal part of the solution:

    $$6.02\times9 = 54.18$$

2. **Multiply the two exponential parts by adding their exponents. Now multiply $10^{23}\times10^{-28}$:**

    $$10^{23}\times10^{-28} = 10^{-5}$$

3. **Write the answer as the product of the numbers you found in Steps 1 and 2.**

    $$54.18\times10^{-5}$$

4. **If the decimal part of the solution is 10 or greater, move the decimal point one place to the left and add 1 to the exponent.**

This method works even when one or both of the exponents are negative numbers. For example, if you follow the preceding series of steps, you find that $\left(6.02\times10^{23}\right)\left(9\times10^{-28}\right)=5.418\times10^{-4}$.

Dividing with Scientific Notation

Here's how to divide two numbers in scientific notation:

1. **Divide the decimal part of the first number by the decimal part of the second number to find the decimal part of the answer.**

2. **To find the power of ten in the answer, subtract the exponent on the second power of ten from the exponent on the first.**

   You're really just dividing the first power of ten by the second.

3. **If the decimal part of the result is less than 1, adjust the result by moving the decimal point one place to the right and subtracting 1 from the exponent.**

Q. Multiply 2.0×10^3 by 4.1×10^4.

A. $\mathbf{8.2 \times 10^7}$. Multiply the two decimal parts:

$$2.0 \times 4.1 = 8.2$$

Then multiply the powers of ten by adding the exponents:

$$10^3 \times 10^4 = 10^{3+4} = 10^7$$

In this case, no adjustment is necessary because the resulting decimal part is less than 10.

Q. Divide 3.4×10^4 by 2.0×10^9.

A. $\mathbf{1.7 \times 10^{-5}}$. Divide the first decimal part by the second:

$$3.4 \div 2.0 = 1.7$$

Then divide the first power of ten by the second by subtracting the exponents:

$$10^4 \div 10^9 = 10^{4-9} = 10^{-5}$$

In this case, no adjustment is necessary because the resulting decimal part isn't less than 1.

YOUR TURN

 13   Multiply 1.5×10^7 by 6.0×10^5.

14   Divide 6.6×10^8 by 1.1×10^3.

Practice Questions Answers and Explanations

1 In each case, write the digit 1 followed by the number of 0s indicated by the exponent:

(a) **10,000**

(b) **10,000,000**

(c) **100,000,000,000,000**

(d) **10,000,000,000,000,000,000,000**

2 In each case, count the number of 0s; then write a power of ten with this number as the exponent.

(a) 10^9

(b) 10^{12}

(c) 10^{16}

(d) 10^{32}

3 Write a decimal beginning with all 0s and ending in 1. The exponent indicates the number of 0s in this decimal (including the leading 0):

(a) **0.1**

(b) **0.00001**

(c) **0.00000000001**

(d) **0.0000000000000001**

4 In each case, count the number of 0s (including the leading 0); then write a power of ten using this number *negated (with the negative sign as the exponent)*:

(a) 10^{-2}

(b) 10^{-6}

(c) 10^{-12}

(d) 10^{-18}

5 Add the exponents and use this sum as the exponent of the answer.

(a) $10^9 \times 10^2 = 10^{9+2} = \mathbf{10^{11}}$

(b) $10^5 \times 10^5 = 10^{5+5} = \mathbf{10^{10}}$

(c) $10^{13} \times 10^{-16} = 10^{13+-16} = \mathbf{10^{-3}}$

(d) $10^{100} \times 10^{21} = 10^{100+21} = \mathbf{10^{121}}$

(e) $10^{-15} \times 10^0 = 10^{-15+0} = \mathbf{10^{-15}}$

(f) $10^{-10} \times 10^{-10} = 10^{-10+-10} = \mathbf{10^{-20}}$

(6) In each case, subtract the first exponent minus the second and use this result as the exponent of the answer.

(a) $10^6 \div 10^4 = 10^{6-4} = \mathbf{10^2}$

(b) $10^{12} \div 10^1 = 10^{12-1} = \mathbf{10^{11}}$

(c) $10^{-7} \div 10^{-7} = 10^{-7-(-7)} = 10^{-7+7} = \mathbf{10^0}$

(d) $10^{18} \div 10^0 = 10^{18-0} = \mathbf{10^{18}}$

(e) $10^{100} \div 10^{-19} = 10^{100-(-19)} = 10^{100+19} = \mathbf{10^{119}}$

(f) $10^{-50} \div 10^{50} = 10^{-50-50} = \mathbf{10^{-100}}$

(7) $\mathbf{2.591 \times 10^3}$. Write 2,591 as a decimal:

2,591.0

To change 2,591.0 to a decimal between 1 and 10, move the decimal point three places to the left and drop the trailing zero:

2.591

Because you moved the decimal point three places, multiply the new decimal by 10^3:

2.591×10^3

You moved the decimal point to the left, so the exponent stays positive. The answer is 2.591×10^3.

(8) $\mathbf{8.7 \times 10^{-2}}$. To change 0.087 to a decimal between 1 and 10, move the decimal point two places to the right and drop the leading zero:

8.7

Because you moved the decimal point two places to the right, multiply the new decimal by 10^{-2}:

8.7×10^{-2}

(9) $\mathbf{1.00000783}$. The decimal 1.00000783 is already between 1 and 10, so no change is needed.

(10) $\mathbf{2.000200002 \times 10^4}$. The number 20,002.00002 is already a decimal. To change it to a decimal between 1 and 10, move the decimal point four places to the left:

2.000200002

Because you moved the decimal point four places, multiply the new decimal by 10^4:

2.000200002×10^4

You moved the decimal point to the left, so the answer is 2.000200002×10^4.

(11) The exponent of a number that's expressed in scientific notation is the magnitude of that number.

    (a) **4**

    (b) **23**

    (c) **−8**

    (d) **1**

(12) To find the magnitude of a number that's expressed in standard notation, convert that number to scientific notation.

    (a) **−1.** $0.8 = 8 \times 10^{-1}$.

    (b) **2.** $538 = 5.38 \times 10^{2}$.

    (c) **9.** $6,000,000,000 = 6 \times 10^{9}$.

    (d) **−6.** $0.000004321 = 4.321 \times 10^{-6}$.

(13) 9.0×10^{12}. Multiply the two decimal parts:

$$1.5 \times 6.0 = 9.0$$

Multiply the two powers of ten:

$$10^{7} \times 10^{5} = 10^{7+5} = 10^{12}$$

In this case, no adjustment is necessary because the decimal is less than 10.

(14) 6.0×10^{5}. Divide the first decimal part by the second:

$$6.6 \div 1.1 = 6.0$$

Divide the first power of ten by the second:

$$10^{8} \div 10^{3} = 10^{8-3} = 10^{5}$$

In this case, no adjustment is necessary because the decimal is greater than 1.

How did you do with the practice problems? When you're ready, the quiz in the next section will test your mastery of the skills in this chapter.

Whaddya Know? Chapter 17 Quiz

The 13 questions in the quiz that follows test your knowledge of the topics covered in this chapter. When you're done, check out the next section for answers and explanations.

1. Write the number 10^8 in standard notation.

2. Multiply: $10^6 \times 10^8$.

3. Write the number 1,000,000 using exponential notation.

4. Change the number to scientific notation: 4,310.

5. Find the order of magnitude of the number 43,000,000.

6. Divide: $\left(6.4 \times 10^8\right) \div \left(8 \times 10^4\right)$.

7. Multiply: $10^{-3} \times 10^{-4}$.

8. Write the number 0.00000001 using exponential notation.

9. Divide: $10^{-11} \div 10^{11}$.

10. Multiply: $\left(4 \times 10^{16}\right) \times \left(6.2 \times 10^{-7}\right)$.

11. Write the number 10^{-8} in standard notation.

12. Change the number to scientific notation: 0.00167.

13. Divide: $10^8 \div 10^{-6}$.

Answers to Chapter 17 Quiz

(1) **100,000,000.** There are eight zeros in the eighth power of 10.

(2) 10^{14}. Add the exponents: $10^6 \times 10^8 = 10^{6+8} = 10^{14}$.

(3) 10^6. There are 6 zeros in the power of 10, so the exponent is 6.

(4) 4.31×10^3. You move the decimal point three places to the left, so the exponent on the ten is 3.

(5) **7.** First write the number using scientific notation. Moving the decimal point 7 places to the left gives you an exponent of 7 on the ten: $43,000,000 = 4.3 \times 10^7$. The order of magnitude is the exponent on the ten.

(6) 8×10^3. $(6.4 \times 10^8) \div (8 \times 10^4)$. Divide 6.4 by 8, and you have $\frac{6.4}{8} = 0.8$. Written in scientific notation, $0.8 = 8 \times 10^{-1}$. Dividing the two powers of 10, you subtract the exponents, giving you $10^8 \div 10^4 = 10^{8-4} = 10^4$. Now multiply the 8×10^{-1} times 10^4 and you have $8 \times 10^{-1} \times 10^4 = 8 \times 10^{-1+4} = 8 \times 10^3$.

(7) 10^{-7}. Add the exponents: $10^{-3} \times 10^{-4} = 10^{-3+(-4)} = 10^{-7}$.

(8) 10^{-8}. You move the decimal point 8 places to the right to get it to the right of the 1.

(9) 10^{-22}. Subtract the exponents: $10^{-11} \div 10^{11} = 10^{-11-11} = 10^{-22}$.

(10) 2.48×10^{10}. Multiply the 4 and the 6.2 first: $4 \times 6.2 = 24.8$. Then change the result to scientific notation by moving the decimal point one place to the left: $24.8 = 2.48 \times 10^1$. Find the product of the two powers of ten by adding the exponents: $10^{16} \times 10^{-7} = 10^{16+(-7)} = 10^9$. Now find the product of the two results and write it in scientific notation: $2.48 \times 10^1 \times 10^9 = 2.48 \times 10^{1+9} = 2.48 \times 10^{10}$.

(11) **0.00000001.** Move the decimal point eight places to the left of the 1; this means adding seven zeros: $10^{-8} = 0.00000001$.

(12) 1.67×10^{-3}. Move the decimal point three places to the right: $0.00167 = 1.67 \times 10^{-3}$.

(13) 10^{14}. Subtract the exponents: $10^8 \div 10^{-6} = 10^{8-(-6)} = 10^{14}$.

» **Discovering differences between the English and metric systems**

» **Estimating and calculating English and metric system conversions**

Chapter **18**

How Much Have You Got? Weights and Measures

This chapter introduces you to *units*, which are items that can be counted, such as apples, coins, or hats. Apples, coins, and hats are easy to count because they're *discrete* — that is, you can easily see where one ends and the next one begins. But not everything is so easy. For example, how do you count water — by the drop? Even if you tried, exactly how big is a drop?

Units of measurement come in handy at this point. A *unit of measurement* allows you to count something that isn't discrete: an amount of a liquid or solid, the distance from one place to another, a length of time, the speed at which you're traveling, or the temperature of the air.

In this chapter, I discuss two important systems of measurement: English and metric. You're probably familiar with the English system already, and you may know more than you think about the metric system. Each of these measurement systems provides a different way to measure distance, volume, weight (or mass), time, and speed. Next, I show you how to estimate metric amounts in English units. Finally, I show how to convert from English units to metric and vice versa.

Understanding Units

Anything that can be counted is a *unit.* That category is a pretty large one because almost anything that you can name can be counted. For now, just understand that all units can be counted, which means that you can apply the Big Four operations to units.

Adding and subtracting units

Adding and subtracting units isn't very different from adding and subtracting numbers. Just remember that you can add or subtract only when the units are the same. For example,

$$3 \text{ chairs} + 2 \text{ chairs} = 5 \text{ chairs}$$
$$4 \text{ oranges} - 1 \text{ orange} = 3 \text{ oranges}$$

What happens when you try to add or subtract different units? Here's an example:

$$3 \text{ chairs} + 2 \text{ tables} = ?$$

The only way you can complete this addition is to make the units the same:

$$3 \text{ pieces of furniture} + 2 \text{ pieces of furniture} = 5 \text{ pieces of furniture}$$

Multiplying and dividing units

You can always multiply and divide units by a *number.* For example, suppose you have four chairs but find that you need twice as many for a party. Here's how you represent this idea in math:

$$4 \text{ chairs} \times 2 = 8 \text{ chairs}$$

Similarly, suppose you have 20 cherries and want to split them among four people. Here's how you represent this idea:

$$20 \text{ cherries} \div 4 = 5 \text{ cherries}$$

But you have to be careful when multiplying or dividing units by units. For example:

$$2 \text{ apples} \times 3 \text{ apples} = ? \text{ WRONG!}$$
$$12 \text{ hats} \div 6 \text{ hats} = ? \text{ WRONG!}$$

Neither of these equations makes any sense. In these cases, multiplying or dividing by units is meaningless.

In many cases, however, multiplying and dividing units is okay. For example, multiplying *units of length* (such as inches, miles, or meters) results in *square units.* For example,

$$3 \text{ inches} \times 3 \text{ inches} = 9 \text{ square inches}$$
$$10 \text{ miles} \times 5 \text{ miles} = 50 \text{ square miles}$$
$$100 \text{ meters} \times 200 \text{ meters} = 20,000 \text{ square meters}$$

You find out more about units of length later in this chapter. Similarly, here are some examples of when dividing units makes sense:

$$12 \text{ slices of pizza} \div 4 \text{ people} = 3 \text{ slices of pizza/person}$$
$$140 \text{ miles} \div 2 \text{ hours} = 70 \text{ miles/hour}$$

In these cases, you read the fraction slash (/) as *per:* slices of pizza *per* person or miles *per* hour. You find out more about multiplying and dividing by units later in this chapter, when I show you how to convert from one unit of measurement to another.

Examining Differences between the English and Metric Systems

The two most common measurement systems today are the *English system* and the *metric system*.

Most Americans learn the units of the English system — for example, pounds and ounces, feet and inches, and so forth — and use them every day. Unfortunately, the English system is awkward for use with math. English units such as inches and fluid ounces are often measured in fractions, which (as you may know from Chapters 10 and 11) can be difficult to work with.

The *metric system* was invented to simplify the application of math to measurement. Metric units are based on the number 10, which makes them much easier to work with. Parts of units are expressed as decimals, which (as Chapter 13 shows you) are much friendlier than fractions.

Yet despite these advantages, the metric system has been slow to catch on in the U.S. Many Americans feel comfortable with English units and are reluctant to part with them. For example, if I ask you to carry a 20-pound bag for one-fourth of a mile, you know what to expect. However, if I ask you to carry a bag weighing 10 kilograms half a kilometer, you may not be sure.

In this section, I show you the basic units of measurement for both the English and metric systems.

If you want an example of the importance of converting carefully, you may want to look to NASA — they kind of lost a Mars orbiter in the late 1990s because an engineering team used English units and NASA used metric to navigate!

Looking at the English system

The *English system of measurement* is most commonly used in the United States (but, ironically, not in England). Although you're probably familiar with most of the English units of measurement, in the following list, I make sure you know the most important ones. I also show you some equivalent values that can help you do conversions from one type of unit to another.

>> **Units of distance:** Distance — also called *length* — is measured in inches (in.), feet (ft.), yards (yd.), and miles (mi.):

12 inches = 1 foot

3 feet = 1 yard

5,280 feet = 1 mile

>> **Units of fluid volume:** Fluid volume (also called *capacity*) is the amount of space occupied by a liquid, such as water, milk, or wine. I discuss volume when I talk about geometry in Chapter 19. Volume is measured in fluid ounces (fl. oz.), cups (c.), pints (pt.), quarts (qt.), and gallons (gal.):

8 fluid ounces = 1 cup

2 cups = 1 pint

2 pints = 1 quart

4 quarts = 1 gallon

Units of fluid volume are typically used for measuring the volume of things that can be poured. The volume of solid objects is more commonly measured in cubic units of distance, such as cubic inches and cubic feet.

>> **Units of weight:** Weight is the measurement of how strongly gravity pulls an object toward Earth. Weight is measured in ounces (oz.), pounds (lb.), and tons.

16 ounces = 1 pound

2,000 pounds = 1 ton

WARNING

Don't confuse *fluid ounces,* which measure volume, with *ounces,* which measure weight. These units are two completely different types of measurements!

>> **Units of time:** Time is hard to define, but everybody knows what it is. Time is measured in seconds, minutes, hours, days, weeks, and years:

60 seconds = 1 minute

60 minutes = 1 hour

24 hours = 1 day

7 days = 1 week

365 days = 1 year

The conversion from days to years is approximate because Earth's daily rotation on its axis and its yearly revolution around the sun aren't exactly synchronized. A year is closer to 365.25 days, which is why leap-years exist.

I left months out of the picture because the definition of a month is imprecise — it can vary from 28 to 31 days.

>> **Unit of speed:** Speed is the measurement of how much time an object takes to move a given distance. The most common unit of speed is miles per hour (mph).

>> **Unit of temperature:** Temperature is measured in proportion to how much heat an object contains. This object can be a glass of water, a turkey in the oven, or the air surrounding your house. Temperature is measured in degrees Fahrenheit (°F).

Table 18-1 summarizes the most commonly used units of English measurement. To use this chart, remember the following rules:

Table 18-1 Commonly Used English Units of Measurement

Measure of . . .	English Units	Conversion Equations
Distance (length)	Inches (in.) Feet (ft.) Yards (yd.) Miles (mi.)	12 inches = 1 foot 3 feet = 1 yard 5,280 feet = 1 mile
Fluid volume (capacity)	Fluid ounces (fl. oz.) Cups (c.) Pints (pt.) Quarts (qt.) Gallons (gal.)	8 fluid ounces = 1 cup 2 cups = 1 pint 2 pints = 1 quart 4 quarts = 1 gallon
Weight	Ounces (oz.) Pounds (lb.) Tons	16 ounces = 1 pound 2,000 pounds = 1 ton
Time	Seconds Minutes Hours Days Weeks Years	60 seconds = 1 minute 60 minutes = 1 hour 24 hours = 1 day 7 days = 1 week 365 days = 1 year
Temperature	Degrees Fahrenheit (°F)	
Speed (rate)	Miles per hour (mph)	

>> When converting from a large unit to a smaller unit, always multiply. For example, 2 pints are in 1 quart, so to convert 10 quarts to pints, multiply by 2:

$$10 \text{ quarts} \times 2 = 20 \text{ pints}$$

>> When converting from a small unit to a larger unit, always divide. For example, 3 feet are in 1 yard, so to convert 12 feet to yards, divide by 3:

$$12 \text{ feet} \div 3 = 4 \text{ yards}$$

When converting from large units to very small ones (for example, from tons to ounces), you may need to multiply more than once. Similarly, when converting from small units to much larger ones (for example, from minutes to days), you may need to divide more than once.

TIP

After doing a conversion, step back and apply a *reasonability test* to your answer — that is, think about whether your answer makes sense. For example, when you convert feet to inches, the number you end up with should be a lot bigger than the number you started with because there are lots of inches in a foot.

EXAMPLE

Q. How many minutes are in a day?

A. **1,440 minutes.** You're going from a larger unit (a day) to a smaller unit (minutes); 60 minutes are in an hour and 24 hours are in a day, so you simply multiply:

$$60 \times 24 = 1,440 \text{ minutes}$$

Therefore, 1,440 minutes are in a day.

Q. 5 pints = _____ fluid ounces.

A. **80 fluid ounces.** 8 fluid ounces are in a cup and 2 cups are in a pint, so

$$8 \times 2 = 16 \text{ fl.oz.}$$

16 fluid ounces are in a pint, so

$$5 \text{ pints} = 5 \times 16 \text{ fluid ounces} = 80 \text{ fl.oz.}$$

Q. If you have 32 fluid ounces, how many pints do you have?

A. **2 pints.** 8 fluid ounces are in a cup, so divide as follows:

$$32 \text{ fluid ounces} \div 8 = 4 \text{ cups}$$

And 2 cups are in a pint, so divide again:

$$4 \text{ cups} \div 2 = 2 \text{ pints}$$

Therefore, 32 fluid ounces equals 2 pints.

Q. 504 hours = _____ weeks.

A. **3 weeks.** 24 hours are in a day, so divide as follows:

$$504 \text{ hours} \div 24 = 21 \text{ days}$$

And 7 days are in a week, so divide again:

$$21 \text{ days} \div 7 = 3 \text{ weeks}$$

Therefore, 504 hours equals 3 weeks.

1   Answer each of the following questions:

   (a) How many inches are in a yard?

   (b) How many hours are in a week?

   (c) How many ounces are in a ton?

   (d) How many cups are in a gallon?

2   Calculate each of the following:

   (a) 7 quarts = _____ cups

   (b) 5 miles = _____ inches

   (c) 3 gallons = _____ fluid ounces

   (d) 4 days = _____ seconds

3   Answer each of the following questions:

   (a) If you have 420 minutes, how many hours do you have?

   (b) If you have 144 inches, how many yards do you have?

   (c) If you have 22,000 pounds, how many tons do you have?

   (d) If you have 256 fluid ounces, how many gallons do you have?

4   Calculate each of the following:

   (a) 168 inches = _____ feet

   (b) 100 quarts = _____ gallons

   (c) 288 ounces = _____ pounds

   (d) 76 cups = _____ quarts

Looking at the metric system

Like the English system, the metric system provides units of measurement for distance, volume, and so on. Unlike the English system, however, the metric system builds these units using a *basic unit* and a set of *prefixes*.

Table 18-2 shows five important basic units in the metric system.

Table 18-2   Five Basic Metric Units

Measure Of	Basic Metric Unit
Distance	Meter
Volume (capacity)	Liter
Mass (weight)	Gram
Time	Second
Temperature	Degrees Celsius (°C)

For scientific purposes, the metric system has been updated to the more rigorously defined *System of International Units (SI)*. Each basic SI unit correlates directly to a measurable scientific process that defines it. In SI, the kilogram (not the gram) is the basic unit of mass, the kelvin is the basic unit of temperature, and the liter is not considered a basic unit. For technical reasons, scientists tend to use the more rigidly defined SI, but most other people use the looser metric system. In everyday practice, you can think of the units in Table 18-2 as basic units.

Table 18-3 shows ten metric prefixes, with the three most commonly used italicized (see Chapter 17 for more information on powers of ten).

Table 18-3   Ten Metric Prefixes

Prefix	Meaning	Number	Power of Ten
Giga-	One billion	1,000,000,000	10^9
Mega-	One million	1,000,000	10^6
Kilo-	*One thousand*	*1,000*	10^3
Hecta-	One hundred	100	10^2
Deca-	Ten	10	10^1
(none)	One	1	10^0
Deci-	One tenth	0.1	10^{-1}
Centi-	*One hundredth*	*0.01*	10^{-2}
Milli-	*One thousandth*	*0.001*	10^{-3}
Micro-	One millionth	0.000001	10^{-6}
Nano-	One billionth	0.000000001	10^{-9}

Large and small metric units are formed by linking a basic unit with a prefix. For example, linking the prefix *kilo-* to the basic unit *meter* gives you the *kilometer*, which means 1,000 meters. Similarly, linking the prefix *milli-* to the basic unit *liter* gives you the *milliliter*, which means 0.001 (one thousandth) of a liter.

Here's a list giving you the basics.

>> **Units of distance:** The basic metric unit of distance is the meter (m). Other common units are millimeters (mm), centimeters (cm), and kilometers (km):

$$1 \text{ kilometer} = 1,000 \text{ meters}$$
$$1 \text{ meter} = 100 \text{ centimeters}$$
$$1 \text{ meter} = 1,000 \text{ millimeters}$$

>> **Units of fluid volume:** The basic metric unit of fluid volume (also called capacity) is the liter (L). Another common unit is the milliliter (mL):

$$1 \text{ liter} = 1,000 \text{ milliliters}$$

Note: One milliliter is equal to 1 cubic centimeter (cc).

» **Units of mass:** Technically, the metric system measures not weight, but mass. *Weight* is the measurement of how strongly gravity pulls an object toward Earth. *Mass,* however, is the measurement of the amount of matter an object has. If you traveled to the moon, your weight would change, so you would feel lighter. But your mass would remain the same, so all of you would still be there. Unless you're planning a trip into outer space or performing a scientific experiment, you probably don't need to know the difference between weight and mass. In this chapter, you can think of them as equivalent, and I use the word *weight* when referring to metric mass.

The basic unit of weight in the metric system is the gram (g). Even more commonly used, however, is the kilogram (kg):

$$1\,\text{kilogram} = 1{,}000\,\text{grams}$$

Note: 1 kilogram of water has a volume of 1 liter.

» **Units of time:** As in the English system, the basic metric unit of time is a second (s). For most purposes, people also use other English units, such as minutes and hours.

For many scientific purposes, the second is the only unit used to measure time. Large numbers of seconds and small fractions of sections are represented with *scientific notation,* which I cover in Chapter 17.

» **Units of speed:** For most purposes, the most common metric unit of speed (also called velocity) is *kilometers per hour* (km/hr.). Another common unit is *meters per second* (m/s).

» **Units of temperature (degrees Celsius or Centigrade):** The basic metric unit of temperature is the *Celsius degree* (°C), also called the *Centigrade degree.* The Celsius scale is set up so that, at sea level, water freezes at 0°C and boils at 100°C.

Scientists often use another unit — the kelvin (K) — to talk about temperature. The degrees are the same size as in Celsius, but 0 K is set at *absolute zero,* the temperature at which atoms don't move at all. Absolute zero is approximately equal to –273.15°C.

EXAMPLE

Q. How many millimeters are in a meter?

A. **1,000.** The prefix *milli–* means one thousandth, so a millimeter is $\frac{1}{1{,}000}$ of a meter. Therefore, a meter contains 1,000 millimeters.

Q. A dyne is an old unit of force, the push or pull on an object. Using what you know about metric prefixes, how many dynes do you think are in 14 teradynes?

A. **14,000,000,000,000 (14 trillion).** The prefix *tera–* means one trillion, so 1,000,000,000,000 dynes are in a teradyne; therefore,

$$14\,\text{teradynes} = 14 \times 1{,}000{,}000{,}000{,}000\,\text{dynes} = 14{,}000{,}000{,}000{,}000\,\text{dynes}$$

YOUR TURN

5   Give the basic metric unit for each type of measurement listed here:

(a) The amount of vegetable oil for a recipe

(b) The weight of an elephant

(c) How much water a swimming pool can hold

(d) How hot a swimming pool is

(e) How long you can hold your breath

(f) Your height

(g) Your weight

(h) How far you can run

 Write down the number or decimal associated with each of the following metric prefixes:

(a) kilo–

(b) milli–

(c) centi–

(d) mega–

(e) micro–

(f) giga–

(g) nano–

(h) no prefix

7 Answer each of the following questions:

(a) How many centimeters are in a meter?

(b) How many milliliters are in a liter?

(c) How many milligrams are in a kilogram?

(d) How many centimeters are in a kilometer?

8 Using what you know about metric prefixes, calculate each of the following:

(a) 75 kilowatts = _____ watts

(b) 12 seconds = _____ microseconds

(c) 7 megatons = _____ tons

(d) 400 gigaHertz = _____ Hertz

Estimating and Converting between the English and Metric Systems

Most Americans use the English system of measurement all the time and have only a passing acquaintance with the metric system. But metric units are being used more commonly as the units for tools, footraces, soft drinks, and many other things. Also, if you travel abroad, you need to know how far 100 kilometers is or how long you can drive on 10 liters of gasoline.

In this section, I show you how to make ballpark estimates of metric units in terms of English units, which can help you feel more comfortable with metric units. I also show you how to convert between English and metric units, which is a common type of math problem.

When I talk about *estimating*, I mean very loose ways of measuring metric amounts using the English units you are familiar with. In contrast, when I talk about *converting*, I mean using an equation to change from one system of units to the other. Neither method is exact, but converting provides a much closer approximation (and takes longer) than estimating.

Estimating in the metric system

One reason people sometimes feel uncomfortable using the metric system is that, when you're not familiar with it, estimating amounts in practical terms is hard. For example, if I tell you that we're going out to a beach that's $\frac{1}{4}$ mile away, you prepare yourself for a short walk. And if I tell you that it's 10 miles away, you head for the car. But what do you do with the information that the beach is 3 kilometers away?

Similarly, if I tell you that the temperature is 85°F, you'll probably wear a bathing suit or shorts. And if I tell you it's 40°F, you'll probably wear a coat. But what do you wear if I tell you that the temperature is 25°C?

In this section, I give you a few rules of thumb to estimate metric amounts. In each case, I show you how a common metric unit compares with an English unit that you already feel comfortable with.

Approximating short distances: 1 meter is about 1 yard (3 feet)

TIP

Here's how to convert meters to feet: 1 meter ≈ 3.28 feet. But for estimating, use the simple rule that 1 meter is about 1 yard (that is, about 3 feet).

By this estimate, a 6-foot man stands about 2 meters tall. A 15-foot room is 5 meters wide. And a football field that's 100 yards long is about 100 meters long. Similarly, a river with a depth of 4 meters is about 12 feet deep. A mountain that's 3,000 meters tall is about 9,000 feet tall. And a child who is only half a meter tall is about a foot and a half tall.

Estimating longer distances and speed

TIP

Here's how to convert kilometers to miles: 1 kilometer ≈ 0.62 mile. For a ballpark estimate, you can remember that 1 kilometer is about $\frac{1}{2}$ mile. By the same token, 1 kilometer per hour is about $\frac{1}{2}$ mile per hour.

This guideline tells you that if you live 2 miles from the nearest supermarket, then you live about 4 kilometers from there. A marathon of 26 miles is about 52 kilometers. And if you run on a treadmill at 6 miles per hour, then you can run at about 12 kilometers per hour. By the same token, a 10-kilometer race is about 5 miles. If the Tour de France is about 4,000 kilometers, then it's about 2,000 miles. And if light travels about 300,000 kilometers per second, then it travels about 150,000 miles per second.

Approximating volume: 1 liter is about 1 quart (1/4 gallon)

Here's how to convert liters to gallons: 1 liter ≈ 0.26 gallon. A good estimate here is that 1 liter is about 1 quart (a gallon consists of about 4 liters).

TIP

Using this estimate, a gallon of milk is 4 quarts, so it's about 4 liters. If you put 10 gallons of gasoline in your tank, it's about 40 liters. In the other direction, if you buy a 2-liter bottle of cola, you have about 2 quarts. If you buy an aquarium with a 100-liter capacity, it holds about 25 gallons of water. And if a pool holds 8,000 liters of water, it holds about 2,000 gallons.

Estimating weight: 1 kilogram is about 2 pounds

TIP

Here's how to convert kilograms to pounds: 1 kilogram ≈ 2.20 pounds. For estimating, figure that 1 kilogram is equal to about 2 pounds.

By this estimate, a 5-kilogram bag of potatoes weighs about 10 pounds. If you can bench-press 70 kilograms, then you can bench-press about 140 pounds. And because a liter of water weighs exactly 1 kilogram, you know that a quart of water weighs about 2 pounds. Similarly, if a baby weighs 8 pounds at birth, they weigh about 4 kilograms. If you weigh 150 pounds, then you weigh about 75 kilograms. And if your New Year's resolution is to lose 20 pounds, then you want to lose about 10 kilograms.

Estimating temperature

The most common reason for estimating temperature in Celsius is in connection with the weather. The formula for converting from Celsius to Fahrenheit is kind of messy:

$$\text{Fahrenheit} = \text{Celsius} \times \frac{9}{5} + 32$$

Instead, use the handy chart in Table 18-4.

Table 18-4   Comparing Celsius and Fahrenheit Temperatures

Celsius (Centigrade)	Description	Fahrenheit
0°	Cold	32°
10°	Cool	50°
20°	Warm	68°
30°	Hot	86°

Any temperature below 0°C is cold, and any temperature over 30°C is hot. Most of the time, the temperature falls in this middling range. So now you know that when the temperature is 6°C, you want to wear a coat. When it's 14°C, you may want a sweater — or at least long sleeves. And when it's 25°C, head for the beach!

Q. While in training for lacrosse over the summer, Devin typically runs 3 miles and drinks 1 gallon of water per day. What are these approximate amounts in metric units?

EXAMPLE

A. 6 kilometers; 4 liters.

9 Elaine can comfortably carry a backpack of up to 30 kilograms a distance of 30 kilometers per day. What do this weight and this distance approximately equal in English units?

 10 A road is 17 miles long and 27 feet wide. What is its approximate length and width in metric units?

 11 An Olympic-size swimming pool is typically 50 meters long and holds 2,500,000 liters of water. What is its approximate length and capacity in English units?

12 Driving home from college, Brayden filled his gas tank with 12.5 gallons of gasoline and then drove it 400 miles. What are these approximate amounts in metric units? How many kilometers per liter did he average on this trip?

 13 Do you think you would feel comfortable wearing a T-shirt with no jacket when the weather is 35°C? What about when the weather is 20°C? And what about 5°C?

Converting units of measurement

Many books give you one formula for converting from English to metric and another for converting from metric to English. People often find this conversion method confusing because they have trouble remembering which formula to use in which direction.

In this section, I show you a simple way to convert between English and metric units that uses only one formula for each type of conversion.

Here's a nice pair that's easy to remember: 16°C is about 61°F.

Understanding conversion factors

When you multiply any number by 1, that number stays the same. For example, $36 \times 1 = 36$. And when a fraction has the same numerator (top number) and denominator (bottom number), that fraction equals 1 (see Chapter 11 for details). So when you multiply a number by a fraction that equals 1, the number stays the same. For example:

$$36 \times \frac{5}{5} = 36$$

If you multiply a measurement by a special fraction that equals 1, you can switch from one unit of measurement to another without changing the value. People call such fractions *conversion factors*.

Take a look at some equations that show how metric and English units are related (all conversions between English and metric units are approximate):

- » $1 \text{ meter} \approx 3.26 \text{ feet}$
- » $1 \text{ kilometer} \approx 0.62 \text{ mile}$

>> 1 liter ≈ 0.26 gallon

>> 1 kilogram ≈ 2.20 pounds

Because the values on each side of the equations are equal, you can create

>> $\dfrac{1 \text{ meter}}{3.26 \text{ feet}}$ or $\dfrac{3.26 \text{ feet}}{1 \text{ meter}}$

>> $\dfrac{1 \text{ kilometer}}{0.62 \text{ mile}}$ or $\dfrac{0.62 \text{ mile}}{1 \text{ kilometer}}$

>> $\dfrac{1 \text{ liter}}{0.26 \text{ gallon}}$ or $\dfrac{0.26 \text{ gallon}}{1 \text{ liter}}$

>> $\dfrac{1 \text{ kilogram}}{2.2 \text{ pounds}}$ or $\dfrac{2.2 \text{ pounds}}{1 \text{ kilogram}}$

When you understand how units of measurement cancel (which I discuss in the next section), you can easily choose which fractions to use to switch between units of measurement.

Canceling units of measurement

When you're multiplying fractions, you can cancel any factor that appears in both the numerator and the denominator (see Chapter 11 for details). Just as with numbers, you can also cancel out units of measurement in fractions. For example, suppose you want to evaluate this fraction:

$$\dfrac{6 \text{ gallons}}{2 \text{ gallons}}$$

You already know that you can cancel out a factor of 2 in both the numerator and the denominator. But you can also cancel out the unit *gallons* in both the numerator and the denominator:

$$= \dfrac{6 \cancel{\text{ gallons}}}{2 \cancel{\text{ gallons}}}$$

So this fraction simplifies as follows:

$$= 3$$

Converting units

When you understand how to cancel out units in fractions and how to set up fractions equal to 1 (see the preceding sections), you have a foolproof system for converting units of measurement.

Suppose you want to convert 7 meters into feet. Using the equation 1 meter = 3.26 feet, you can make a fraction out of the two values, as follows:

$$\dfrac{1 \text{ meter}}{3.26 \text{ feet}} = 1 \text{ or } \dfrac{3.26 \text{ feet}}{1 \text{ meter}} = 1$$

Both fractions equal 1 because the numerator and the denominator are equal. So you can multiply the quantity you're trying to convert (7 meters) by one of these fractions without changing it. Remember that you want the meters unit to cancel out. You already have the word *meters* in the numerator (to make this clear, place 1 in the denominator), so use the fraction that puts *1 meter* in the denominator:

$$\frac{7\,\text{meters}}{1} \times \frac{3.26\,\text{feet}}{1\,\text{meter}}$$

Now cancel out the unit that appears in both the numerator and the denominator:

$$= \frac{7\,\cancel{\text{meters}}}{1} \times \frac{3.26\,\text{feet}}{1\,\cancel{\text{meters}}}$$

At this point, the only value in the denominator is 1, so you can ignore it. And the only unit left is *feet*, so place it at the end of the expression:

$$= 7 \times 3.26\,\text{feet}$$

Now do the multiplication (Chapter 13 shows how to multiply decimals):

$$= 22.82\,\text{feet}$$

It may seem strange that the answer appears with the units already attached, but that's the beauty of this method: When you set up the right expression, the answer just appears.

Converting between English and Metric Units

To convert between English units and metric units, use the four conversion equations shown in the first column of Table 18-5.

The remaining columns in Table 18-5 show conversion factors (fractions) that you multiply by to convert from metric units to English or from English units to metric. To convert from one unit to another, multiply by the conversion factor and cancel out any unit that appears in both the numerator and denominator.

Table 18-5   Conversion Factors for English and Metric Units

Conversion Equation	English-to-Metric	Metric-to-English
1 meter ≈ 3.26 feet	$\dfrac{1\text{m}}{3.26\text{ ft.}}$	$\dfrac{3.26\text{ ft.}}{1\text{m}}$
1 kilometer ≈ 0.62 mile	$\dfrac{1\text{km}}{0.62\text{mi.}}$	$\dfrac{0.62\text{mi.}}{1\text{km}}$
1 liter ≈ 0.26 gallon	$\dfrac{1\text{L}}{0.26\text{ gal.}}$	$\dfrac{0.26\text{ gal.}}{1\text{ L}}$
1 kilogram ≈ 2.20 pounds	$\dfrac{1\text{kg}}{2.20\text{ lbs.}}$	$\dfrac{2.20\text{ lbs.}}{1\text{kg}}$

Always use the conversion factor that has units you're converting *from* in the *denominator*. For example, to convert from miles to kilometers, use the conversion factor that has miles in the denominator: that is, $\frac{1\text{km}}{0.62\text{mi}}$.

Sometimes, you may want to convert between units for which no direct conversion factor exists. In these cases, set up a *conversion chain* to convert via one or more intermediate units. For instance, to convert centimeters into inches, you may go from centimeters to meters to feet to inches. When the conversion chain is set up correctly, every unit cancels out except for the unit that you're converting to. You can set up a conversion chain of any length to solve a problem.

With a long conversion chain, it's sometimes helpful to take an extra step and turn the whole chain into a single fraction. Place all the numerators above a single fraction bar and all the denominators below it, keeping multiplication signs between the numbers.

A chain can include conversion factors built from any conversion equation in this chapter. For example, you know from the section "Looking at the English system" that 2 pints = 1 quart, so you can use the following two fractions:

$$\frac{2\text{pt.}}{1\text{qt.}} \quad \frac{1\text{qt.}}{2\text{pt.}}$$

Similarly, you know from the section "Looking at the metric system" that 1 kilogram = 1,000 grams, so you can use these two fractions:

$$\frac{1\text{kg}}{1,000\text{g}} \quad \frac{1,000\text{g}}{1\text{kg}}$$

Q. Convert 5 kilometers to miles.

A. **3.1 miles.** To convert from kilometers, multiply 5 kilometers by the conversion factor with kilometers in the denominator:

$$5\text{km} \times \frac{0.62\text{mi.}}{1\text{km}}$$

Now you can cancel the unit *kilometer* in both the numerator and denominator:

$$= 5\,\cancel{\text{km}} \times \frac{0.62\text{mi.}}{1\,\cancel{\text{km}}}$$

Calculate the result:

$$= 5 \times \frac{0.62\text{mi.}}{1} = 3.1\text{mi.}$$

Notice that when you set up the conversion correctly, you don't have to think about the unit — it changes from kilometers to miles automatically!

Q. Convert 21 grams to pounds.

A. **0.0462 pound.** You don't have a conversion factor to convert grams to pounds directly, so set up a conversion chain that makes the following conversion:

grams → kilograms → pounds

To convert grams to kilograms, use the equation 1,000 g = 1 kg. Multiply by the fraction with kilograms in the numerator and grams in the denominator:

$$21\,g \times \frac{1\,kg}{1,000\,g}$$

To convert kilograms to pounds, use the equation 1 kg = 2.2 lb. and multiply by the fraction with pounds in the numerator and kilograms in the denominator:

$$= 21\,g \times \frac{1\,kg}{1,000\,g} \times \frac{2.2\,lb.}{1\,kg}$$

Cancel *grams* and *kilograms* in both the numerator and denominator:

$$= 21\,g \times \frac{1\,\cancel{kg}}{1,000\,g} \times \frac{2.2\,lb.}{1\,\cancel{kg}}$$

Finally, calculate the result:

$$= 21 \times \frac{1}{1,000} \times \frac{2.2\,lb.}{1}$$
$$= 0.021 \times 2.2\,lbs = 0.0462\,lb.$$

After doing the conversion, step back and apply a reasonability test to your answers. Does each answer make sense? For example, when you convert grams to kilograms, the number you end up with should be a lot smaller than the number you started with because each kilogram contains lots of grams.

Q. For a science project, Anita observed that an ant could walk a distance of 1 foot in 3.8 seconds. How many centimeters per second is that, to the nearest hundredth of a centimeter?

A. **8.07 centimeters per second.** Use the conversion equation, 1 foot = 3.8 seconds. Then set up a conversion chain with centimeters in the numerator and seconds in the denominator.

$$\frac{1\,\cancel{ft.}}{3.8\,sec.} \times \frac{1\,\cancel{m}}{3.26\,\cancel{ft.}} \times \frac{100\,cm}{1\,\cancel{m}} \approx 8.07\frac{cm}{sec.}$$

YOUR TURN

14 Convert 8 kilometers to miles.

15 If you weigh 72 kilograms, what's your weight in pounds to the nearest whole pound?

16 If you're 1.8 meters tall, what's your height in inches to the nearest whole inch?

17 Change 100 cups to liters, rounded to the nearest whole liter.

 18 To the nearest whole gram, how many grams are in a ton?

 19 Cathy has decided to cut back to drinking only 5 cups of coffee per day. How many liters of coffee is that, to the nearest tenth of a liter?

20 If 1 milliliter of water weighs exactly 1 gram, what is the weight of the water in an Olympic-size swimming pool that holds 660,000 gallons of water, to the nearest whole kilogram?

21 In 1973, Secretariat ran the 12-furlong Belmont Stakes in 144 seconds — a record that still stands to this day. If 1 furlong equals $\frac{1}{8}$ of a mile, what was Secretariat's average speed in kilometers per hour, to the nearest whole number of units? (*Hint:* Use the conversion equation 12 furlongs = 144 seconds. Then set up a conversion chain with miles in the numerator and seconds in the denominator.)

Practice Questions Answers and Explanations

(1) All these questions ask you to convert a large unit to a smaller one, so use multiplication.

(a) **36 inches.** 12 inches are in a foot and 3 feet are in a yard, so

$$3 \times 12 = 36 \, \text{in.}$$

(b) **168 hours.** 24 hours are in a day and 7 days are in a week, so

$$24 \times 7 = 168 \, \text{hours}$$

(c) **32,000 ounces.** 16 ounces are in a pound and 2,000 pounds are in a ton, so

$$16 \times 2,000 = 32,000 \, \text{oz.}$$

(d) **16 cups.** 2 cups are in a pint, 2 pints are in a quart, and 4 quarts are in a gallon, so

$$2 \times 2 \times 4 = 16 \, \text{c.}$$

(2) All these problems ask you to convert a large unit to a smaller one, so use multiplication:

(a) **28 cups.** 2 cups are in a pint and 2 pints are in a quart, so

$$2 \times 2 = 4 \, \text{c.}$$

So 4 cups are in a quart; therefore,

$$7 \, \text{qt.} = 7 \times 4 \, \text{c.} = 28 \, \text{c.}$$

(b) **316,800 inches.** 12 inches are in a foot and 5,280 feet are in a mile, so

$$12 \times 5,280 = 63,360 \, \text{in.}$$

63,360 inches are in a mile, so

$$5 \, \text{mi.} = 5 \times 63,360 \, \text{in.} = 316,800 \, \text{in.}$$

(c) **384 fluid ounces.** 8 fluid ounces are in a cup, 2 cups are in a pint, 2 pints are in a quart, and 4 quarts are in a gallon, so

$$8 \times 2 \times 2 \times 4 = 128 \, \text{fl.oz.}$$

Therefore, 128 fluid ounces are in a gallon, so

$$3 \, \text{gal.} = 3 \times 128 \, \text{fl.oz.} = 384 \, \text{fl.oz.}$$

(d) 345,600 seconds. 60 seconds are in a minute, 60 minutes are in an hour, and 24 hours are in a day, so

$$60 \times 60 \times 24 = 86,400 \, \text{seconds}$$

(e) 86,400 seconds are in a day, so

$$4 \, \text{days} = 4 \times 86,400 \, \text{seconds} = 345,600 \, \text{seconds}$$

③ All these problems ask you to convert a smaller unit to a larger one, so use division:

(a) 7 hours. There are 60 minutes in an hour, so divide by 60:

$$420 \, \text{minutes} \div 60 = 7 \, \text{hours}$$

(b) 4 yards. There are 12 inches in a foot, so divide by 12:

$$144 \, \text{in.} \div 12 = 12 \, \text{ft.}$$

There are 3 feet in a yard, so divide by 3:

$$12 \, \text{ft.} \div 3 = 4 \, \text{yd.}$$

(c) 11 tons. There are 2,000 pounds in a ton, so divide by 2,000:

$$22,000 \, \text{lb} \div 2,000 = 11 \, \text{tons}$$

(d) 2 gallons. There are 128 fluid ounces in a gallon, so divide by 128: 256 oz./128 oz. = 2.

④ All these problems ask you to convert a smaller unit to a larger one, so use division:

(a) 14 feet. There are 12 inches in a foot, so divide by 12:

$$168 \, \text{in.} \div 12 = 14 \, \text{ft.}$$

(b) 25 gallons. There are 4 quarts in a gallon, so divide by 4:

$$100 \, \text{qt.} \div 4 = 25 \, \text{gal.}$$

(c) 18 pounds. There are 16 ounces in a pound, so divide by 16:

$$288 \, \text{oz.} \div 16 = 18 \, \text{lbs.}$$

(d) 19 quarts. There are 2 cups in a pint, so divide by 2:

$$76 \, \text{cups} \div 2 = 38 \, \text{pints}$$

There are 2 pints in a quart, so divide by 2:

$$38 \, \text{pints} \div 2 = 19 \, \text{quarts}$$

(5) Note that you want only the base unit, without any prefixes such as *kilo, milli,* and so forth.

(a) liters

(b) grams

(c) liters

(d) degrees Celsius (Centigrade)

(e) seconds

(f) meters

(g) grams

(h) meters

(6)

(a) **1,000** $\left(\text{one thousand or } 10^3\right)$

(b) **0.001** $\left(\text{one thousandth or } 10^{-3}\right)$

(c) **0.01** $\left(\text{one hundredth or } 10^{-2}\right)$

(d) **1,000,000** $\left(\text{one million or } 10^6\right)$

(e) **0.000001** $\left(\text{one millionth or } 10^{-6}\right)$

(f) **1,000,000,000** $\left(\text{one billion or } 10^9\right)$

(g) **0.000000001** $\left(\text{one billionth or } 10^{-9}\right)$

(h) **1** $\left(\text{one or } 10^0\right)$

(7)

(a) **100 centimeters.**

(b) **1,000 milliliters.**

(c) **1,000,000 milligrams.** 1,000 milligrams are in a gram and 1,000 grams are in a kilogram, so

$$1,000 \times 1,000 = 1,000,000 \, \text{mg}$$

Therefore, 1,000,000 milligrams are in a kilogram.

(d) **100,000 centimeters.** 100 centimeters are in a meter and 1,000 meters are in a kilometer, so

$$100 \times 1,000 = 100,000 \, \text{cm}$$

Therefore, 100,000 centimeters are in a kilometer.

(8)

(a) **75,000 watts.** The prefix *kilo-* means *one thousand,* so 1,000 watts are in a kilowatt; therefore,

$$75 \, \text{kilowatts} = 75 \times 1,000 \, \text{watts} = 75,000 \, \text{watts}$$

(b) **12,000,000 microseconds.** The prefix *micro-* means *one millionth*, so a micro-second is a millionth of a second. Therefore, 1,000,000 microseconds are in a second. Thus,

$$12 \text{ seconds} = 12 \times 1,000,000 \text{ microseconds} = 12,000,000 \text{ microseconds}$$

(c) **7,000,000 tons.** The prefix *mega-* means *one million*, so 1,000,000 tons are in a megaton; therefore,

$$7 \text{ megatons} = 7 \times 1,000,000 \text{ tons} = 7,000,000 \text{ tons}$$

(d) **400,000,000,000 Hertz.** The prefix *giga-* means *one billion*, so 1,000,000,000 Hertz are in a gigaHertz; thus,

$$400 \text{ gigaHertz} = 400 \times 1,000,000,000 \text{ Hertz} = 400,000,000,000 \text{ Hertz}$$

9 **60 pounds; 15 miles.**

10 **34 kilometers; 9 meters.**

11 **150 feet; 625,000 gallons.**

12 **50 liters; 800 kilometers; 16 kilometers per liter.** $800 \div 50 = 16$.

13 **Yes; yes; no.** A temperature of 35°C is very hot (95°F); 20°C is warm (68°F); 5°C is between cool and cold (41°F).

14 **4.96 miles.** To convert kilometers to miles, multiply by the conversion fraction with miles in the numerator and kilometers in the denominator:

$$8 \text{ km} \times \frac{0.62 \text{ mi.}}{1 \text{ km}}$$

Cancel the unit *kilometer* in both the numerator and denominator:

$$= 8 \cancel{\text{ km}} \times \frac{0.62 \text{ mi.}}{1 \cancel{\text{ km}}}$$

Now calculate the result:

$$= 8 \times 0.62 \text{ mi.} = 4.96 \text{ mi.}$$

15 **158 pounds.** To convert kilograms to pounds, multiply by the conversion factor with pounds in the numerator and kilograms in the denominator:

$$72 \text{ kg} \times \frac{2.2 \text{ lb.}}{1 \text{ kg}}$$

Cancel the unit *kilogram* in both the numerator and denominator:

$$= 72 \, \cancel{kg} \times \frac{2.2 \, \text{lb.}}{1 \, \cancel{kg}}$$

$$= 72 \times 2.2 \, \text{lb.}$$

Multiply to find the answer:

$$= 158.4 \, \text{lb.}$$

Round to the nearest whole pound:

$$\approx 158 \, \text{lb.}$$

(16) **70 inches.** You don't have a conversion factor to change meters to inches directly, so set up a conversion chain as follows:

$$\text{meters} \rightarrow \text{feet} \rightarrow \text{inches}$$

Convert meters to feet with the fraction that puts meters in the denominator:

$$1.8 \, \text{m} \times \frac{3.26 \, \text{ft.}}{1 \, \text{m}}$$

Convert the feet to inches with the conversion factor that has feet in the denominator:

$$= 1.8 \, \text{m} \times \frac{3.26 \, \text{ft.}}{1 \, \text{m}} \times \frac{12 \, \text{in.}}{1 \, \text{ft.}}$$

Cancel the units *meters* and *feet* in both the numerator and denominator:

$$= 1.8 \, \cancel{m} \times \frac{3.26 \, \cancel{\text{ft.}}}{1 \, \cancel{m}} \times \frac{12 \, \text{in.}}{1 \, \cancel{\text{ft.}}}$$

Multiply to find the answer:

$$= 70.416 \, \text{in.}$$

Round the answer to the nearest whole inch:

$$\approx 70 \, \text{in.}$$

(17) **24 liters.** You don't have a conversion factor to change cups to liters directly, so set up a conversion chain:

$$\text{cups} \rightarrow \text{pints} \rightarrow \text{quarts} \rightarrow \text{gallons} \rightarrow \text{liters}$$

Convert cups to pints:

$$100\,c. \times \frac{1\,pt.}{2\,c.}$$

Convert pints to quarts:

$$=100\,c. \times \frac{1\,pt.}{2\,c.} \times \frac{1\,qt.}{2\,pt.}$$

Convert quarts to gallons:

$$=100\,c. \times \frac{1\,pt.}{2\,c.} \times \frac{1\,qt.}{2\,pt.} \times \frac{1\,gal.}{4\,qt.}$$

Convert gallons to liters:

$$=100\,c. \times \frac{1\,pt.}{2\,c.} \times \frac{1\,qt.}{2\,pt.} \times \frac{1\,gal.}{4\,qt.} \times \frac{1\,L}{0.26\,gal.}$$

Now all units *except* liters cancel out:

$$=100\,\cancel{c.} \times \frac{1\,\cancel{pt.}}{2\,\cancel{c.}} \times \frac{1\,\cancel{qt.}}{2\,\cancel{pt.}} \times \frac{1\,\cancel{gal.}}{4\,\cancel{qt.}} \times \frac{1\,L}{0.26\,\cancel{gal.}}$$

$$=100 \times \frac{1}{2} \times \frac{1}{2} \times \frac{1}{4} \times \frac{1\,L}{0.26}$$

To avoid confusion, set this chain up as a single fraction:

$$\frac{100 \times 1 \times 1 \times 1 \times 1\,L}{2 \times 2 \times 4 \times 0.26}$$

$$=\frac{100\,L}{4.16}$$

Use decimal division to find the answer to at least one decimal place:

$$\approx 24.04\,L$$

Round to the nearest whole liter:

$$\approx 24\,L$$

(18) **909,091 grams.**

$$1 \text{ ton} \times \frac{2{,}000 \text{ lbs.}}{1 \text{ ton}} \times \frac{1 \text{ kg}}{2.2 \text{ lbs.}} \times \frac{1{,}000 \text{ g}}{1 \text{ kg}} \approx 909{,}091 \text{ g}$$

19 **1.2 liters.**

$$5 \text{ c.} \times \frac{1 \text{ qt.}}{4 \text{ c.}} \times \frac{1 \text{ gal.}}{4 \text{ qt.}} \times \frac{1 \text{ L}}{0.26 \text{ gal.}} \approx 1.2 \text{ L}$$

20 **2,538,462 kilograms.**

$$660{,}000 \text{ gal.} \times \frac{1 \text{ L}}{0.26 \text{ gal.}} \times \frac{1{,}000 \text{ mL}}{1 \text{ L}} \times \frac{1 \text{ g}}{1 \text{ mL}} \times \frac{1 \text{ kg}}{1{,}000 \text{ g}} \approx 2{,}538{,}462 \text{ kg}$$

21 **60 km per hour.**

$$\frac{1 \text{ km}}{0.62 \text{ mi.}} \times \frac{1 \text{ mi.}}{8 \text{ furlongs}} \times \frac{12 \text{ furlongs}}{144 \text{ sec.}} \times \frac{60 \text{ sec.}}{1 \text{ min.}} \times \frac{60 \text{ min.}}{1 \text{ hr.}} \approx 60 \frac{\text{km}}{\text{hr.}}$$

Ready to test your skills? Then check out the chapter quiz in the next section.

Whaddya Know? Chapter 18 Quiz

The following 12-question quiz will give you a chance to test your knowledge of weights and measures. When you're done, you'll find solutions and explanations in the next section.

 1 How many yard sticks would it take — laid end to end — to cover a mile?

 2 You are using a gallon milk jug to fill a pool. If it will take 78 kiloliters of water to fill the pool, then how many gallon jugs will you need?

3 If Billy weighs 150 pounds, how many ounces is that?

 4 It's 40 weeks until your next birthday. How many minutes do you have to wait?

5 You want to make 80 one-cup servings of eggnog. How many gallons of eggnog will you need?

6 How many hours in April?

7 You have 6,000 seconds to finish the test. How many hours and minutes is that?

8 You caught a whale and were able to prepare 633,600 servings of whale meat that weighed 8 ounces each. How much did the whale weigh in kilograms?

9 Your 12-inch ruler also measures how many centimeters?

10 You are driving your American-made vehicle in Europe and see that the speed limit is 120. This is in kilometers per hour. What is this equivalent to on your speedometer (in miles per hour)?

11 George is preparing for the 1,500-meter freestyle swim competition. His pool is measured in feet, and he swims 4,900 feet during each practice. How close is this to the 1,500 meters?

12 You work 40 hours per week and take 4 weeks of vacation every year. If you never miss a day of work, then how many hours are you working each year?

Answers to Chapter 18 Quiz

1. **1,760.** There are 5,280 feet in one mile, and there are three feet in one yard. You will divide.

$$\frac{5,280 \text{ feet}}{1 \text{ mile}} \times \frac{1 \text{ yard}}{3 \text{ feet}} = \frac{5,280 \text{ feet}}{1 \text{ mile}} \times \frac{1 \text{ yard}}{3 \text{ feet}} = \frac{5,280}{3} = 1,760 \text{ yards}/1 \text{ mile}$$

2. **20,280.** One kiloliter is equal to 1,000 liters, and a liter is about 0.26 gallon.

$$\frac{78 \text{ kiloliters}}{1} \times \frac{1,000 \text{ liters}}{1 \text{ kiloliter}} \times \frac{0.26 \text{ gallon}}{1 \text{ liter}}$$
$$= \frac{78 \text{ kiloliters}}{1} \times \frac{1,000 \text{ liters}}{1 \text{ kiloliter}} \times \frac{0.26 \text{ gallon}}{1 \text{ liter}} = 20,280 \text{ gallon jugs}$$

3. **2,400.** There are 16 ounces in a pound, so you multiply.

$$\frac{150 \text{ pounds}}{1} \times \frac{16 \text{ ounces}}{1 \text{ pound}}$$
$$= \frac{150 \text{ pounds}}{1} \times \frac{16 \text{ ounces}}{1 \text{ pound}} = 150 \times 16 = 2400 \text{ ounces}$$

4. **403,200.** There are 7 days in a week, 24 hours in a day, and 60 minutes in an hour. Multiply!

$$\frac{40 \text{ weeks}}{1} \times \frac{7 \text{ days}}{1 \text{ week}} \times \frac{24 \text{ hours}}{1 \text{ day}} \times \frac{60 \text{ minutes}}{1 \text{ hour}}$$
$$= \frac{40 \text{ weeks}}{1} \times \frac{7 \text{ days}}{1 \text{ week}} \times \frac{24 \text{ hours}}{1 \text{ day}} \times \frac{60 \text{ minutes}}{1 \text{ hour}}$$
$$= 40 \times 7 \times 24 \times 60 = 403,200 \text{ minutes}$$

5. **5.** There are 2 cups in a pint, 2 pints in a quart, and 4 quarts in a gallon. You multiply and divide.

$$\frac{80 \text{ cups}}{1} \times \frac{1 \text{ pint}}{2 \text{ cups}} \times \frac{1 \text{ quart}}{2 \text{ pints}} \times \frac{1 \text{ gallon}}{4 \text{ quarts}}$$
$$= \frac{80 \text{ cups}}{1} \times \frac{1 \text{ pint}}{2 \text{ cups}} \times \frac{1 \text{ quart}}{2 \text{ pint}} \times \frac{1 \text{ gallon}}{4 \text{ quarts}}$$
$$= \frac{80}{2 \times 2 \times 4} = \frac{80}{16} = 5 \text{ gallons}$$

6) **720.** There are 30 days in April and 24 hours in a day. Multiply.

$$\frac{30 \text{ days}}{1} \times \frac{24 \text{ hours}}{1 \text{ day}} = \frac{30 \text{ days}}{1} \times \frac{24 \text{ hours}}{1 \text{ day}} = 30 \times 24 = 720 \text{ hours}$$

7) **1 hour, 40 minutes.** There are 60 seconds in a minute and 60 minutes in an hour. Divide to determine the number of hours. If it doesn't come out even, then change the remainder to minutes.

$$\frac{6{,}000 \text{ seconds}}{1} \times \frac{1 \text{ minute}}{60 \text{ seconds}} \times \frac{1 \text{ hour}}{60 \text{ minutes}}$$
$$= \frac{6{,}000 \text{ seconds}}{1} \times \frac{1 \text{ minute}}{60 \text{ seconds}} \times \frac{1 \text{ hour}}{60 \text{ minutes}}$$
$$= \frac{6{,}000}{60 \times 60} = \frac{6{,}000}{3600} = 1\frac{2}{3} \text{ hours}$$

Multiply $\frac{2}{3}$ times 60 to get the number of minutes: $\frac{2}{3} \times 60 = 40$. You have 1 hour and 40 minutes.

8) **144,000.** First multiply the 633,600 times 8 to get the total number of ounces. There are 16 ounces in a pound and about 2.20 pounds in a kilogram. You'll be dividing.

$$633{,}600 \times 8 = 5{,}068{,}800 \text{ ounces}$$

$$\frac{5{,}068{,}800 \text{ ounces}}{1} \times \frac{1 \text{ pound}}{16 \text{ ounces}} \times \frac{1 \text{ kilogram}}{2.2 \text{ pounds}}$$
$$= \frac{5{,}068{,}800 \text{ ounces}}{1} \times \frac{1 \text{ pound}}{16 \text{ ounces}} \times \frac{1 \text{ kilogram}}{2.2 \text{ pounds}}$$
$$= \frac{5{,}068{,}800}{16 \times 2.2} = \frac{5{,}068{,}800}{35.2} = 144{,}000 \text{ kilograms}$$

9) **30.67.** There are about 3.26 feet in one meter, and there are 100 centimeters in a meter.

$$\frac{3.26 \text{ feet}}{1 \text{ meter}} \times \frac{1 \text{ meter}}{100 \text{ centimeters}} = \frac{3.26 \text{ feet}}{100 \text{ centimeters}}$$

You want just one foot, so divide by 3.26.

$$\frac{3.26 \text{ feet}}{100 \text{ centimeters}} \div \frac{3.26}{3.26} \approx \frac{1 \text{ foot}}{30.67 \text{ centimeters}}$$

10) **74.4.** One kilometer is equivalent to about 0.62 miles per hour.

$$\frac{120 \text{ kilometers}}{1} \times \frac{0.62 \text{ mile}}{1 \text{ kilometer}} = 74.4 \text{ miles per hour}$$

(11) **About 3 feet more.** One meter is about 3.26 feet. Divide the number of feet by 3.26.

$$\frac{4900 \text{ feet}}{1} \times \frac{1 \text{ meter}}{3.26 \text{ feet}} = \frac{4900}{3.26} \approx 1503 \text{ meters}$$

(12) **1,920.** There are 52 weeks in a year, so you are working 52 − 4 = 48 weeks.

$$\frac{48 \text{ weeks}}{1} \times \frac{40 \text{ hours}}{1 \text{ week}} = 48 \times 40 = 1,920 \text{ hours}$$

Chapter **19**

Getting the Picture with Geometry

Geometry is the mathematics of figures such as squares, circles, triangles, and lines. Because geometry is the math of physical space, it's one of the most useful areas of math. Geometry comes into play when measuring rooms or walls in your house, the area of a circular garden, the volume of water in a pool, or the shortest distance across a rectangular field.

Although geometry is usually a year-long course in high school, you may be surprised by how quickly you can pick up what you need to know about basic geometry. Much of what you discover in a geometry course is how to write geometric proofs, which you don't need for algebra — or trigonometry, or even calculus.

In this chapter, I give you a quick and practical overview of geometry. First, I show you four important concepts in plane geometry: points, lines, angles, and shapes. Then I give you the basics on geometric shapes, from flat circles to solid cubes. Finally, I discuss how to measure geometric shapes by finding the area and perimeter of two-dimensional forms and the volume and surface area of some geometric solids.

Of course, if you want to know more about geometry, the ideal place to look beyond this chapter is *Geometry For Dummies*, 3rd Edition, by Mark Ryan (published by Wiley)!

Getting on the Plane: Points, Lines, Angles, and Shapes

Plane geometry is the study of figures on a two-dimensional surface — that is, on a *plane*. You can think of the plane as a piece of paper with no thickness at all. Technically, a plane doesn't end at the edge of the paper — it continues forever.

In this section, I introduce you to four important concepts in plane geometry: points, lines, angles, and shapes (squares, circles, triangles, and so forth).

Making some points

A *point* is a location on a plane. It has no size or shape. Although in reality a point is too small to be seen, you can represent it visually in a drawing by using a dot.

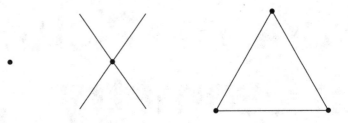

When two lines intersectthey share a single point. Additionally, each corner of a polygon is a point. (Keep reading for more on lines and polygons.)

Knowing your lines

A *line* — also called a *straight line* — is pretty much what it sounds like; it marks the shortest distance between two points, but it extends infinitely in both directions. It has length but no width, making it a one-dimensional (1-D) figure.

Given any two points, you can draw exactly one line that passes through both of them. In other words, two points *determine* a line.

When two lines intersect, they share a single point. When two lines don't intersect, they are *parallel*, which means that they remain the same distance from each other everywhere. A good visual aid for parallel lines is a set of railroad tracks. In geometry, you draw a line with arrows at both ends. Arrows on either end of a line mean that the line goes on forever (as you can see in Chapter 1, where I discuss the number line).

A *line segment* is a piece of a line that has endpoints, as shown here.

A *ray* is a piece of a line that starts at a point and extends infinitely in one direction, kind of like a laser. It has one endpoint and one arrow.

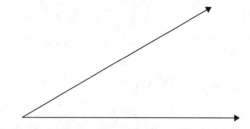

Figuring the angles

An *angle* is formed when two rays extend from the same point.

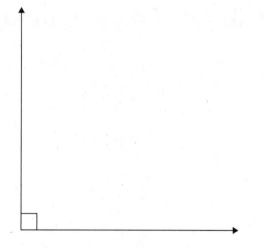

Angles are typically used in carpentry to measure the corners of objects. They're also used in navigation to indicate a sudden change in direction. For example, when you're driving, it's common to distinguish when the angle of a turn is "sharp" or "not so sharp."

The sharpness of an angle is usually measured in *degrees.* The most common angle is the *right angle* — the angle at the corner of a square — which is a 90° (90-degree) angle:

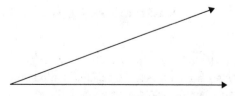

Angles that have fewer than 90° — that is, angles that are sharper than a right angle — are called *acute angles,* like this one:

Angles that measure greater than 90° — that is, angles that aren't as sharp as a right angle — are called *obtuse angles,* as shown here:

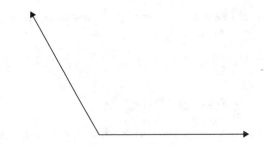

When an angle is exactly 180°, it forms a straight line and is called a *straight angle*.

Shaping things up

A shape is any closed geometrical figure that has an inside and an outside. Circles, squares, triangles, and larger polygons are all examples of shapes.

Much of plane geometry focuses on different types of shapes. In the next section, I show you how to identify a variety of shapes. Later in this chapter, I show you how to measure these shapes.

Getting in Shape: Polygon (and Non-Polygon) Basics

You can divide shapes into two basic types: polygons and non-polygons. A *polygon* has all straight sides, and you can easily classify polygons by the number of sides they have:

Polygon	Number of Sides
Triangle	3
Quadrilateral	4
Pentagon	5
Hexagon	6
Heptagon	7
Octagon	8

Any shape that has at least one curved edge is a *non-polygon*. The most common non-polygon is the circle.

REMEMBER The area of a shape — the space inside — is usually measured in square units, such as square inches (in.²), square meters (m²), or square kilometers (km²). If a problem mixes units of measurement, such as inches and feet, you have to convert to one or the other before doing the math (for more on conversions, see Chapter 18).

Closed Encounters: Shaping Up Your Understanding of 2-D Shapes

REMEMBER

A *shape* is any closed two-dimensional (2-D) geometrical figure that has an inside and an out-side, separated by the *perimeter* (boundary) of the shape. The area of a shape is the measurement of the size inside that shape.

A few shapes that you're probably familiar with include the square, rectangle, and triangle. However, many shapes don't have names, as you can see in Figure 19-1.

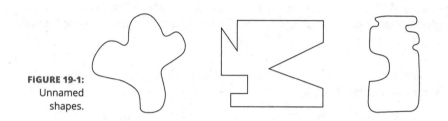

FIGURE 19-1: Unnamed shapes.

Measuring the perimeter and area of shapes is useful for a variety of applications, from land surveying (to get information about a parcel of land that you're measuring) to sewing (to figure out how much material you need for a project). In this section, I introduce you to a variety of geometric shapes. Later in the chapter, I show you how to find the perimeter and area of each, but for now, I just acquaint you with them.

Polygons

A *polygon* is any shape whose sides are all straight. Every polygon has three or more sides (if it had fewer than three, it wouldn't really be a shape at all). Following are a few of the most common polygons.

Triangles

The most basic shape with straight sides is the *triangle*, a three-sided polygon. You find out all about triangles when you study trigonometry (and what better place to begin than *Trigonometry For Dummies*, 2nd Edition, by Mary Jane Sterling [Wiley]). Triangles are classified on the basis of their sides and angles. Take a look at the differences (and see Figure 19-2).

>> **Equilateral:** An *equilateral triangle* has three sides that are all the same length and three angles that all measure 60°.

>> **Isosceles:** An *isosceles triangle* has two sides that are the same length and two equal angles.

>> **Scalene:** *Scalene triangles* have three sides that are all different lengths and three angles that are all unequal.

>> **Right:** A *right triangle* has one right angle. It may be isosceles or scalene.

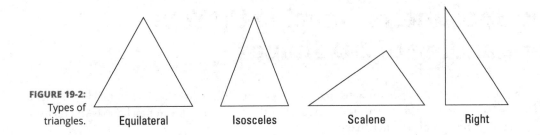

FIGURE 19-2:
Types of triangles.

Equilateral   Isosceles   Scalene   Right

Quadrilaterals

A *quadrilateral* is any shape that has four straight sides. Quadrilaterals are one of the most common shapes you see in daily life. If you doubt this statement, look around and notice that most rooms, doors, windows, and tabletops are quadrilaterals. Here I introduce you to a few common quadrilaterals (Figure 19-3 shows you what they look like).

>> **Square:** A *square* has four right angles and four sides of equal length; also, both pairs of opposite sides (sides directly across from each other) are parallel.

>> **Rectangle:** Like a square, a *rectangle* has four right angles and two pairs of opposite sides that are parallel. Unlike the square, however, although opposite sides are equal in length, sides that share a corner — *adjacent* sides — may have different lengths.

>> **Rhombus:** Imagine starting with a square and collapsing it as if its corners were hinges. This shape is called a *rhombus*. All four sides are equal in length, and both pairs of opposite sides are parallel.

>> **Parallelogram:** Imagine starting with a rectangle and collapsing it as if the corners were hinges. This shape is a *parallelogram* — both pairs of opposite sides are equal in length, and both pairs of opposite sides are parallel.

>> **Trapezoid:** The *trapezoid*'s only important feature is that at least two opposite sides are parallel.

>> **Kite:** A kite is a quadrilateral with two pairs of adjacent sides that are the same length.

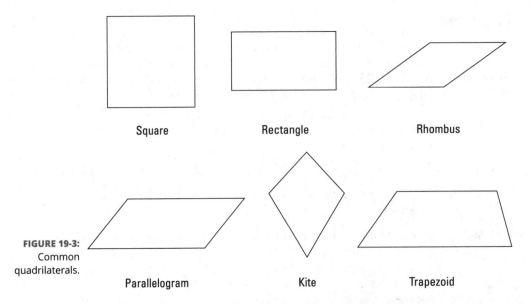

Square   Rectangle   Rhombus

FIGURE 19-3:
Common quadrilaterals.

Parallelogram   Kite   Trapezoid

A quadrilateral can fit into more than one of these categories. For example, every parallelogram (with two sets of parallel sides) is also a trapezoid (with at least one set of parallel sides). Every rectangle and rhombus is also both a parallelogram and a trapezoid. And every square is also all five other types of quadrilaterals. In practice, however, it's common to identify a quadrilateral as descriptively as possible — that is, use the *first* word in the list that accurately describes it.

Polygons on steroids — larger polygons

A polygon can have any number of sides. Polygons with more than four sides aren't as common as triangles and quadrilaterals, but they're still worth knowing about. Larger polygons come in two basic varieties: regular and irregular.

A *regular polygon* has equal sides and equal angles. The most common with more than four sides are regular pentagons (five sides), regular hexagons (six sides), and regular octagons (eight sides). See Figure 19-4.

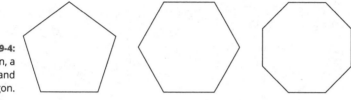

FIGURE 19-4:
A pentagon, a hexagon, and an octagon.

Every other polygon is an *irregular polygon* (see Figure 19-5).

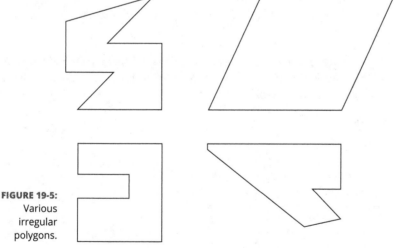

FIGURE 19-5:
Various irregular polygons.

Circles

A circle is the set of all points that are a constant distance from the circle's center. The distance from any point on the circle to its center is called the *radius* of the circle. The distance from any point on the circle straight through the center to the other side of the circle is called the *diameter* of the circle.

Unlike polygons, a circle has no straight edges. The ancient Greeks — who invented much of geometry as we know it today — thought that the circle was the most perfect geometric shape.

Squaring Off with Quadrilaterals

Any shape with four sides is a *quadrilateral*. Quadrilaterals include squares, rectangles, rhombuses, parallelograms, and trapezoids, plus a host of more irregular shapes. In this section, I show you how to find the area (A) and in some cases the perimeter (P) of these five basic types of quadrilaterals.

A *square* has four right angles and four equal sides. To find the area and perimeter of a square, use the following formulas, where s stands for the length of a side (see Figure 19-6):

$$A = s^2$$
$$P = 4 \times s$$

Square

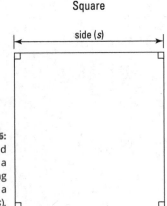

FIGURE 19-6:
The area and perimeter of a square using the length of a side (*s*).

A *rectangle* has four right angles and opposite sides that are equal. The long side of a rectangle is called the *length*, and the short side is called the *width*. To find the area and perimeter of a rectangle, use the following formulas, where l stands for the length of a side and w stands for the width (see Figure 19-7):

$$A = l \times w$$
$$P = 2 \times (l + w)$$

Rectangle

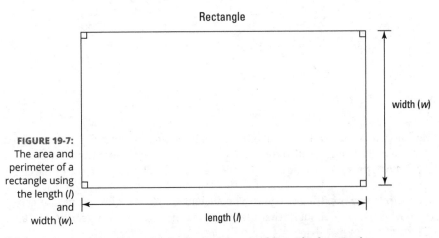

FIGURE 19-7:
The area and perimeter of a rectangle using the length (*l*) and width (*w*).

A *rhombus* resembles a collapsed square. It has four equal sides, but its four angles aren't necessarily right angles. Similarly, a *parallelogram* resembles a collapsed rectangle. Its opposite sides are equal, but its four angles aren't necessarily right angles. To find the area of a rhombus or a parallelogram, use the following formula, where b stands for the length of the base (either the bottom or top side) and h stands for the height (the shortest distance between the two bases); also see Figure 19-8:

$$A = b \times h$$

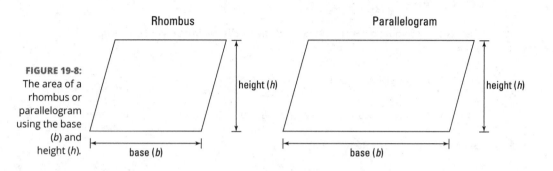

FIGURE 19-8:
The area of a rhombus or parallelogram using the base (*b*) and height (*h*).

A *trapezoid* is a quadrilateral whose only distinguishing feature is that it has two parallel bases (top side and bottom side). To find the area of a trapezoid, use the following formula, where b_1 and b_2 stand for the lengths of the two bases and h stands for the height (the shortest distance between the two bases); also see Figure 19-9:

$$A = \frac{1}{2} \times \left(b_1 + b_2 \right) \times h$$

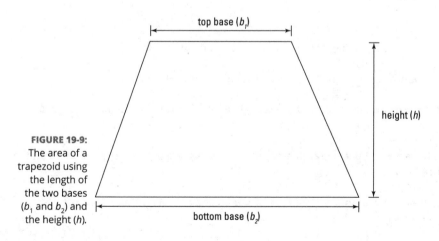

FIGURE 19-9:
The area of a trapezoid using the length of the two bases (*b₁* and *b₂*) and the height (*h*).

Q. Find the area and perimeter of a square with a side that measures 5 inches.

A. The area is 25 square inches, and the perimeter is 20 inches.

$$A = s^2 = (5 \text{ in.})^2 = 25 \text{ in.}^2$$
$$P = 4 \times s = 4 \times 5 \text{ in.} = 20 \text{ in.}$$

Q. Find the area and perimeter of a rectangle with a length of 9 centimeters and a width of 4 centimeters.

A. The area is 36 square centimeters, and the perimeter is 26 centimeters.

$$A = l \times w = 9 \text{ cm} \times 4 \text{ cm} = 36 \text{ cm}^2$$
$$P = 2 \times (l + w) = 2 \times (9 \text{ cm} + 4 \text{ cm}) = 2 \times 13 \text{ cm} = 26 \text{ cm}$$

Q. Find the area of a parallelogram with a base of 4 feet and a height of 3 feet.

A. The area is 12 square feet.

$$A = b \times h = 4 \text{ ft.} \times 3 \text{ ft.} = 12 \text{ ft.}^2$$

Q. Find the area of a trapezoid with bases of 3 meters and 5 meters and a height of 2 meters.

A. The area is 8 square meters.

$$A = \frac{1}{2} \times (b_1 + b_2) \times h$$
$$= \frac{1}{2} \times (3 \text{ m} + 5 \text{ m}) \times 2 \text{ m}$$
$$= \frac{1}{2} \times (8 \text{ m}) \times 2 \text{ m} = 8 \text{ m}^2$$

---

1 What are the area and perimeter of a square with a side of 9 miles?

2 Find the area and perimeter of a square with a side of 31 centimeters.

3 Figure out the area and perimeter of a rectangle with a length of 10 inches and a width of 5 inches.

4 Determine the area and perimeter of a rectangle that has a length of 23 kilometers and a width of 19 kilometers.

5 What's the area of a rhombus with a base of 9 meters and a height of 6 meters?

6 Figure out the area of a parallelogram with a base of 17 yards and a height of 13 yards.

7 Write down the area of a trapezoid with bases of 6 feet and 8 feet and a height of 5 feet.

8 What's the area of a trapezoid that has bases of 15 millimeters and 35 millimeters and a height of 21 millimeters?

Making a Triple Play with Triangles

Any shape with three straight sides is a *triangle.* To find the area of a triangle, use the following formula, in which b is the length of the base (one side of the triangle) and h is the height of the triangle (the shortest distance from the base to the opposite corner); also see Figure 19-10:

$$A = \frac{1}{2} \times b \times h$$

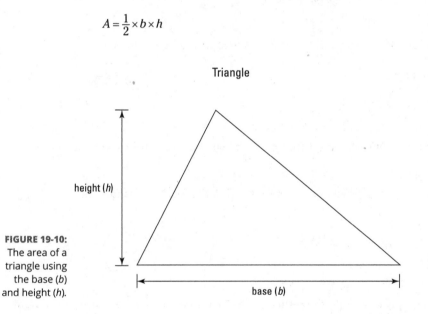

FIGURE 19-10:
The area of a triangle using the base (*b*) and height (*h*).

Any triangle that has a 90-degree angle is called a *right triangle.* The right triangle is one of the most important shapes in geometry. In fact, *trigonometry,* which is devoted entirely to the study of triangles, begins with a set of key insights and observations about right triangles.

The longest side of a right triangle (*c*) is called the *hypotenuse,* and the two short sides (*a* and *b*) are each called *legs.* The most important right triangle formula — called the *Pythagorean Theorem* — allows you to find the length of the hypotenuse given only the length of the legs:

$$a^2 + b^2 = c^2$$

Figure 19-11 shows this theorem in action.

Right Triangle

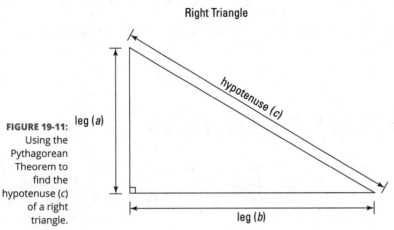

FIGURE 19-11:
Using the
Pythagorean
Theorem to
find the
hypotenuse (c)
of a right
triangle.

leg (a)

hypotenuse (c)

leg (b)

EXAMPLE

Q. Find the area of a triangle with a base of 5 meters and a height of 6 meters.

A. **15 square meters.**

$$A = \frac{1}{2} \times 5 \text{ m} \times 6 \text{ m} = 15 \text{ m}^2$$

Q. Find the hypotenuse of a right triangle with legs that are 6 inches and 8 inches in length.

A. **10 inches.** Use the Pythagorean Theorem to find the value of c as follows:

$$a^2 + b^2 = c^2$$
$$6^2 + 8^2 = c^2$$
$$36 + 64 = c^2$$
$$100 = c^2$$

So when you multiply c by itself, the result is 100. Therefore, c = 10 in., because 10 × 10 = 100.

YOUR TURN

9 What's the area of a triangle with a base of 7 centimeters and a height of 4 centimeters?

10 Find the area of a triangle with a base of 10 kilometers and a height of 17 kilometers.

11 Figure out the area of a triangle with a base of 2 feet and a height of 33 inches.

12 Discover the hypotenuse of a right triangle whose two legs measure 3 miles and 4 miles.

13 What's the hypotenuse of a right triangle with two legs measuring 5 millimeters and 12 millimeters?

14 Calculate the hypotenuse of a right triangle with two legs measuring 8 feet and 15 feet.

Getting Around with Circle Measurements

A *circle* is the set of all points that are a constant distance from a point inside it. Here are a few terms that are handy when talking about circles (see Figure 19-12):

>> The *center* (*c*) of a circle is the point that's the same distance from any point on the circle itself.

>> The *radius* (*r*) of a circle is the distance from the center to any point on the circle.

>> The *diameter* (*d*) of a circle is the distance from any point on the circle through the center to the opposite point on the circle.

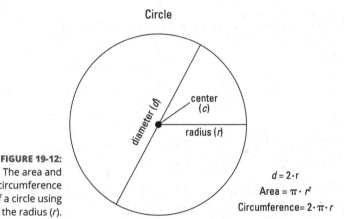

Circle

diameter (*d*)

center (*c*)

radius (*r*)

$d = 2 \cdot r$

Area $= \pi \cdot r^2$

Circumference $= 2 \cdot \pi \cdot r$

FIGURE 19-12: The area and circumference of a circle using the radius (*r*).

To find the area (*A*) of a circle, use the following formula:

$$A = \pi \times r^2$$

The symbol π is called *pi* (pronounced *pie*). It's a decimal that goes on forever, so you can't get an exact value for π. However, the number 3.14 is a good approximation of π that you can use when solving problems that involve circles. (Note that when you use an approximation, the $\approx$ symbol replaces the = sign in problems.)

The perimeter of a circle has a special name: the *circumference* (*C*). The formulas for the circumference of a circle also include π:

$$C = 2 \times \pi \times r$$
$$C = \pi \times d$$

These circumference formulas say the same thing because, as you can see in Figure 19-12, the diameter of a circle is always twice the radius of that circle. That gives you the following formula:

$$d = 2 \times r$$

Q. What's the diameter of a circle that has a radius of 3 inches?

A. **6 inches.**

$$d = 2 \times r = 2 \times 3 \text{ in.} = 6 \text{ in.}$$

Q. What's the approximate area of a circle that has a radius of 10 millimeters?

A. **314 square millimeters.**

$$A = \pi \times r^2$$
$$\approx 3.14 \times (10 \text{ mm})^2$$
$$\approx 3.14 \times 100 \text{ mm}^2$$
$$\approx 314 \text{ mm}^2$$

Q. What's the approximate circumference of a circle that has a radius of 4 feet?

A. **25.12 feet.**

$$C = 2\pi \times r$$
$$\approx 2 \times 3.14 \times 4 \text{ ft.}$$
$$\approx 25.12 \text{ ft.}$$

 YOUR TURN

15 What's the approximate area and circumference of a circle that has a radius of 3 kilometers?

16 Figure out the approximate area and circumference of a circle that has a radius of 12 yards.

17 Write down the approximate area and circumference of a circle with a diameter of 52 centimeters.

18 Find the approximate area and circumference of a circle that has a diameter of 86 inches.

Taking a Trip to Another Dimension: Solid Geometry

Solid geometry is the study of shapes in *space* — that is, the study of shapes in three dimensions. A *solid* is the spatial (three-dimensional, or 3-D) equivalent of a shape. Every solid has an *inside* and an *outside* separated by the surface of the solid. Here, I introduce you to a variety of solids.

The many faces of polyhedrons

A *polyhedron* is the three-dimensional equivalent of a polygon. As you may recall from earlier in the chapter, a polygon is a shape that has only straight sides. Similarly, a polyhedron is a solid that has only straight edges and flat faces (that is, faces that are polygons).

The most common polyhedron is the *cube* (see Figure 19-13). As you can see, a cube has 6 flat faces that are polygons — in this case, all the faces are square — and 12 straight edges. Additionally, a cube has eight *vertexes*, or *vertices* (corners). Later in this chapter, I show you how to measure the surface area and volume of a cube.

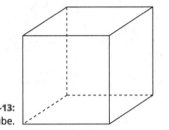

FIGURE 19-13:
A typical cube.

Figure 19-14 shows a few common polyhedrons (or polyhedra).

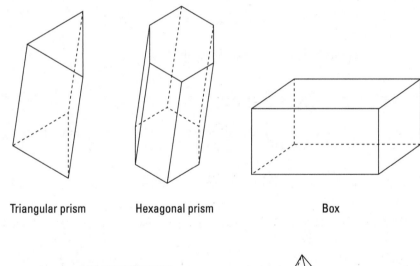

Triangular prism   Hexagonal prism   Box

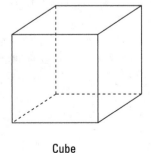 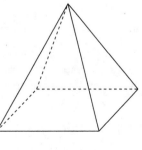

FIGURE 19-14:
Common
polyhedrons.

Cube   Pyramid

Later in this chapter, I show you how to measure each of these polyhedrons to determine its volume — that is, the amount of space contained inside its surface.

One special set of polyhedrons is called the *five regular solids* (see Figure 19-15). Each regular solid has identical faces that are regular polygons. Notice that a cube is a type of regular solid. Similarly, the tetrahedron is a pyramid with four faces that are equilateral triangles.

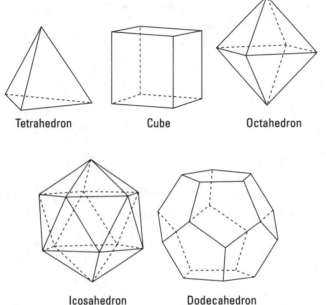

FIGURE 19-15:
The five
regular solids.

Tetrahedron   Cube   Octahedron

Icosahedron   Dodecahedron

3-D shapes with curves

Many solids aren't polyhedrons because they contain at least one curved surface. Here are a few of the most common of these types of solids (also see Figure 19-16).

>> **Sphere:** A *sphere* is the solid, or three-dimensional, equivalent of a circle. A ball is a perfect visual aid for a sphere.

>> **Cylinder:** A *cylinder* has a circular base and extends vertically from the plane. A good visual aid for a cylinder is a can of soup.

>> **Cone:** A *cone* is a solid with a round base that extends vertically to a single point. A good visual aid for a cone is an ice-cream cone.

In the next section, I show you how to measure a sphere and a cylinder to determine their volume — that is, the amount of space contained within.

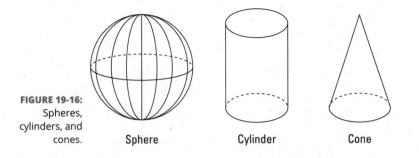

FIGURE 19-16:
Spheres,
cylinders, and
cones.

Sphere   Cylinder   Cone

Building Solid Measurement Skills

Solids take you into the real world, the third dimension. One of the simplest solids is the *cube*, a solid with six identical square faces. To find the volume of a cube, use the following formula, where *s* is the length of the side of any one face (check out Figure 19-17):

$$V = s^3$$

Cube

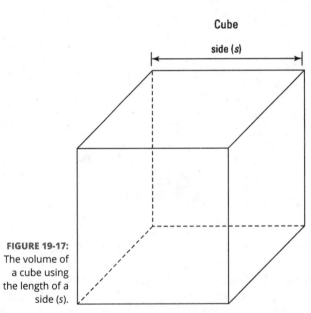

side (*s*)

FIGURE 19-17:
The volume of
a cube using
the length of a
side (*s*).

A *box* (also called a *rectangular solid*) is a solid with six rectangular faces. To find the volume of a box, use the following formula, where *l* is the length, *w* is the width, and *h* is the height (see Figure 19-18):

$$V = l \times w \times h$$

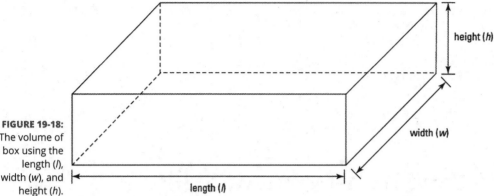

FIGURE 19-18:
The volume of
a box using the
length (*l*),
width (*w*), and
height (*h*).

A *prism* is a solid with two identical bases and a constant cross-section — that is, whenever you slice a prism parallel to the bases, the cross-section is the same size and shape as the bases. A *cylinder* is a solid with two identical circular bases and a constant cross-section. To find the volume of a prism or cylinder, use the following formula, where A_b is the area of the base and *h* is the height (see Figure 19-19). You can find the area formulas throughout this chapter:

$$V = A_b \times h$$

Prism   Cylinder

FIGURE 19-19:
The volume of
a prism or
cylinder using
the area of the
base (A_b) and
the height (*h*).

height (*h*)   area of base (A_b)   height (*h*)   area of base (A_b)

A *pyramid* is a solid that has a base that's a polygon (a shape with straight sides), with straight lines that extend from the sides of the base to meet at a single point. Similarly, a *cone* is a solid that has a base that's a circle, with straight lines extending from every point on the edge of the base to meet at a single point. The formula for the volume of a pyramid is the same as for the volume of a cone. In this formula, illustrated in Figure 19-20, A_b is the area of the base, and *h* is the height:

$$V = \frac{1}{3} \times A_b \times h$$

A *sphere* is a perfectly round solid, such as a ball. In this formula, illustrated in Figure 19-21, *r* is the radius — that is, the distance from the center to any point on the sphere:

$$V = \frac{4}{3} \times \pi \times r^3$$

Pyramid   Cone

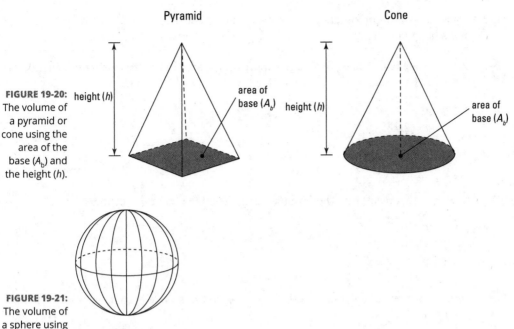

FIGURE 19-20: height (*h*)   area of   height (*h*)   area of
The volume of   base (*A_b*)   base (*A_b*)
a pyramid or
cone using the
area of the
base (*A_b*) and
the height (*h*).

FIGURE 19-21:
The volume of
a sphere using
the radius (*r*).   Sphere

REMEMBER

Volume measurements are usually in cubic units, such as cubic centimeters $\left(cm^3\right)$ or cubic feet $\left(ft.^3\right)$.

EXAMPLE

Q. What's the volume of a cube with a side that measures 4 centimeters?

A. **64 cubic centimeters.**

$$V = s^3 = (4 \text{ cm})^3 = 4 \text{ cm} \times 4 \text{ cm} \times 4 \text{ cm} = 64 \text{ cm}^3$$

Q. Calculate the volume of a box with a length of 7 inches, a width of 4 inches, and a height of 2 inches.

A. **56 cubic inches.**

$$V = l \times w \times h = 7 \text{ in.} \times 4 \text{ in.} \times 2 \text{ in.} = 56 \text{ in.}^3$$

Q. Find the volume of a prism with a base that has an area of 6 square centimeters and a height of 3 centimeters.

A. **18 cubic centimeters.**

$$V = A_b \times h = 6 \text{ cm}^2 \times 3 \text{ cm} = 18 \text{ cm}^3$$

Q. Figure out the approximate volume of a cylinder with a base that has a radius of 2 feet and a height of 8 feet.

A. **100.48 cubic feet.** To begin, find the approximate area of the base using the formula for the area of a circle:

$$A_b = \pi \times r^2$$
$$\approx 3.14 \times (2 \text{ ft.})^2$$
$$\approx 3.14 \times 4 \text{ ft.}^2$$
$$\approx 12.56 \text{ ft.}^2$$

Now plug this result into the formula for the volume of a prism/cylinder:

$$V = A_b \times h$$
$$= 12.56 \text{ ft.}^2 \times 8 \text{ ft.}$$
$$= 100.48 \text{ ft.}^3$$

Q. Find the volume of a pyramid with a square base whose side is 10 inches and with a height of 6 inches.

A. **200 cubic inches.** First, find the area of the base using the formula for the area of a square:

$$A_b = s^2 = (10 \text{ in.})^2 = 100 \text{ in.}^2$$

Now plug this result into the formula for the volume of a pyramid/cone:

$$V = \frac{1}{3} \times A_b \times h$$
$$= \frac{1}{3} \times 100 \text{ in.}^2 \times 6 \text{ in.}$$
$$= 200 \text{ in.}^3$$

Q. Find the approximate volume of a cone with a base that has a radius of 2 meters and with a height of 3 meters.

A. **12.56 cubic meters.** First, find the approximate area of the base using the formula for the area of a circle:

$$A_b = \pi \times r^2$$
$$\approx 3.14 \times (2 \text{ m})^2$$
$$\approx 3.14 \times 4 \text{ m}^2$$
$$= 12.56 \text{ m}^2$$

Now plug this result into the formula for the volume of a pyramid/cone:

$$V = \frac{1}{3} \times A_b \times h$$
$$= \frac{1}{3} \times 12.56 \text{ m}^2 \times 3 \text{ m}$$
$$= 12.56 \text{ m}^3$$

Find the volume of a cube that has a side of 19 meters.

 Figure out the volume of a box with a length of 18 centimeters, a width of 14 centimeters, and a height of 10 centimeters.

21 Figure out the approximate volume of a cylinder whose base has a radius of 7 millimeters and whose height is 16 millimeters.

22 Find the approximate volume of a cone whose base has a radius of 3 inches and whose height is 8 inches.

23 What is the volume of a ball that has a radius of 4 feet?

Solving Geometry Word Problems

Some geometry word problems present you with a picture. In other cases, you have to draw a picture yourself. Sketching figures is always a good idea because it can usually give you an idea of how to proceed. The following sections present you with both types of problems. (To solve these word problems, you need some of the geometry formulas I discuss in the previous section.)

Working from words and images

Sometimes you have to interpret a picture to solve a word problem. Read the problem carefully, recognize shapes in the drawing, pay attention to labels, and use whatever formulas you have to help you answer the question. In this problem, you get to work with a picture.

> Mr. Dennis is a farmer with two teenage sons. He gave them a rectangular piece of land with a creek running through it diagonally, as shown in Figure 19-22. The elder boy took the larger area, and the teenaged boy took the smaller. What is the area of each boy's land in square feet?

To find the area of the smaller triangular plot, use the formula for the area of a triangle, where A is the area, b is the base, and h is the height:

$$A = \frac{1}{2}(b \times h)$$

The whole piece of land is a rectangle, so you know that the corner the triangle shares with the rectangle is a right angle. Therefore, you know that the sides labeled 200 feet and 250 feet are the base and height. Find the area of this plot by plugging the base and height into the formula:

$$A = \frac{200 \text{ feet} \times 250 \text{ feet}}{2}$$

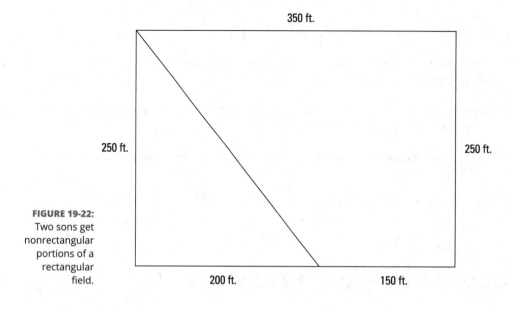

FIGURE 19-22:
Two sons get nonrectangular portions of a rectangular field.

350 ft.

250 ft.　　　　　　　　　　　　　　　　　　250 ft.

200 ft.　　　　　　　　　150 ft.

To make this calculation a little easier, notice that you can cancel a factor of 2 from the numerator and denominator:

$$A = \frac{\cancel{200}^{100} \text{ feet} \times 250 \text{ feet}}{\cancel{2}_{1}} = 25,000 \text{ square feet}$$

The shape of the remaining area is a trapezoid. You can find its area by using the formula for a trapezoid, but there's an easier way. Because you know the area of the triangular plot, you can use this word equation to find the area of the trapezoid:

area of trapezoid = area of whole plot − area of triangle

To find the area of the whole plot, remember the formula for the area of a rectangle. Plug its length and width into the formula:

$$A = \text{length} \times \text{width}$$
$$A = 350 \text{ ft.} \times 250 \text{ ft.}$$
$$A = 87,500 \text{ square ft.}^2$$

Now just substitute the numbers that you know into the word equation you set up:

$$\text{Area of trapezoid} = 87,500 \text{ square feet} - 25,000$$
$$= 62,500 \text{ square feet}$$

So the area of the elder boy's land is 62,500 square feet, and the area of the younger boy's land is 25,000 square feet.

Breaking out those sketching skills

Geometry word problems may not make much sense until you draw some pictures. Here's an example of a geometry problem without a picture provided:

> In Elmwood Park, the flagpole is due south of the swing set and exactly 20 meters due west of the tree house. If the area of the triangle made by the flagpole, the swing set, and the tree house is 150 square meters, what is the distance from the swing set to the tree house?

This problem is bound to be confusing until you draw a picture of what it's telling you. Start with the first sentence, depicted in Figure 19-23. As you can see, I've drawn a right triangle whose corners are the swing set (S), the flagpole (F), and the tree house (T). I've also labeled the distance from the flagpole to the tree house as 20 meters.

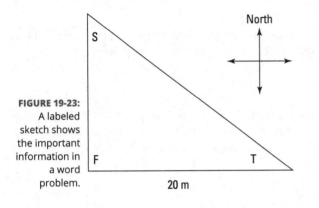

North

FIGURE 19-23: A labeled sketch shows the important information in a word problem.

20 m

The next sentence tells you the area of this triangle:

$$A = 150\,\text{m}^2$$

Now you're out of information, so you need to remember anything you can from geometry. Because you know the area of the triangle, you may find the formula for the area of a triangle helpful:

$$A = \frac{1}{2}(b \times h)$$

Here b is the base and h is the height. In this case, you have a right triangle, so the base is the distance from F to T, and the height is the distance from S to F. So you already know the area of the triangle, and you also know the length of the base. Fill in the equation:

$$150 = \frac{1}{2}(20 \times h)$$

You can now solve this equation for h. Start by simplifying:

$$150 = 10 \times h$$
$$15 = h$$

Now you know that the height of the triangle is 15 meters, so you can add this information to your picture (see Figure 19-24).

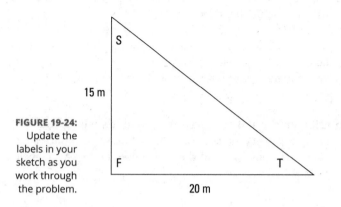

FIGURE 19-24:
Update the labels in your sketch as you work through the problem.

To solve the problem, though, you still need to find out the distance from S to T. Because this is a right triangle, you can use the Pythagorean Theorem to figure out the distance:

$$a^2 + b^2 = c^2$$

Remember that a and b are the lengths of the short sides, and c is the length of the longest side, called the *hypotenuse*. (See the previous section for more on the Pythagorean Theorem.) You can substitute numbers into this formula and solve, as follows:

$$15^2 + 20^2 = c^2$$
$$225 + 400 = c^2$$
$$625 = c^2$$
$$\sqrt{625} = \sqrt{c^2}$$
$$25 = c$$

So the distance from the swing set to the tree house is 25 meters.

Q. The long side of a rectangular field is 150 meters, and the diagonal length from one corner of the field to the opposite corner measures 170 meters. What is the area of the field?

EXAMPLE

A. **12,000 square meters.** Plug in the length of the field into the Pythagorean Theorem as a and the diagonal length as c, and solve for b:

$$a^2 + b^2 = c^2$$
$$150^2 + b^2 = 170^2$$
$$22,500 + b^2 = 28,900$$
$$b^2 = 6,400$$
$$b = 80$$

Thus, the field has a width of 80 meters and a length of 150 meters:

$$A = 150 \times 80 = 12,000$$

24 Becca and Sarah are sisters who share a room that measures 15 feet by 16 feet. Last year, they had a big fight and drew a diagonal line down the middle of the room, giving each girl the triangular half of the room containing her bed and desk. Can you find the area of each triangular half of the room?

25 A guard dog is kept in a circular area of approximately 1,017 square feet, limited by a chain attached to a stake in the ground. What is the length of the chain, to the nearest foot? (Use 3.14 to approximate the value of π.)

26 If a rectangular room has a width of 13 feet and a perimeter of 64 feet, what is its area?

27 The Great Pyramid at Giza has a height of 147 meters and a square base that measures 230 meters on each side. What is its volume?

28 A sign outside of a bowling alley includes a statue of a large bowling ball hitting ten pins. If the height of the bowling ball is 3.2 meters, what is its volume in cubic meters, to the nearest thousandth. (Use 3.14 to approximate the value of π.)

29 A small can of soup has a height of 4 inches and a radius of 1.5 inches, while a large can has a height of 6 inches and a radius of 2 inches. By volume, how much greater is the capacity of the large can than that of the small can? (Use 3.14 to approximate the value of π.)

30 A room has a height of 11 feet and a length of 40 feet. If the volume of the room measures 13,200 cubic feet, what is the distance from one corner of the floor to the opposite corner of the floor?

31 A circular room in a museum has a diameter of 8 yards. At the center of the room is a large square pedestal whose edge measures 3 yards. To the nearest whole square yard, what is the area of the floor not covered by the pedestal? (Use 3.14 to approximate the value of π.)

Practice Questions Answers and Explanations

(1) **Area is 81 square miles; perimeter is 36 miles.** Use the formulas for the area and perimeter of a square:

$$A = s^2 = (9\,\text{mi.})^2 = 81\,\text{mi.}^2$$
$$P = 4 \times s = 4 \times 9\,\text{mi.} = 36\,\text{mi.}$$

(2) **Area is 961 square centimeters; perimeter is 124 centimeters.** Plug in 31 cm for s in the formulas for the area and perimeter of a square.

$$A = s^2 = (31\,\text{cm})^2 = 961\,\text{cm}^2$$
$$P = 4 \times s = 4 \times 31\,\text{cm} = 124\,\text{cm}$$

(3) **Area is 50 square inches; perimeter is 30 inches.** Plug the length and width into the area and perimeter formulas for a rectangle:

$$A = l \times w = 10\,\text{in.} \times 5\,\text{in.} = 50\,\text{in.}^2$$
$$P = 2 \times (l + w) = 2 \times (10\,\text{in.} + 5\,\text{in.}) = 30\,\text{in.}$$

(4) **Area is 437 square kilometers; perimeter is 84 kilometers.** Use the rectangle area and perimeter formulas:

$$A = l \times w = 23\,\text{km} \times 19\,\text{km} = 437\,\text{km}^2$$
$$P = 2 \times (l + w) = 2 \times (23\,\text{km} + 19\,\text{km}) = 84\,\text{km}$$

(5) **54 square meters.** Use the parallelogram/rhombus area formula:

$$A = b \times h = 9\,\text{m} \times 6\,\text{m} = 54\,\text{m}^2$$

(6) **221 square yards.** Use the parallelogram/rhombus area formula:

$$A = b \times h = 17\,\text{yd.} \times 13\,\text{yd.} = 221\,\text{yd.}^2$$

(7) **35 square feet.** Plug your numbers into the trapezoid area formula:

$$A = \frac{1}{2} \times (b_1 + b_2) \times h$$
$$= \frac{1}{2} \times (6\,\text{ft.} + 8\,\text{ft.}) \times 5\,\text{ft.}$$
$$= \frac{1}{2} \times 14\,\text{ft.} \times 5\,\text{ft.}$$
$$= 35\,\text{ft.}^2$$

(8) **525 square millimeters.** Use the trapezoid area formula:

$$A = \frac{1}{2} \times (b_1 + b_2) \times h$$
$$= \frac{1}{2} \times (15\,\text{mm} + 35\,\text{mm}) \times 21\,\text{mm}$$
$$= \frac{1}{2} \times 50\,\text{mm} \times 21$$
$$= 525\,\text{mm}^2$$

(9) **14 square centimeters.** Use the triangle area formula:

$$A = \frac{1}{2} \times b \times h = \frac{1}{2} \times 7\,\text{cm} \times 4\,\text{cm} = 14\,\text{cm}^2$$

(10) **85 square kilometers.** Plug in the numbers for the base and height of the triangle:

$$A = \frac{1}{2} \times b \times h = \frac{1}{2} \times 10\,\text{km} \times 17\,\text{km} = 85\,\text{km}^2$$

(11) **396 square inches.** First, convert feet to inches. Twelve inches are in 1 foot:

$$2\,\text{ft.} = 24\,\text{in.}$$

Now use the area formula for a triangle:

$$A = \frac{1}{2} \times b \times h$$
$$= \frac{1}{2} \times 24\,\text{in.} \times 33\,\text{in.}$$
$$= 396\,\text{in.}^2$$

Note: If you instead converted from inches to feet, the answer 2.75 square feet is also correct.

(12) **5 miles.** Use the Pythagorean Theorem to find the value of c as follows:

$$a^2 + b^2 = c^2$$
$$3^2 + 4^2 = c^2$$
$$9 + 16 = c^2$$
$$25 = c^2$$

When you multiply c by itself, the result is 25. Therefore,

$$c = 5\,\text{mi.}$$

(13) **13 millimeters.** Use the Pythagorean Theorem to find the value of c:

$$a^2 + b^2 = c^2$$
$$5^2 + 12^2 = c^2$$
$$25 + 144 = c^2$$
$$169 = c^2$$

When you multiply c by itself, the result is 169. The hypotenuse is longer than the longest leg, so c has to be greater than 12. Use trial and error, starting with 13:

$$13^2 = 169$$

Therefore, the hypotenuse is 13 mm.

(14) **17 feet.** Use the Pythagorean Theorem to find the value of c:

$$a^2 + b^2 = c^2$$
$$8^2 + 15^2 = c^2$$
$$64 + 225 = c^2$$
$$289 = c^2$$

When you multiply c by itself, the result is 289. The hypotenuse is longer than the longest leg, so c has to be greater than 15. Use trial and error, starting with 16:

$$16^2 = 256$$
$$17^2 = 289$$

Therefore, the hypotenuse is 17 ft.

(15) **Approximate area is 28.26 square kilometers; approximate circumference is 18.84 kilometers.** Use the area formula for a circle to find the area:

$$A = \pi \times r^2$$
$$\approx 3.14 \times 3^2$$
$$= 3.14 \times 9$$
$$= 28.26$$

Use the circumference formula to find the circumference:

$$C = 2 \times \pi \times r$$
$$\approx 2 \times 3.14 \times 3$$
$$= 18.84$$

16 **Approximate area is 452.16 square yards; approximate circumference is 75.36 yards.** Use the area formula for a circle to find the area:

$$A = \pi \times r^2$$
$$\approx 3.14 \times 12^2$$
$$= 3.14 \times 144$$
$$= 452.16$$

Use the circumference formula to find the circumference:

$$C = 2 \times \pi \times r$$
$$\approx 2 \times 3.14 \times 12$$
$$= 75.36$$

17 **Approximate area is 2,122.64 square centimeters; approximate circumference is 163.28 centimeters.** The diameter is 52 cm, so the radius is half of that, which is 26 cm. Use the area formula for a circle to find the area:

$$A = \pi \times r^2$$
$$\approx 3.14 \times (26 \text{ cm})^2$$
$$= 3.14 \times 676 \text{ cm}^2$$
$$= 2,122.64 \text{ cm}^2$$

Use the circumference formula to find the circumference:

$$C = 2\pi \times r$$
$$\approx 2 \times 3.14 \times 26 \text{ cm}$$
$$= 163.28 \text{ cm}$$

18 **Approximate area is 5,805.86 square inches; approximate circumference is 270.04 inches.** The diameter is 86 in., so the radius is half of that, which is 43 in. Use the area formula for a circle to find the area:

$$A = \pi \times r^2$$
$$\approx 3.14 \times (43 \text{ in.})^2$$
$$= 3.14 \times 1,849 \text{ in.}^2$$
$$= 5,805.86 \text{ in.}^2$$

Use the circumference formula to find the circumference:

$$C = 2\pi \times r$$
$$\approx 2 \times 3.14 \times 43 \text{ in.}$$
$$= 270.04 \text{ in.}$$

19 **6,859 cubic meters.** Substitute 19 m for s in the cube volume formula:

$$V = s^3 = (19 \text{ m})^3 = 6{,}859 \text{ m}^3$$

20 **2,520 cubic centimeters.** Use the box area formula:

$$V = l \times w \times h$$
$$= 18 \text{ cm} \times 14 \text{ cm} \times 10 \text{ cm} = 2{,}520 \text{ cm}^3$$

21 **Approximately 2,461.76 cubic millimeters.** First, use the area formula for a circle to find the area of the base:

$$A_b = \pi \times r^2$$
$$\approx 3.14 \times (7 \text{ mm})^2$$
$$= 3.14 \times 49 \text{ mm}^2$$
$$= 153.86 \text{ mm}^2$$

Plug this result into the formula for the volume of a prism/cylinder:

$$V = A_b \times h$$
$$= 153.86 \text{ mm}^2 \times 16 \text{ mm} = 2{,}461.76 \text{ mm}^3$$

22 **Approximately 75.36 cubic inches.** Use the area formula for a circle to find the area of the base:

$$A_b = \pi \times r^2$$
$$\approx 3.14 \times (3 \text{ in.})^2$$
$$= 3.14 \times 9 \text{ in.}^2$$
$$= 28.26 \text{ in.}^2$$

Plug this result into the formula for the volume of a pyramid/cone:

$$V = \frac{1}{3} \times A_b \times h = \frac{1}{3} \times 28.26 \text{ in.}^2 \times 8 \text{ in.} = 75.36 \text{ in.}^3$$

23 **Approximately 269.95 cubic feet.** Plug the radius into the formula for the volume of a sphere:

$$V = \frac{4}{3} \times \pi \times r^3$$
$$\approx \frac{4}{3} \times \pi \times (4 \text{ ft.})^3$$
$$\approx \frac{4}{3} \times 3.14 \times 64 \text{ ft.}^3$$
$$\approx 267.95 \text{ ft.}^3$$

24 **120 square feet.** Plug the length and width of the room into the formula for the area of a triangle:

$$A = \frac{1}{2}bh = \frac{1}{2} \times 15 \times 16 = 120$$

25 **18 feet.** Plug 1,017 into the formula for the area of a circle, and solve for the radius:

$$A = \pi r^2$$
$$1{,}017 = 3.14r^2$$
$$323.88 \approx r^2$$
$$18 \approx r$$

26 **247 square feet.** Find the length of the room using the formula for the perimeter of a rectangle:

$$P = 2(l + w)$$
$$64 = 2(l + 13)$$
$$64 = 2l + 26$$
$$38 = 2l$$
$$19 = l$$

So, the length of the room is 19 feet and the width is 13 feet. Plug these values into the formula for the area of a rectangle:

$$A = 19 \times 13 = 247$$

27 **2,592,100 cubic meters.** Plug the length of the side and the height of the pyramid into the formula for the volume of a pyramid:

$$V = \frac{1}{3}s^2h = \frac{1}{3}230^2(147) = 2{,}592{,}100$$

28 **17.149 cubic meters.** The height of the bowling ball is 3.2 meters, so its radius is 1.6 meters. Plug this value into the formula for the volume of a sphere:

$$V = \frac{4}{3}\pi r^3 = \frac{4}{3}(3.14)(1.6)^3 \approx 17.149$$

29 **47.1 cubic inches.** Begin by finding the volume of the two cans, using the formula for the volume of a cylinder:

$$\text{Volume of small can} = \pi r^2 h = (3.14)(1.5^2)(4) = 28.26$$
$$\text{Volume of large can} = \pi r^2 h = (3.14)(2^2)(6) = 75.36$$

Now, subtract the volume of the large can minus that of the small can:

$$75.36 - 28.26 = 47.1$$

(30) **50 feet.** To begin, plug the room's volume, height, and length into the formula for the volume of a box and calculate the width:

$$V = lwh$$
$$13,200 = (40)w(11)$$
$$13,200 = 440w$$
$$30 = w$$

Now, plug the width and the length into the Pythagorean Theorem as a and b and solve for c:

$$a^2 + b^2 = c^2$$
$$30^2 + 40^2 = c^2$$
$$900 + 1,600 = c^2$$
$$2,500 = c^2$$
$$50 = c$$

(31) **41.24 square yards.** The room is a circle with a diameter of 8 yards, so its radius is 4:

$$\text{Area of room} = \pi r^2 = \pi\left(4^2\right) = 16\pi \approx 50.24$$

The pedestal is a square with a side of 3 yards:

$$\text{Area of pedestal} = s^2 = 3^2 = 9$$

Now, subtract the area of the room minus the area of the pedestal:

$$50.24 - 9 = 41.24$$

Ready to test your geometry skills? Then take the chapter quiz in the next section.

Whaddya Know? Chapter 19 Quiz

This quiz includes 20 questions that test the geometry skills covered in this chapter. When you're finished, flip to the next section for answers and complete explanations.

1. A right triangle has two legs measuring 7 feet and 24 feet. What is the measure of the hypotenuse?

2. A square has sides measuring 12 inches. What is its area?

3. A circle has a radius measuring 10 yards. What is its approximate area?

4. A rectangle measures 7 feet by 14 feet. What is its perimeter?

5. A trapezoid has two parallel sides that measure 6 yards and 12 yards. The height measured between those parallel sides is 4 yards. What is the area of the trapezoid?

6. Charlie and Clair both left home at the same time riding their bikes. Charlie rode due north and Clair rode due east. Two hours later, Charlie was 10 miles from home and Clair was 24 miles from home. Assuming their routes stayed on that 90-degree angle for the entire rides, how far apart were they?

7. A cube has edges measuring 8 feet. What is its volume?

8. A triangle measures 16 feet high from its base, which measures 10 feet. What is the triangle's area?

9. What is the volume of a rectangular prism whose length, width, and height are 2.3 feet, 1.2 feet, and 5 feet, respectively?

10. A sphere has a radius measuring 9 inches. What is its approximate volume?

11. The perimeter of a square is 40 yards. What is the area of that square?

12. A cylindrical can stands 6 inches tall. Its circular base has a radius of 2 inches. What is its approximate volume?

13. A rectangular parcel of land measures 100 kilometers by 430 kilometers. What is its area?

14. An irrigation system sprinkler waters a circular area with a radius of 1,600 feet. If the sprinkler head is in the middle of a square piece of land, then approximately how much of the land in the square does not get watered?

15. A triangle has a base measuring 10 yards and a height of 11 feet. What is its area?

16. A square has sides each measuring 13 feet. What is its perimeter?

17. A circle has a radius of 6 inches. What is its approximate circumference?

18 Stella wants to paint her room. She'll save enough money to purchase the paint and do it herself! Her room measures 18 feet by 20 feet, and the height of the room is 9 feet. If she's just painting the four walls, then what is the square footage she has to cover? (Don't take into account any windows or the door.)

19 A rhombus measures 6 inches on each side and has a height between the parallel sides of 4.5 inches. What is the area of the rhombus?

20 A cone is 4 feet tall, and its radius is 1 foot. What is its approximate volume?

Answers to Chapter 19 Quiz

(1) **25 ft.** The Pythagorean Theorem states that the legs, a and b, and the hypotenuse, c, are related by $a^2 + b^2 = c^2$.

$$7^2 + 24^2 = c^2$$
$$49 + 576 = c^2$$
$$625 = c^2$$
$$25 = c$$

(2) **144 in.²** The formula for the area of a square is $A = s^2$.

$$A = (12 \text{ in.})^2 = 144 \text{ in.}^2$$

(3) **314 yd.²** The area of a circle is found with $A = \pi \times r^2$.

$$A \approx 3.14 \times (10 \text{ yd.})^2 = 3.14 \times (100 \text{ yd.}^2) = 314 \text{ yd.}^2$$

(4) **42 ft.** The formula for the perimeter of a rectangle is $P = 2 \times (l + w)$.

$$P = 2 \times (7 \text{ ft.} + 14 \text{ ft.}) = 2 \times (21 \text{ ft.}) = 42 \text{ ft.}$$

(5) **36 yd.²** The formula for the area of a trapezoid is $A = \frac{1}{2} \times (b_1 + b_2) \times h$.

$$A = \frac{1}{2} \times (6 \text{ yd.} + 12 \text{ yd.}) \times 4 \text{ yd.} = \frac{1}{2} \times (18 \text{ yd.}) \times 4 \text{ yd.} = 36 \text{ yd.}^2$$

(6) **26 miles.** The two routes form the two legs of a right triangle. You solve for the hypotenuse for their distance apart.

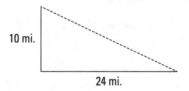

10 mi.

24 mi.

$$a^2 + b^2 = c^2$$
$$10^2 + 24^2 = c^2$$
$$100 + 576 = c^2$$
$$676 = c^2$$
$$26 = c$$

7 **512 ft.³** To find the volume of a cube, use $V = s^3$.

$$V = (8 \text{ ft.})^3 = 512 \text{ ft.}^3$$

8 **80 ft.²** The area of a triangle is found with $A = \frac{1}{2} \times b \times h$.

$$A = \frac{1}{2} \times 10 \text{ ft.} \times 16 \text{ ft.} = 80 \text{ ft.}^2$$

9 **13.8 ft.³** The volume of a rectangular prism is found with $V = l \times w \times h$.

$$V = 2.3 \text{ ft.} \times 1.2 \text{ ft.} \times 5 \text{ ft.} = 13.8 \text{ ft.}^3$$

10 **3,052.08 in.³** To find the volume of a sphere, use $V = \frac{4}{3} \times \pi \times r^3$.

$$V \approx \frac{4}{3} \times 3.14 \times (9 \text{ in.})^3 = 3,052.08 \text{ in.}^3$$

11 **100 yd.²** The perimeter of a square is found with $P = 4 \times s$. If the perimeter is 40 yards, then $40 \text{ yd.} = 4 \times s$. This means that $s = 10$ yd. Using the formula for the area of a square, $A = (10 \text{ yd.})^2 = 100 \text{ yd.}^2$

12 **75.36 in.³** The volume of a cylinder is found with $V = A_b \times h$, and you find the area of the base with $A_b = \pi \times r^2$. First, finding the area of the base:

$$A_b \approx 3.14 \times (2 \text{ in.})^2 = 3.14 \times 4 \text{ in.}^2 = 12.56 \text{ in.}^2$$

And the volume is: $V = 12.56 \text{ in.}^2 \times 6 \text{ in.} = 75.36 \text{ in.}^3$

13 **43,000 km².** The area of a rectangle is found with $A = l \times w$.

$$A = 100 \text{ km} \times 430 \text{ km} = 43,000 \text{ km}^2$$

14 **2,201,600 ft.²** Find the area of the square and subtract the area of the circle to find the amount of land left unwatered.

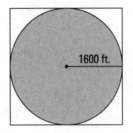

1600 ft.

The length of a side of the square will be twice the radius of the circle.

$$A = s^2$$
$$A = (2 \times 1600 \text{ ft.})^2$$
$$A = (3200 \text{ ft.})^2 = 10,240,000 \text{ ft.}^2$$

The area of the circle is found with $A = \pi \times r^2$.

$$A = \pi \times r^2$$
$$A \approx 3.14 \times (1,600 \text{ ft.})^2$$
$$A = 3.14 \times 2,560,000 \text{ ft.}^2 = 8,038,400 \text{ ft.}^2$$

Finding the difference between the areas: $10,240,000 \text{ ft.}^2 - 8,038,400 \text{ ft.}^2 = 2,201,600 \text{ ft.}^2$

⑮ **165 ft.²** First, change the measure of the base to feet. Because 1 yard = 3 feet, 10 yd. $= (10 \times 3 \text{ ft.}) = 30$ ft. The area of a triangle is found with $A = \frac{1}{2} \times b \times h$.

$$A = \frac{1}{2} \times 30 \text{ ft.} \times 11 \text{ ft.} = 165 \text{ ft.}^2$$

⑯ **52 ft.** The perimeter of a square is found with $P = 4 \times s$.

$$P = 4 \times 13 \text{ ft.} = 52 \text{ ft.}$$

⑰ **37.68 in.** The circumference of a circle is found with $C = 2\pi \times r$.

$$C \approx 2(3.14) \times 6 \text{ in.} = 6.28 \times 6 \text{ in.} = 37.68 \text{ in.}$$

⑱ **684 ft.²** Determine the area of each wall.

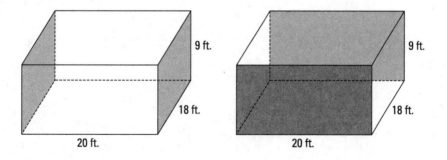

To find the area of a rectangle, you use $A = l \times w$.

There are two walls measuring 9 by 18 feet, and two walls measuring 9 by 20 feet.

The total area: $A = 2(9 \times 18) + 2(9 \times 20) = 2(162) + 2(180) = 684$ square feet.

(19) **27 in.²** You find the area of a rhombus with $A = b \times h$.

$A = 6 \text{ in.} \times 4.5 \text{ in.} = 27 \text{ in.}^2$

(20) **4.19 ft.³** You find the volume of a cone with $V = \frac{1}{3} \times A_b \times h$ where the area of the base is the area of a circle, $A_b = \pi \times r^2$.

First, finding the area of the base, $A_b \approx 3.14 \times (1 \text{ ft.})^2 = 3.14 \text{ ft.}^2$

And the volume is: $V \approx \frac{1}{3} \times (3.14 \text{ ft.}^2) \times (4 \text{ ft.}) = \frac{12.56 \text{ ft.}^3}{3} \approx 4.19 \text{ ft.}^3$

Chapter **20**

Figuring Your Chances: Statistics and Probability

Statistics and probability are two of the most important and widely used applications of math. They're applicable to virtually every aspect of the real world — business, biology, city planning, politics, meteorology, and many more areas of study. Even physics, once thought to be devoid of uncertainty, now relies on probability.

In this chapter, I give you a basic understanding of these two mathematical ideas. First, I introduce you to statistics and the important distinction between qualitative and quantitative data. I show you how to work with both types of data to find meaningful answers. Then I give you the basics of probability. I show you how the probability that an event will occur is always a number from 0 to 1 — that is, usually a fraction, decimal, or percent. After that, I demonstrate how to build this number by counting both favorable outcomes and possible outcomes. Finally, I put these ideas to work by showing you how to calculate the probability of tossing coins.

Gathering Data Mathematically: Basic Statistics

Statistics is the science of gathering and drawing conclusions from data, which is information that's measured objectively in an unbiased, reproducible way.

An individual *statistic* is a conclusion drawn from this data. Here are some examples:

>> The average working person drinks 3.7 cups of coffee every day.

>> Only 52% of students who enter law school actually graduate.

>> The cat is the most popular pet in the United States.

>> In the last year, the cost of a high-definition TV dropped by an average of $575.

Statisticians do their work by identifying a population that they want to study: working people, law students, pet owners, buyers of electronics, whoever. Because most populations are far too large to work with, a statistician collects data from a smaller, randomly selected sample of this population. Much of statistics concerns itself with gathering data that's reliable and accurate. You can read all about this idea in *Statistics For Dummies*, 2nd Edition, by Deborah J. Rumsey (Wiley).

In this section, I give you a short introduction to the more mathematical aspects of statistics.

Understanding differences between qualitative and quantitative data

Data — the information used in statistics — can be either qualitative or quantitative. *Qualitative data* divides a data set (the pool of data that you've gathered) into discrete chunks based on a specific attribute. For example, in a class of students, qualitative data can include

>> Each child's gender

>> Their favorite color

>> Whether they own at least one pet

>> How they get to and from school

REMEMBER

You can identify qualitative data by noticing that it links an attribute — that is, a quality — to each member of the data set. For example, four attributes of Emma are that she's female, her favorite color is green, she owns a dog, and she walks to school.

On the other hand, *quantitative data* provides numerical information — that is, information about quantities, or amounts. For example, quantitative data on this same classroom of students can include the following:

>> Each child's height in inches

>> Each child's weight in pounds

>> The number of siblings each child has

>> The number of words each child spelled correctly on the most recent spelling test

REMEMBER

You can identify quantitative data by noticing that it links a number to each member of the data set. For example, Carlos is 55 inches tall, weighs 68 pounds, has three siblings, and spelled 18 words correctly on his last spelling test.

Working with qualitative data

Qualitative data usually divides a sample into discrete chunks. As my sample — which is purely fictional — I use 25 children in Sister Elena's fifth-grade class. For example, suppose all 25 children in Sister Elena's class answer the three yes/no questions in Table 20-1.

Table 20-1   Sister Elena's Fifth-Grade Survey

Question	Yes	No
Are you an only child?	5	20
Do you own any pets?	14	11
Do you take the bus to school?	16	9

The students also answer the question, "What is your favorite color?" with the results in Table 20-2.

Table 20-2   Favorite Colors in Sister Elena's Class

Color	Number of Students	Color	Number of Students
Blue	8	Orange	1
Red	6	Yellow	1
Green	5	Gold	1
Purple	3		

Even though the information that each child provided is non-numerical, you can handle it numerically by counting how many students made each response and working with these numbers.

Given this information, you can now make informed statements about the students in this class just by reading the charts. For instance,

>> Exactly 20 children have at least one brother or sister.

>> Nine children don't take the bus to school.

>> Only one child's favorite color is yellow.

Playing the percentages

You can make more sophisticated statistical statements about qualitative data by finding out the percentage of the sample that has a specific attribute. Here's how you do so:

1. **Write a statement that includes the number of members who share that attribute and the total number in the sample.**

   Suppose you want to know what percentage of students in Sister Elena's class are only children. The chart tells you that 5 students have no siblings, and you know that 25 kids are in the class. So you can begin to answer this question as follows:

   Five out of 25 children are only children.

2. **Rewrite this statement, turning the numbers into a fraction:**

$$\frac{\text{Number who share attribute}}{\text{Number in sample}} = \frac{5}{25}$$

   In the example, $\frac{5}{25}$ of the children are only children.

3. **Turn the fraction into a percent, using the method I show you in Chapter 14.**

   You find that $\frac{5}{25} = \frac{1}{5} = 0.2$, so 20% of the children are only children.

Similarly, suppose you want to find out what percentage of children take the bus to school. This time, the chart tells you that 16 children take the bus, so you can write this statement:

   Sixteen out of 25 children take the bus to school.

Now rewrite the statement as follows:

   $\frac{16}{25}$ of the children take the bus to school.

Finally, turn this fraction into a percent: $16 \div 25 = 0.64$, which equals 64%, so

   64% of the children take the bus to school.

Getting into the mode

The *mode* tells you the most popular answer to a statistical question. For example, in the poll of Sister Elena's class (see Tables 20-1 and 20-2), the mode groups are children who

>> Have at least one brother or sister (20 students)

>> Own at least one pet (14 students)

>> Take the bus to school (16 students)

>> Chose blue as their favorite color (8 students)

REMEMBER

When a question divides a data set into two parts (as with all yes/no questions), the mode group represents more than half of the data set. But when a question divides a data set into more than two parts, the mode doesn't necessarily represent more than half of the data set.

For example, 14 children own at least one pet, and the other 11 children don't own one. So the mode group — children who own a pet — is more than half the class. But 8 of the 25 children chose blue as their favorite color. So even though this is the mode group, fewer than half the class chose this color.

With a small sample, you may have more than one mode — for example, perhaps the number of students who like red is equal to the number who like blue. However, getting multiple modes isn't usually an issue with a larger sample because it becomes less likely that exactly the same number of people will have the same preference.

EXAMPLE

Q. When 25 children were asked, "Who usually helps you with your homework?" their answers were as follows:

Mother: 14

Father: 5

Older sibling: 3

Grandparent: 2

Aunt or uncle: 1

(a) Which answer is the mode?

(b) What percent of the children answered "Father"?

(c) What percent of the children named somebody other than a parent?

A.

(a) **Mother.** The most popular answer was "Mother."

(b) **20%.** Five children out of 25 answered "Father," so divide 5 by 25 $\left(5 \div 25 = 0.2 = 20\%\right)$.

(c) **24%.** Six children out of 25 named somebody other than a parent $\left(3 + 2 + 1 = 6\right)$, so divide 6 by 25 $\left(6 \div 25 = 0.24 = 24\%\right)$.

YOUR TURN

1 When 48 people were asked, "If you could go anywhere, where would you most like to take your next vacation?" their answers were as follows:

Europe: 16

North America: 14

Asia: 9

South America: 4

Someplace else: 5

(a) What is the mode answer?

(b) What percentage of people chose Asia?

(c) What percentage of people chose either North America or South America?

(d) What percentage of people chose a place other than Europe?

2 When 80 people were asked their opinion of a politician, they answered as follows:

Very favorable: 16

Somewhat favorable: 30

Somewhat unfavorable: 22

Very unfavorable: 8

No opinion: 4

(a) What is the mode answer?

(b) What percentage of people answered either "Very favorable" or "Somewhat favorable"?

(c) What percentage of people answered either "Somewhat favorable" or "Somewhat unfavorable"?

(d) What percentage of people had no opinion?

Working with quantitative data

Quantitative data assigns a numerical value to each member of the sample. As my sample — again, fictional — I use five members of Sister Elena's basketball team. Suppose that the information in Table 20-3 has been gathered about each team member's height and most recent spelling test.

Table 20-3   Height and Spelling Test Scores

Student	Height in Inches	Number of Words Spelled Correctly
Carlos	55	18
Dwight	60	20
Patrick	59	14
Tyler	58	17
William	63	18

In this section, I show you how to use this information to find the mean and median for both sets of data. Both terms refer to ways to calculate the average value in a quantitative data set. An *average* gives you a general idea of where most individuals in a data set fall so you know what kinds of results are standard. For example, the average height of Sister Elena's fifth-grade class is probably less than the average height of the Los Angeles Lakers. As I show you in the sections that follow, an average can be misleading in some cases, so knowing when to use the mean versus the median is important.

Finding the mean

The mean is the most commonly used average. In fact, when most people use the word *average*, they're referring to the mean. Here's how you find the mean of a set of data:

1. **Add up all the numbers in that set.**

   For example, to find the average height of the five team members, first add up all their heights:

   $$55 + 60 + 59 + 58 + 63 = 295$$

2. **Divide this result by the total number of members in that set.**

   Divide 295 by 5 (that is, by the total number of boys on the team):

   $$295 \div 5 = 59$$

   So the mean height of the boys on Sister Elena's team is 59 inches.

This procedure is summed up (so to speak) in a simple formula:

$$\text{Mean} = \frac{\text{Sum of values}}{\text{Number of values}}$$

You can use this formula to find the mean number of words that the boys spelled correctly. To do this, plug the number of words that each boy spelled correctly into the top part of the formula, and then plug the number of boys in the group into the bottom part:

$$\text{Mean} = \frac{18 + 20 + 14 + 17 + 18}{5}$$

Now simplify to find the result:

$$= \frac{87}{5} = 17.4$$

As you can see, when you divide, you end up with a decimal in your answer. If you round to the nearest whole word, the mean number of words that the five boys spelled correctly is about 17 words. (For more information about rounding, see Chapter 3.)

The mean can be misleading when you have a strong skew in data — that is, when the data has many low values and a few very high ones, or vice versa.

For example, suppose that the president of a company tells you, "The average salary in my company is $200,000 a year!" But on your first day at work, you find out that the president's salary is $19,010,000 and each of his 99 employees earns $10,000. To find the mean, first plug the total salaries ($19,010,000 for the president plus $10,000 for each of 99 employees) into the top of the formula. Next, plug the number of employees (100) into the bottom:

$$\text{Mean} = \frac{\$19,010,000 + (\$10,000 \times 99)}{100}$$

Now calculate:

$$= \frac{\$19,010,000 + \$990,000}{100} = \frac{\$20,000,000}{100} = \$200,000$$

So the president didn't lie. However, the skew in salaries resulted in a misleading mean.

Finding the median

When data values are skewed (when a few very high or very low numbers differ significantly from the rest of the data), the median can give you a more accurate picture of what's standard. Here's how to find the median of a set of data:

1. **Arrange the set from lowest to highest.**

   To find the median height of the boys in Table 20-3, arrange their five heights in order from lowest to highest.

   55   58   <u>59</u>   60   63

2. **Choose the middle number.**

   The middle value, 59 inches, is the median average height.

To find the median number of words that the boys spelled correctly (refer to Table 20-3), arrange their scores in order from lowest to highest:

14   17   <u>18</u>   18   20

This time, the middle value is 18, so 18 is the median score.

REMEMBER

If you have an even number of values in the data set, put the numbers in order and find the mean of the *two middle numbers* in the list (see the preceding section for details on the mean). For instance, consider the following:

2   3   <u>5</u>   <u>7</u>   9   11

The two center numbers are 5 and 7. Add them together to get 12, and then divide by 2 to get their mean. The median in this list is 6.

Now recall the company president who makes $19,010,000 a year and his 99 employees who each earn $10,000. Here's how this data looks:

10,000   10,000   10,000...   10,000   19,010,000

As you can see, if you wrote out all 100 salaries, the center numbers would obviously both be 10,000. The median salary is $10,000, and this result is much more reflective of what you'd probably earn if you worked at this company.

EXAMPLE

Q. Ten basketball players were each given five free throws. In order, the number of successful free throws were as follows:

2, 3, 4, 1, 4, 2, 5, 3, 1, 2

(a) What is the mean number of successful free throws, rounded to the nearest tenth?

(b) What is the median number of free throws?

A.

(a) **2.7.** To find the mean, find the sum of successful free throws $(2+3+4+1+4+2+5+3+1+2=27)$ and divide this number by 10 $(27 \div 10 = 2.7)$.

(b) **2.5.** To find the median, arrange the list in order from smallest to largest and find the two middle numbers:

1, 1, 2, 2, 2, 3, 3, 4, 4, 5

Take the average of the two middle numbers $(2+3=5; \ 5 \div 2 = 2.5)$.

**YOUR
TURN**

3 Thirteen children took a pop quiz scored on a scale from 0 to 10. Their scores, in order from least to greatest, were as follows:

4, 6, 6, 7, 8, 8, 8, 8, 9, 9, 9, 10, 10

(a) What is the mean score on the quiz, rounded to the nearest tenth?

(b) What is the median score on the quiz?

4 Sixteen people were asked how many cups of coffee they drink each day. Their answers, in the order that they were asked, were as follows:

3, 0, 1, 2, 2, 3, 3, 2, 0, 5, 2, 2, 2, 1, 0, 1

(a) What is the mean number of cups of coffee, rounded to the nearest tenth?

(b) What is the median number of cups of coffee?

Looking at Likelihoods: Basic Probability

Probability is the mathematics of deciding how likely an event is to occur. For example,

» What's the likelihood that the lottery ticket I bought will win?

» What's the likelihood that my new car will need repairs before the warranty runs out?

» What's the likelihood that more than 100 inches of snow will fall in Manchester, New Hampshire, this winter?

Probability has a wide variety of applications in insurance, weather prediction, biological sciences, and even physics. You can read all about the details of probability in *Probability For Dummies*, by Deborah J. Rumsey (Wiley). In this section, I give you a little taste of this fascinating subject.

Figuring the probability

The *probability* that an event will occur is a fraction whose numerator (top number) and denominator (bottom number) are as follows (for more on fractions, flip to Chapter 10):

$$\text{Probability} = \frac{\text{Target outcomes}}{\text{Total outcomes}}$$

In this case, the number of *target outcomes* (or *successes*) is simply the number of outcomes in which the event you're examining does happen. In contrast, the number of *total outcomes* (or *sample space*) is the number of outcomes that *can* happen.

For example, suppose you want to know the probability that a tossed coin will land heads up. Notice that there are two total outcomes (heads or tails), but only one of these outcomes is the target — the outcome in which heads comes up. To find the probability of this event, make a fraction as follows:

$$\text{Probability} = \frac{1}{2}$$

So the probability that the coin will land heads up is $\frac{1}{2}$.

So what's the probability that, when you roll a die, the number 3 will land face up? To figure this one out, notice that there are *six* total outcomes $(1,2,3,4,5,$ and $6)$, but in only *one* of these does 3 land face up. To find the probability of this outcome, make a fraction as follows:

$$\text{Probability} = \frac{1}{6}$$

So the probability that the number 3 will land face up is $\frac{1}{6}$.

And what's the probability that, if you pick a card at random from a deck, it'll be an ace? To figure this out, notice that there are 52 total outcomes (one for each card in the deck), but in only 4 of these do you pick an ace. So

$$\text{Probability} = \frac{4}{52}$$

So the probability that you'll pick an ace is $\frac{4}{52}$, which reduces to $\frac{1}{13}$ (see Chapter 10 for more on reducing fractions).

Probability is always a number from 0 to 1. When the probability of an outcome is 0, the outcome is *impossible*. When the probability of an outcome is 1, the outcome is *certain*.

Oh, the possibilities! Counting outcomes with multiple coins

Although the basic probability formula isn't difficult, sometimes finding the numbers to plug into it can be tricky. One source of confusion is in counting the number of outcomes, both target and total. In this section, I focus on tossing coins.

When you flip a coin, you can generally get two total outcomes: heads or tails. When you flip two coins at the same time — say, a penny and a nickel — you can get four total outcomes:

Outcome	Penny	Nickel
#1	Heads	Heads
#2	Heads	Tails
#3	Tails	Heads
#4	Tails	Tails

When you flip three coins at the same time — say, a penny, a nickel, and a dime — eight outcomes are possible:

Outcome	Penny	Nickel	Dime
#1	Heads	Heads	Heads
#2	Heads	Heads	Tails
#3	Heads	Tails	Heads
#4	Heads	Tails	Tails
#5	Tails	Heads	Heads
#6	Tails	Heads	Tails
#7	Tails	Tails	Heads
#8	Tails	Tails	Tails

Notice the pattern: Every time you add a coin, the number of total outcomes doubles. So if you flip six coins, here's how many total outcomes you have:

$$2 \times 2 \times 2 \times 2 \times 2 \times 2 = 64$$

The number of total outcomes equals the number of outcomes per coin (2) raised to the number of coins (6): Mathematically, you have $2^6 = 64$.

TIP

Here's a handy formula for calculating the number of outcomes when you're flipping, shaking, or rolling multiple coins, dice, or other objects at the same time:

$$\text{Total outcomes} = \text{Number of outcomes per object}^{\text{Number of objects}}$$

Suppose you want to find the probability that six tossed coins will all fall heads up. To do this, you want to build a fraction, and you already know that the denominator — the number of total outcomes — is 64. Only one outcome is the target outcome, so the numerator is 1:

$$\text{Probability} = \frac{1}{64}$$

So the probability that six tossed coins will all fall heads up is $\frac{1}{64}$.

Here's a more subtle question: What's the probability that exactly five out of six tossed coins will all fall heads up? Again, you're building a fraction, and you already know that the denominator is 64. To find the numerator (target outcomes), think about it this way: If the first coin falls tails up, then all the rest must fall heads up. If the second coin falls tails up, then again all the rest must fall heads up. This is true of all six coins, so you have six target outcomes:

$$\text{Probability} = \frac{6}{64}$$

Therefore, the probability that exactly five out of six coins will fall heads up is $\frac{6}{64}$, which reduces to $\frac{3}{32}$ (see Chapter 10 for more on reducing fractions).

EXAMPLE

Q. Suppose that you pick a random card from a deck of 52 cards:

   (a) How many possible outcomes are there?

   (b) What is the probability that the card is a heart? (*Hint:* There are 13 hearts in the deck.)

A.

   (a) **52.** The deck has 52 cards, so there are 52 possible outcomes.

   (b) $\frac{1}{4}$. There are a total of 52 possible outcomes, and 13 target outcomes:

$$\frac{\text{Target outcomes}}{\text{Total outcomes}} = \frac{13}{52} = \frac{1}{4}$$

YOUR TURN

5 How many possible outcomes are there in each of the following cases?

   (a) Flipping 4 coins.

   (b) Flipping 8 coins.

   (c) Rolling a pair of dice.

   (d) Rolling 4 dice.

If you roll a six-sided die with the numbers 1 through 6 on it, what is the probability of each of the following outcomes?

(a) Rolling a 6.

(b) Rolling either a 2 or a 3.

(c) Rolling an odd number.

(d) Rolling any number but a 1.

7 If you flip three coins, what is the probability of each of the following outcomes?

(a) All three coins are heads.

(b) Exactly one coin is heads and exactly two are tails.

(c) At least one coin is heads and at least one coin is tails.

(d) At least two coins are heads and at least two coins are tails.

8 Suppose you pick one card at random from a deck of 52 cards. What is the probability of each of the following outcomes?

(a) The card is the ace of spades.

(b) The card is one of the four jacks.

(c) The card is either the jack of diamonds or the queen of spades.

(d) The card is not one of the 13 clubs.

Practice Questions Answers and Explanations

1

(a) **Europe.** The mode answer is the most popular, which is Europe.

(b) **18.75%.** Nine people out of 48 chose Asia, so divide 9 by 48 $(9 \div 48 = 0.1875 = 18.75\%)$.

(c) **37.5%.** Eighteen people $(14 + 4 = 18)$ out of 48 chose either North America or South America, so divide 18 by 48 $(18 \div 48 = 0.375 = 37.5\%)$.

(d) $66\frac{2}{3}\%$. Thirty-two people out of 48 chose a place other than Europe $(48 - 16 = 32)$, so divide 32 by 48:

$$\frac{32}{48} = \frac{2}{3} = 66\frac{2}{3}\%$$

2

(a) **Somewhat favorable.** The mode answer is the most popular, which is "Somewhat favorable."

(b) **57.5%.** Forty-six people $(30 + 16 = 46)$ answered either "Very favorable" or "Somewhat favorable," so divide 46 by 80 $(46 \div 80 = 0.575 = 57.5\%)$.

(c) **65%.** Fifty-two people $(22 + 30 = 52)$ answered either "Somewhat favorable" or "Somewhat unfavorable," so divide 52 by 80 $(52 \div 80 = 0.65 = 65\%)$.

(d) **5%.** Four people answered "No opinion," so divide 4 by 80 $(4 \div 80 = 0.05 = 5\%)$.

3

(a) **7.8.** To find the mean score, add the 13 scores $(4 + 6 + 6 + 7 + 8 + 8 + 8 + 8 + 9 + 9 + 9 + 10 + 10 = 102)$ and divide their sum by 13 $(102 \div 13 \approx 7.8)$.

(b) **8.** The median score is the middle number in the list:

4, 6, 6, 7, 8, 8, **8**, 8, 9, 9, 9, 10, 10

4

(a) **1.8.** To find the mean score, add the 16 numbers $(3 + 0 + 1 + 2 + 2 + 3 + +3 + 2 + 0 + 5 + 2 + 2 + 2 + 1 + 0 + 1 = 29)$ and divide their sum by 16 $(29 \div 16 = 1.8125 \approx 1.8)$.

(b) **2.** To find the median, arrange the answers in order from smallest to largest and find the two middle numbers:

0, 0, 0, 1, 1, 1, 2, 2, 2, 2, 2, 2, 3, 3, 3, 5

So the median is 2.

(5)

(a) **16.** There are 4 events, each with 2 possible outcomes $\left(2^4 = 16\right)$.

(b) **256.** There are 8 events, each with 2 possible outcomes $\left(2^8 = 256\right)$.

(c) **36.** There are 2 events, each with 6 possible outcomes $\left(6^2 = 36\right)$.

(d) **1,296.** There are 4 events, each with 6 possible outcomes $\left(6^4 = 1,296\right)$.

(6) In each case, there are a total of 6 possible outcomes.

(a) $\dfrac{1}{6}$. There is 1 target outcome:

$$\frac{\text{Target outcomes}}{\text{Total outcomes}} = \frac{1}{6}$$

(b) $\dfrac{1}{3}$. There are 2 target outcomes:

$$\frac{\text{Target outcomes}}{\text{Total outcomes}} = \frac{2}{6} = \frac{1}{3}$$

(c) $\dfrac{1}{2}$. There are 3 target outcomes:

$$\frac{\text{Target outcomes}}{\text{Total outcomes}} = \frac{3}{6} = \frac{1}{2}$$

(d) $\dfrac{5}{6}$. There are 5 target outcomes:

$$\frac{\text{Target outcomes}}{\text{Total outcomes}} = \frac{5}{6}$$

(7) In each case, there are a total of 8 target outcomes.

(a) $\dfrac{1}{8}$. There is one target outcome:

$$\frac{\text{Target outcomes}}{\text{Total outcomes}} = \frac{1}{8}$$

(b) $\dfrac{3}{8}$. There are three target outcomes (see the table in the section, "Oh, the possibilities! Counting outcomes with multiple coins"):

$$\frac{\text{Target outcomes}}{\text{Total outcomes}} = \frac{3}{8}$$

(c) $\dfrac{3}{4}$. There are only 2 outcomes that are not target outcomes (3 heads and 3 tails), so there are 6 target outcomes:

$$\frac{\text{Target outcomes}}{\text{Total outcomes}} = \frac{6}{8} = \frac{3}{4}$$

(d) 0. There are 0 target outcomes:

$$\frac{\text{Target outcomes}}{\text{Total outcomes}} = \frac{0}{8} = 0$$

(8) In each case, the total number of possible outcomes is 52.

(a) $\frac{1}{52}$. There is only one target outcome:

$$\frac{\text{Target outcomes}}{\text{Total outcomes}} = \frac{1}{52}$$

(b) $\frac{1}{13}$. There are four target outcomes:

$$\frac{\text{Target outcomes}}{\text{Total outcomes}} = \frac{4}{52} = \frac{1}{13}$$

(c) $\frac{1}{26}$. There are two target outcomes:

$$\frac{\text{Target outcomes}}{\text{Total outcomes}} = \frac{2}{52} = \frac{1}{26}$$

(d) $\frac{3}{4}$. There are 13 clubs and 39 other cards, so there are 39 target outcomes:

$$\frac{\text{Target outcomes}}{\text{Total outcomes}} = \frac{39}{52} = \frac{3}{4}$$

If you're ready to test your skills a bit more, take the following chapter quiz that incorporates all the chapter topics.

Whaddya Know? Chapter 20 Quiz

Quiz time! Complete each problem to test your knowledge on the various topics covered in this chapter. You can then find the solutions and explanations in the next section.

 1 A group of 30 travelers was asked: "What world city would you most like to visit?" Their responses were:

Rome: 3

Barcelona: 4

Amsterdam: 7

London: 5

Berlin: 2

Paris: 9

What percent of the group favored Rome?

2 A group of ten friends in a retirement community were discussing their families. They found that they had the following numbers of grandchildren: 1, 1, 3, 4, 4, 4, 6, 7, 10, 10. What is the mean number of grandchildren among the group?

3 You're rolling a 12-sided (dodecahedron) die with the 12 months of the year written on the sides. When you roll the die, what is the probability that you'll roll the month when you were born?

4 Using a standard deck of 52 playing cards, when you draw a card at random, what is the probability that you'll draw a red king?

5 You're playing Monopoly and are sitting on Park Place. You're two steps away from Boardwalk, which has a hotel on it. When you roll your pair of dice, what is the probability that you won't land on Boardwalk?

6 You kept track of the average daily temperature for a week. The temperatures were: 67, 72, 73, 71, 67, 65, 61. What is the median temperature?

7 You're rolling a 12-sided (dodecahedron) die with the 12 months of the year written on the sides. When you roll the die, what is the probability that you'll roll a month whose name starts with a J?

 8 A group of ten friends in a retirement community were discussing their families. They found that they had the following numbers of grandchildren: 1, 1, 3, 4, 4, 4, 6, 7, 10, 10. What is the median number of grandchildren per person?

9 You kept track of the average daily temperature for a week. The temperatures were: 67, 72, 73, 71, 67, 65, 61. What is the mean temperature?

10 A group of 30 travelers was asked: "What world city would you most like to visit?" Their responses were:

Rome: 3

Barcelona: 4

Amsterdam: 7

London: 5

Berlin: 2

Paris: 9

What percent of the people did **not** respond with Rome, Amsterdam, or London?

11 A group of ten friends in a retirement community were discussing their families. They found that they had the following numbers of grandchildren: 1, 1, 3, 4, 4, 4, 6, 7, 10, 10. What is the mode number of grandchildren among the group?

12 Using a standard deck of 52 playing cards, when you draw a card at random, what is the probability that you'll draw either a 2 or a 3?

13 You kept track of the average daily temperature for a week. The temperatures were: 67, 72, 73, 71, 67, 65, 61. What is the mode temperature?

14 You're playing Monopoly and are sitting on Park Place. You're four steps away from passing Go. When you roll your two dice, what is the probability that you will pass or land on Go?

15 Your test scores are: 89, 89, 90, 90, 92, 98, 100, 100. What is the median test score?

Answers to Chapter 20 Quiz

(1) **10%.** Divide the number of people who chose Rome (3) by 30.

$$\frac{3}{30} = 0.10 = 10\%$$

(2) **5.** Add the numbers together and divide by 10.

$$1+1+3+4+4+4+6+7+10+10 = 50$$
$$\frac{50}{10} = 5$$

(3) $8\frac{1}{3}\%$. Divide 1 by 12.

$$\frac{1}{12} = 0.08333... = 8\frac{1}{3}\%$$

(4) **3.8%.** There are 2 red kings in the deck. Divide 2 by 52.

$$\frac{2}{52} = \frac{1}{26} \approx 0.038 = 3.8\%$$

(5) $97\frac{2}{9}\%$. To land on Boardwalk, you have to roll a 2. There are 35 other possible rolls. So divide 35 by 36.

$$\frac{35}{36} = 0.97222... = 97\frac{2}{9}\%$$

(6) **67.** First, put the temperatures in order from smallest to largest: 61, 65, 67, 67, 71, 72, 73. The middle temperature is the fourth one: 67.

(7) **25%.** There are three months starting with J. Divide 3 by 12.

$$\frac{3}{12} = \frac{1}{4} = 25\%$$

(8) **4.** Among the ten people, the two middle numbers are 4 and 4.

(9) **68.** First, find the sum of the temperatures; then divide by 7.

$$67+72+73+71+67+65+61 = 476$$
$$\frac{476}{7} = 68$$

(10) **50%.** Find the sum of those choosing the cities that aren't Rome, Amsterdam, or London – that is, Barcelona, Berlin, and Paris – and then divide that sum by 30.

$$4 + 2 + 9 = 18$$
$$\frac{15}{30} = \frac{1}{2} = 50\%$$

(11) **4.** The most frequent number is 4.

(12) **15.4%.** There are four 2's and four 3's. Divide 8 by 52.

$$\frac{8}{52} = \frac{2}{13} \approx 0.1538 \approx 15.4\%$$

(13) **67.** The 67 appears twice, which is the most frequent temperature.

(14) $91\frac{2}{3}\%$. You pass Go if you roll 4, 5, 6, 7, 8, 9, 10, 11, or 12. That means: don't roll 2 or 3. There is one way to roll 2 and two ways to roll 3, leaving 33 ways to roll a number that will get you past Go. Divide 33 by 36.

$$\frac{33}{36} = \frac{11}{12} = 0.91666.... = 91\frac{2}{3}\%$$

(15) **91.** There are eight test scores, so the middle two scores are 90 and 92. The mean of 90 and 92 is 91.

» Understanding subsets and the
empty set

» Knowing the basic operations on
sets, including union and
intersection

Chapter **21**

Setting Things Up with Basic Set Theory

A *set* is just a collection of things. But in their simplicity, sets are profound. At the deepest level, set theory is the foundation for everything in math.

Set theory provides a way to talk about collections of numbers, such as even numbers, prime numbers, or counting numbers, with ease and clarity. It also gives rules for performing calculations on sets that become useful in higher math. For these reasons, set theory becomes more important the higher up you go in the math food chain — especially when you begin writing mathematical proofs. Studying sets can also be a nice break from the usual math stuff you work with.

In this chapter, I show you the basics of set theory. First, I show you how to define sets and their elements and how you can tell when two sets are equal. I also show you the simple idea of a set's cardinality. Next, I discuss subsets and the all-important empty set ($\emptyset$). After that, I discuss four operations on sets: union, intersection, relative complement, and complement.

Understanding Sets

A *set* is a collection of things, in any order. These things can be buildings, earmuffs, lightning bugs, numbers, qualities of historical figures, names you call your little brother, whatever.

You can define a set in a few main ways.

> » **Placing a list of the elements of the set in braces** { }: You can simply list everything that belongs in the set. When the set is too large, you use an ellipsis (...) to indicate elements of the set not mentioned. For example, to list the set of numbers from 1 to 100, you can write $\{1, 2, 3, \ldots 100\}$. To list the set of all the counting numbers, you can write $\{1, 2, 3, \ldots\}$.
>
> » **Using a verbal description:** If you use a verbal description of what the set includes, make sure the description is clear and unambiguous so you know exactly what's in the set and what isn't. For instance, the set of the four seasons is pretty clear-cut, but you may run into some debate on the set of words that describe my cooking skills because different people have different opinions.
>
> » **Writing a mathematical rule (set-builder notation):** In later algebra, you can write an equation that tells people how to calculate the numbers that are part of a set. Check out *Algebra II For Dummies*, by Mary Jane Sterling (John Wiley & Sons, Inc.), for details.

Sets are usually identified with capital letters to keep them distinct from variables in algebra, which are usually small letters. (Chapter 22 talks about using variables.)

The best way to understand sets is to begin working with them. For example, here I define three sets:

$A = \{$Empire State Building, Eiffel Tower, Roman Colosseum$\}$

$B = \left\{ \begin{array}{l} \text{Albert Einstein's intelligence, Marilyn Monroe's talent, Joe DiMaggio's athletic ability,} \\ \text{Sen. Joseph McCarthy's ruthlessness} \end{array} \right\}$

$C = $ the four seasons of the year

Set A contains three tangible objects: famous works of architecture. Set B contains four intangible objects: attributes of famous people. And set C also contains intangible objects: the four seasons. Set theory allows you to work with either tangible or intangible objects, provided that you define your set properly. In the following sections, I show you the basics of set theory.

Elementary, my dear: Considering what's inside sets

The things contained in a set are called *elements* (also known as *members*). Consider the first two sets I define in the section intro:

$A = \{$Empire State Building, Eiffel Tower, Roman Colosseum$\}$

$B = \left\{ \begin{array}{l} \text{Albert Einstein's intelligence, Marilyn Monroe's talent,} \\ \text{Joe DiMaggio's athletic ability, Sen. Joseph McCarthy's ruthlessness} \end{array} \right\}$

The Eiffel Tower is an element of A, and Marilyn Monroe's talent is an element of B. You can write these statements using the symbol $\in$, which means "is an element of":

Eiffel Tower $\in A$

Marilyn Monroe's talent $\in B$

However, the Eiffel Tower is not an element of B. You can write this statement using the symbol $\notin$, which means "is not an element of":

Eiffel Tower $\notin B$

These two symbols become more common as you move higher in your study of math. The following sections discuss what's inside those braces and how some sets relate to each other.

Cardinality of sets

The *cardinality* of a set is just a fancy word for the number of elements in that set.

When A is {Empire State Building, Eiffel Tower, Roman Colosseum}, it has three elements, so the cardinality of A is three. Set B, which is {Albert Einstein's intelligence, Marilyn Monroe's talent, Joe DiMaggio's athletic ability, Sen. Joseph McCarthy's ruthlessness}, has four elements, so the cardinality of B is four.

Equal sets

REMEMBER

If two sets list or describe the exact same elements, the sets are equal (you can also say they're *identical* or *equivalent*). The order of elements in the sets doesn't matter. Similarly, an element may appear twice in one set, but only the distinct elements need to match.

Suppose I define some sets as follows:

$C =$ the four seasons of the year
$D = \{$spring, summer, fall, winter$\}$
$E = \{$fall, spring, summer, winter$\}$
$F = \{$summer, summer, summer, spring, fall, winter, winter, summer$\}$

Set C gives a clear rule describing a set. Set D explicitly lists the four elements in C. Set E lists the four seasons in a different order. And set F lists the four seasons with some repetition. Thus, all four sets are equal. As with numbers, you can use equals signs to show that sets are equal:

$C = D = E = F$

Subsets

When all the elements of one set are completely contained in a second set, the first set is a subset of the second. For example, consider these sets:

$C = \{$spring, summer, fall, winter$\}$
$G = \{$spring, summer, fall$\}$

As you can see, every element of G is also an element of C, so G is a subset of C. The symbol for subset is $\subset$, so you can write the following:

$G \subset C$

Every set is a subset of itself. This idea may seem odd until you realize that all the elements of any set are obviously contained in that set.

Empty sets

The *empty set* — also called the *null set* — is a set that has no elements:

$$H = \{\}$$

As you can see, I define H by listing its elements, but I haven't listed any, so H is empty. The symbol $\varnothing$ is used to represent the empty set. So $H = \varnothing$.

You can also define an empty set by using a rule. For example,

$$I = \text{types of roosters that lay eggs}$$

Clearly, roosters are male and, therefore, can't lay eggs, so this set is empty.

TIP

You can think of $\varnothing$ as nothing. And because nothing is always nothing, there's only one empty set. All empty sets are equal to each other, so in this case, $H = I$.

Furthermore, $\varnothing$ is a subset of every other set (the preceding section discusses subsets), so the following statements are true:

$$\varnothing \subset A$$
$$\varnothing \subset B$$
$$\varnothing \subset C$$

This concept makes sense when you think about it. Remember that $\varnothing$ has no elements, so technically, every element in $\varnothing$ is in every other set.

Sets of numbers

One important use of sets is to define sets of numbers. As with all other sets, you can do so either by listing the elements or by verbally describing a rule that clearly tells you what's included in the set and what isn't. For example, consider the following sets:

$$J = \{1,2,3,4,5\}$$
$$K = \{2,4,6,8,10,\ldots\}$$
$$L = \text{the set of counting numbers}$$

My definitions of J and K list their elements explicitly. Because K is infinitely large, you need to use an ellipsis (. . .) to show that this set goes on forever. The definition of L is a description of the set in words.

Performing Operations on Sets

In arithmetic, the Big Four operations (adding, subtracting, multiplying, and dividing) allow you to combine numbers in various ways (see Chapters 3 and 4 for more information). Set theory also has four important operations: union, intersection, relative complement, and complement. You'll see more of these operations as you move on in your study of math.

Here are definitions for three sets of numbers:

$$P = \{1,7\}$$
$$Q = \{4,5,6\}$$
$$R = \{2,4,6,8,10\}$$

In this section, I use these three sets and a few others to discuss the four set operations and show you how they work. (*Note:* Within equations, I relist the elements, replacing the names of the sets with their equivalent in braces. Therefore, you don't have to flip back and forth to look up what each set contains.)

Union: Combined elements

The union of two sets is the set of their *combined* elements. For example, the union of $\{1,2\}$ and $\{3,4\}$ is $\{1,2,3,4\}$. The symbol for this operation is $\cup$, so

$$\{1,2\} \cup \{3,4\} = \{1,2,3,4\}$$

Similarly, here's how to find the union of P and Q:

$$P \cup Q = \{1,7\} \cup \{4,5,6\} = \{1,4,5,6,7\}$$

When two sets have one or more elements in common, these elements appear only once in their union set. For example, consider the union of Q and R. In this case, the elements 4 and 6 are in both sets, but each of these numbers appears once in their union:

$$Q \cup R = \{4,5,6\} \cup \{2,4,6,8,10\} = \{2,4,5,6,8,10\}$$

The union of any set with itself is itself:

$$P \cup P = P$$

Similarly, the union of any set with $\varnothing$ (see the earlier section, "Empty sets") is itself:

$$P \cup \varnothing = P$$

Intersection: Elements in common

The intersection of two sets is the set of their common elements (the elements that appear in both sets). For example, the intersection of $\{1,2,3\}$ and $\{2,3,4\}$ is $\{2,3\}$. The symbol for this operation is $\cap$. You can write the following:

$$\{1,2,3\} \cap \{2,3,4\} = \{2,3\}$$

Similarly, here's how to write the intersection of Q and R:

$$Q \cap R = \{4,5,6\} \cap \{2,4,6,8,10\} = \{4,6\}$$

When two sets have no elements in common, their intersection is the empty set ($\varnothing$):

$$P \cap Q = \{1,7\} \cap \{4,5,6\} = \varnothing$$

The intersection of any set with itself is itself:

$$P \cap P = P$$

But the intersection of any set with $\varnothing$ is $\varnothing$:

$$P \cap \varnothing = \varnothing$$

Q. Suppose L = the set of all even integers between 1 and 9, M = the set of all positive integers less than 7, and $N = \{3,6,9,12\}$.

EXAMPLE

    (a) What elements are in L?

    (b) What elements are in M?

    (c) What is $(L \cup M) \cap N$?

A.

    (a) **2, 4, 6, and 8.**

    (b) **1, 2, 3, 4, 5, and 6.**

    (c) $\{3,6\}$. To begin, find the union of the two sets inside the parentheses:

$$L \cup M = \{1,2,3,4,5,6,8\}$$

Next, find the intersection of this set and N:

$$(L \cup M) \cap N = \{3,6\}$$

---

1 Suppose O = the set of all odd integers between 0 and 10, and P = the set of all integers greater than 6 and less than 14.

YOUR TURN

    (a) What elements are in O?

    (b) What elements are in P?

(c) What is $O \cup P$?

(d) What is $O \cap P$?

 2 Let $R = \{1,2,3,4,5,6,7,8,9,10\}$, $S = \{2,4,6,8,10\}$, and $T = \{2,3,5,7\}$.

(a) What is $(R \cup S) \cap T$?

(b) What is $(R \cap S) \cup T$?

(c) What is $R \cup (S \cap T)$?

(d) What is $(R \cap S) \cup (S \cap T)$?

Relative complement: Subtraction (sorta)

The relative complement of two sets is an operation similar to subtraction. The symbol for this operation is the negative sign ($-$). Starting with the first set, you remove every element that appears in the second set to arrive at their relative complement. For example,

$$\{1,2,3,4,5\} - \{1,2,5\} = \{3,4\}$$

Similarly, here's how to find the relative complement of R and Q. Both sets share a 4 and a 6, so you have to remove those elements from R:

$$R - Q = \{2,4,6,8,10\} - \{4,5,6\} = \{2,8,10\}$$

Note that the reversal of this operation gives you a different result. This time, you remove the shared 4 and 6 from Q:

$$Q - R = \{4,5,6\} - \{2,4,6,8,10\} = \{5\}$$

REMEMBER

Like subtraction in arithmetic, the relative complement is not a commutative operation. In other words, order is important. (See Chapter 3 for more on commutative and non-commutative operations.)

Complement: Feeling left out

The complement of a set is everything that isn't in that set. Because "everything" is a difficult concept to work with, you first have to define what you mean by "everything" as the universal set (U). For example, suppose you define the universal set like this:

$$U = \{0,1,2,3,4,5,6,7,8,9\}$$

Now, here are a couple of sets to work with:

$$M = \{1,3,5,7,9\}$$
$$N = \{6\}$$

The complement of each set is the set of every element in U that isn't in the original set:

$$U - M = \{0,1,2,3,4,5,6,7,8,9\} - \{1,3,5,7,9\} = \{0,2,4,6,8\}$$
$$U - N = \{0,1,2,3,4,5,6,7,8,9\} - \{6\} = \{0,1,2,3,4,5,7,8,9\}$$

The complement is closely related to the relative complement (see the preceding section). Both operations are similar to subtraction. The main difference is that the complement is *always* subtraction of a set from U, but the relative complement is subtraction of a set from any other set.

The symbol for the complement is $'$, so you can write the following:

$$M' = \{0,2,4,6,8\} \qquad\qquad N' = \{0,1,2,3,4,5,7,8,9\}$$

EXAMPLE

Q. Let $X = \{1,3,5,7\}, Y = \{2,3,4\}$, and $Z = \{4,5,6,7,8\}$

(a) What is $X - Y$?

(b) What is $Y - X$?

(c) What is $(Z - Y) \cup X$?

A.

(a) $\{1,5,7\}$.

(b) $\{2,4\}$.

(c) $\{1,3,5,6,7,8\}$. To begin, find the relative complement of Z and Y:

$$Z - Y = \{5,6,7,8\}$$

Next, find the union of this set and X:

$$(Z - Y) \cup X = \{1,3,5,6,7,8\}$$

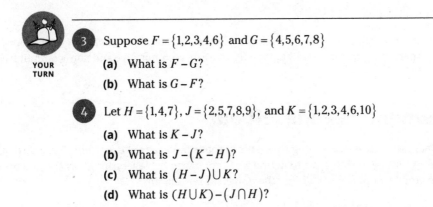

YOUR TURN

3 Suppose $F = \{1,2,3,4,6\}$ and $G = \{4,5,6,7,8\}$

(a) What is $F - G$?

(b) What is $G - F$?

4 Let $H = \{1,4,7\}, J = \{2,5,7,8,9\}$, and $K = \{1,2,3,4,6,10\}$

(a) What is $K - J$?

(b) What is $J - (K - H)$?

(c) What is $(H - J) \cup K$?

(d) What is $(H \cup K) - (J \cap H)$?

Practice Questions Answers and Explanations

1

(a) 1,3,5,7, and 9.

(b) 7,8,9,10,11,12, and 13.

(c) $\{1,3,5,7,8,9,10,11,12,13\}$. The union includes all elements from either set.

(d) $\{7,9\}$. The intersection includes every element in both sets.

2

(a) $\{2,3,5,7\}$. Begin by solving the union inside the parentheses:

$$R \cup S = \{1,2,3,4,5,6,7,8,9,10\}$$

Now, find the intersection of this set with T:

$$(R \cup S) \cap T = \{2,3,5,7\}$$

(b) $\{2,3,4,5,6,7,8,10\}$. Begin by solving the intersection inside the parentheses:

$$R \cap S = \{2,4,6,8,10\}$$

Now, find the union of this set with T:

$$(R \cap S) \cup T = \{2,3,4,5,6,7,8,10\}$$

(c) $\{1,2,3,4,5,6,7,8,9,10\}$. Begin by solving the intersection inside the parentheses:

$$S \cap T = \{2\}$$

Now, find the union of this set with R:

$$R \cup (S \cap T) = \{1,2,3,4,5,6,7,8,9,10\}$$

(d) $\{2,4,6,8,10\}$. Begin by solving the intersections inside both sets of parentheses:

$$R \cap S = \{2,4,6,8,10\}$$
$$S \cap T = \{2\}$$

Now, find the union of these two sets:

$$(R \cap S) \cup (S \cap T) = \{2,4,6,8,10\}$$

3

(a) $\{1,2,3\}$.

(b) $\{5,7,8\}$.

4

(a) $\{1,3,4,6,10\}$.

(b) $\{5,7,8,9\}$. To begin, find the relative complement of K and H:

$$K - H = \{2,3,6,10\}$$

Next, find the relative complement of J and this set:

$$J - (K - H) = \{5,7,8,9\}$$

(c) $\{1,2,3,4,6,10\}$. To begin, find the relative complement of H and J:

$$H - J = \{1,4\}$$

Next, find the union of this set and K:

$$(H - J) \cup K = \{1,2,3,4,6,10\}$$

(d) $\{1,2,3,4,6,10\}$. To begin, find the union of H and K and the intersection of J and H:

$$H \cup K = \{1,2,3,4,6,7,10\}$$
$$J \cap H = \{7\}$$

Next, find the relative complement of these two sets:

$$(H \cup K) - (J \cap H) = \{1,2,3,4,6,10\}$$

Now, see how you do on the chapter quiz in the next section.

Whaddya Know? Chapter 21 Quiz

This 12-question quiz incorporates all the skills you learn in this chapter. When you're done, check out the next section for complete answers and explanations to every question. Use the following sets to answer Questions 1 to 10:

$$U = \{0,1,2,3,4,5,6,7,8,9,10,11,12,13,14,15,16,17,18,19,20\}$$
$$A = \{0,3,6,9,12,15\}, B = \{0,4,8,12,16\}, C = \{1,3,5,7,9\}, D = \{0,5,10,15\}$$

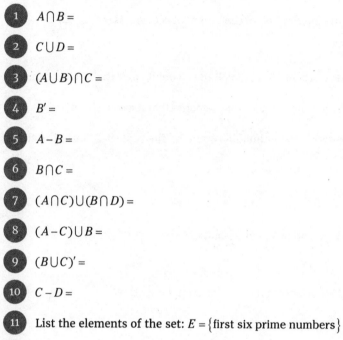

1 $A \cap B =$

2 $C \cup D =$

3 $(A \cup B) \cap C =$

4 $B' =$

5 $A - B =$

6 $B \cap C =$

7 $(A \cap C) \cup (B \cap D) =$

8 $(A - C) \cup B =$

9 $(B \cup C)' =$

10 $C - D =$

11 List the elements of the set: $E = \{\text{first six prime numbers}\}$

12 List the elements of the set: $S = \{\text{all the states in the US whose names start with } Y\}$

Answers to Chapter 21 Quiz

1. $\{0,12\}$. The two sets share the elements 0 and 12.

2. $\{0,1,3,5,7,9,10,15\}$. The union contains all the elements of both sets.

3. $\{3,9\}$. First find the union of sets A and B.

$$(A\cup B)\cap C = \{0,3,4,6,8,9,12,15,16\}\cap C$$

Then find the intersection of that union of sets with set C.

$$= \{3,9\}$$

4. $\{1,2,3,5,6,7,9,10,11,13,14,15\}$. Find all the elements of U that are not in set B.

5. $\{3,6,9,15\}$. Remove from set A all the elements that A and B share.

6. $\{\ \} = \varnothing$. The two sets have nothing in common. Their intersection is the empty set.

7. $\{0,3,9\}$. First find the two intersections of sets.

$$(A\cap C)\cup(B\cap D) = \{3,9\}\cup\{0\}$$

Now find the union of the two results.

$$\{3,9\}\cup\{0\} = \{0,3,9\}$$

8. $\{0,4,6,8,12,15,16\}$. First find the relative complement.

$$(A-C)\cup B = \{0,6,12,15\}\cup B$$

Now find the union of the result and set B.

$$\{0,6,12,15\}\cup B = \{0,4,6,8,12,15,16\}$$

9. $\{2,6,10,11,13,14,15\}$. First find the union of sets B and C.

$$(B\cup C)' = \{0,1,3,4,5,7,8,9,12,16\}'$$

Now find the complement of the union of the sets.

$$\{0,1,3,4,5,7,8,9,12,16\}' = \{2,6,10,11,13,14,15\}$$

10. $\{1,3,7,9\}$. Remove from set C all the elements that C and D share.

11. $\{2,3,5,7,11,13\}$

12. $\{\ \} = \varnothing$. There are no states whose names start with Y, so the set is the empty set.

7

The X-Files: Introduction to Algebra

In This Unit . . .

» **Understanding how a variable such as** *x* **stands for a number**

» **Using substitution to evaluate an algebraic expression**

» **Identifying and rearranging the terms in any algebraic expression**

» **Simplifying algebraic expressions**

Chapter 22

Working with Algebraic Expressions

You never forget your first love, your first car, or your first *x*. Unfortunately for some folks, remembering their first *x* in algebra is similar to remembering their first love who stood them up at the prom or their first car that broke down someplace in Mexico.

The most well-known fact about algebra is that it uses letters — like *x* — to represent numbers. So if you have a traumatic *x*-related tale, all I can say is that the future will be brighter than the past.

What good is algebra? That question is a common one, and it deserves a decent answer. Algebra is used for solving problems that are just too difficult for ordinary arithmetic. And because number crunching is so much a part of the modern world, algebra is everywhere (even if you don't see it): architecture, engineering, medicine, statistics, computers, business, chemistry, physics, biology, and, of course, higher math. Anywhere that numbers are useful, algebra is there. That fact is why virtually every college and university insists that you leave (or enter) with at least a passing familiarity with algebra.

In this chapter, I introduce (or reintroduce) you to that elusive little fellow, Mr. X, in a way that's bound to make him seem a little friendlier. Then I show you how *algebraic expressions* are similar to and different from the arithmetic expressions that you're used to working with. (For a refresher on arithmetic expressions, see Chapter 6.)

Seeing How X Marks the Spot

In math, x stands for a number — any number. Any letter that you use to stand for a number is a *variable*, which means that its value can *vary* — that is, its value is uncertain. In contrast, a number in algebra is often called a *constant* because its value is *fixed*.

Sometimes you have enough information to find out the identity of x. For example, consider the following:

$$2 + 2 = x$$

Obviously, in this equation, x stands for the number 4. But other times, what the number x stands for stays shrouded in mystery. For example:

$$x > 5$$

In this inequality, x stands for some number greater than 5 — maybe 6, maybe $7\frac{1}{2}$, maybe 542.002.

Expressing Yourself with Algebraic Expressions

In Chapter 6, I introduce you to *arithmetic expressions*: strings of numbers and operators that can be evaluated or placed on one side of an equation. For example:

$$2 + 3 \qquad\qquad 7 \times 1.5 - 2 \qquad\qquad 2^4 - |-4| - \sqrt{400}$$

In this chapter, I introduce you to another type of mathematical expression: the algebraic expression. An *algebraic expression* is any string of mathematical symbols that can be placed on one side of an equation and that includes at least one variable.

Here are a few examples of algebraic expressions:

$$5x \qquad\qquad -5x + 2 \qquad\qquad x^2 y - xy^2 + \frac{z}{3} - xyz + 1$$

REMEMBER

As you can see, the difference between arithmetic and algebraic expressions is simply that an algebraic expression includes at least one variable.

In this section, I show you how to work with algebraic expressions. First, I demonstrate how to evaluate an algebraic expression by substituting the values of its variables. Then I show you how to separate an algebraic expression into one or more terms, and I walk through how to identify the coefficient and the variable part of each term.

Evaluating Algebraic Expressions

REMEMBER

To evaluate an algebraic expression, you need to know the numerical value of every variable. For each variable in the expression, substitute, or plug in, the number that it stands for and then evaluate the expression.

In Chapter 6, I show you how to evaluate an arithmetic expression. Briefly, this means finding the value of that expression as a single number (flip to Chapter 6 for more on evaluating arithmetic expressions using PEMDAS).

Knowing how to evaluate arithmetic expressions comes in handy for evaluating algebraic expressions. For example, suppose you want to evaluate the following expression:

$$4x - 7$$

Note that this expression contains the variable x, which is unknown, so the value of the whole expression is also unknown.

An algebraic expression can have any number of variables, but you usually don't work with expressions that have more than two or maybe three, at the most. You can use any letter as a variable, but x, y, and z tend to get a lot of mileage.

Suppose in this case that $x = 2$. To evaluate the expression, substitute 2 for x everywhere it appears in the expression:

$$4(2) - 7$$

After you make the substitution, you're left with an arithmetic expression, so you can finish your calculations to evaluate the expression:

$$= 8 - 7 = 1$$

So given $x = 2$, the algebraic expression $4x - 7 = 1$.

Now suppose you want to evaluate the following expression, where $x = 4$:

$$2x^2 - 5x - 15$$

Again, the first step is to substitute 4 for x everywhere this variable appears in the expression:

$$= 2(4^2) - 5(4) - 15$$

Now evaluate according to the order of operations (PEMDAS) explained in Chapter 6. You do exponents first, so begin by evaluating the exponent 4^2, which equals 4×4:

$$= 2(16) - 5(4) - 15$$

Now proceed to the multiplication, moving from left to right:

$$= 32 - 5(4) - 15$$
$$= 32 - 20 - 15$$

Then evaluate the subtraction, again from left to right:

$$= 12 - 15$$
$$= -3$$

So given $x = 4$, the algebraic expression $2x^2 - 5x - 15 = -3$.

You aren't limited to expressions of only one variable when using substitution. As long as you know the value of every variable in the expression, you can evaluate algebraic expressions with any number of variables. For example, suppose you want to evaluate this expression:

$$3x^2 + 2xy - xyz$$

To evaluate it, you need the values of all three variables:

$$x = 3$$
$$y = -2$$
$$z = 5$$

The first step is to substitute the equivalent value for each of the three variables wherever you find them:

$$3(3^2) + 2(3)(-2) - (3)(-2)(5)$$

Now use the rules for order of operations (PEMDAS) from Chapter 6. Begin by evaluating the exponent 3^2:

$$= 3(9) + 2(3)(-2) - (3)(-2)(5)$$

Next, evaluate the multiplication from left to right (if you need to know more about the rules for multiplying negative numbers, check out Chapter 4):

$$= 27 + (-12) - (-30)$$

Now all that's left is addition and subtraction. Evaluate from left to right, remembering the rules for adding and subtracting negative numbers in Chapter 4:

$$= 15 - (-30) = 15 + 30 = 45$$

So given the three values for x, y, and z, the algebraic expression $3x^2 + 2xy - xyz = 45$.

For practice, copy this expression and the three values on a separate piece of paper, close the book, and see whether you can substitute and evaluate on your own to get the same answer.

TIP

Q. Evaluate the algebraic expression $x^2 - 2x + 5$ when $x = 3$.

EXAMPLE **A.** **8.** First, substitute 3 for x in the expression:

$$x^2 - 2x + 5 = 3^2 - 2(3) + 5$$

Evaluate the expression using the order of operations (PEMDAS). Start by evaluating the power:

$$= 9 - 2(3) + 5$$

Next, evaluate the multiplication:

$$= 9 - 6 + 5$$

Finally, evaluate the addition and subtraction from left to right:

$$= 3 + 5 = 8$$

Q. Evaluate the algebraic expression $3x^2 + 2xy^4 - y^3$ when $x = 5$ and $y = 2$.

A. **227.** Plug in 5 for x and 3 for y in the expression:

$$3x^2 + 2xy^4 - y^3 = 3(5^2) + 2(5)(2^4) - 2^3$$

Evaluate the expression using the order of operations (PEMDAS). Start by evaluating the three powers:

$$= 3(25) + 2(5)(16) - 8$$

Next, evaluate the multiplication:

$$= 75 + 160 - 8$$

Finally, evaluate the addition and subtraction from left to right:

$$= 235 - 8 = 227$$

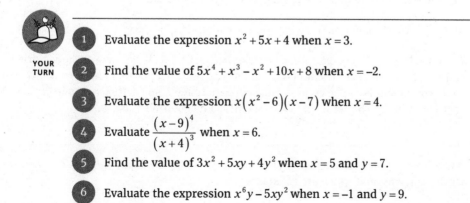

YOUR TURN

1. Evaluate the expression $x^2 + 5x + 4$ when $x = 3$.

2. Find the value of $5x^4 + x^3 - x^2 + 10x + 8$ when $x = -2$.

3. Evaluate the expression $x(x^2 - 6)(x - 7)$ when $x = 4$.

4. Evaluate $\dfrac{(x-9)^4}{(x+4)^3}$ when $x = 6$.

5. Find the value of $3x^2 + 5xy + 4y^2$ when $x = 5$ and $y = 7$.

6. Evaluate the expression $x^6y - 5xy^2$ when $x = -1$ and $y = 9$.

Knowing the Terms

Algebraic expressions begin to make more sense when you understand how they're put together, and the best way to understand this is to take them apart and know what each part is called. Every algebraic expression can be broken into one or more terms. A *term* is any chunk of symbols set off from the rest of the expression by either an addition or subtraction sign. For example,

>> The expression $-7x + 2xy$ has two terms: $-7x$ and $2xy$.

>> The expression $x^4 - \frac{x^2}{5} - 2x + 2$ has four terms: $x^4, \frac{-x^2}{5}, -2x$, and 2.

>> The expression $8x^2y^3z^4$ has only one term: $8x^2y^3z^4$.

REMEMBER

When you separate an algebraic expression into terms, group the plus or negative sign with the term that immediately follows it. Then you can rearrange terms in any order you like. After the terms are separated, you can drop the plus signs.

When a term doesn't have a variable, it's called a *constant*. (*Constant* is just a fancy word for number when you're talking about terms in an expression.) When a term has a variable, it's called an *algebraic term*. Every algebraic term has two parts:

>> The *coefficient* is the signed numerical part of a term — that is, the number and the sign (+ or –) that go with that term. Typically, the coefficients 1 and –1 are dropped from a term, so when a term appears to have no coefficient, its coefficient is either 1 or –1, depending on the sign of that term. The coefficient of a constant is the constant itself.

>> The *variable part* is everything else other than the coefficient.

REMEMBER

When two terms have the same variable part, they're called *like terms*. For two terms to be like, both the letters and their exponents have to be exact matches. For example, the terms $2x$, $-12x$, and $834x$ are all like terms because their variable parts are all x. Similarly, the terms $17xy$, $1,000xy$, and $0.625xy$ are all like terms because their variable parts are all xy.

EXAMPLE

Q. In the expression $3x^2 - 2x - 9$, identify which terms are algebraic and which are constants.

A. $3x^2$ **and** $-2x$ **are both algebraic terms, and** -9 **is a constant.**

Q. Identify the coefficient of every term in the expression $2x^4 - 5x^3 + x^2 - x - 9$.

A. **2, – 5, 1, – 1, and – 9.** The expression has five terms: $2x^4, -5x^3, x^2, -x, -9$. The coefficient of $2x^4$ is 2, and the coefficient of $-5x^3$ is -5. The term x^2 appears to have no coefficient, so its coefficient is 1. The term $-x$ also appears to have no coefficient, so its coefficient is -1. The term -9 is a constant, so its coefficient is -9.

YOUR TURN

7 Write down all the terms in the expression $7x^2yz - 10xy^2z + 4xyz^2 + y - z + 2$. Which are algebraic terms, and which are constants?

8. Identify the coefficient of every term in $-2x^5 + 6x^4 + x^3 - x^2 - 8x + 17$.

9. Name every coefficient in the expression $-x^3y^3z^3 - x^2y^2z^2 + xyz - x$.

10. In the expression $12x^3 + 7x^2 - 2x - x^2 - 8x^4 + 99x + 99$, identify any sets of like terms.

Making the commute: Rearranging your terms

When you understand how to separate an algebraic expression into terms, you can go one step further by rearranging the terms in any order you like. Each term moves as a unit, kind of like a group of people carpooling to work together — everyone in the car stays together for the whole ride.

For example, suppose you begin with the expression $-5x + 2$. You can rearrange the two terms of this expression without changing its value. Notice that each term's sign stays with that term, although dropping the plus sign at the beginning of an expression is customary:

$$= 2 - 5x$$

Rearranging terms in this way doesn't affect the value of the expression because addition is *commutative* — that is, you can rearrange things that you're adding without changing the answer. (See Chapter 3 for more on the commutative property of addition.)

For example, suppose $x = 3$. Then the original expression and its rearrangement evaluate as follows (using the rules that I outline earlier in the section, "Evaluating Algebraic Expressions"):

$$-5x + 2 \qquad 2 - 5x$$
$$= -5(3) + 2 \qquad = 2 - 5(3)$$
$$= -15 + 2 \qquad = 2 - 15$$
$$= -13 \qquad = -13$$

Rearranging expressions in this way becomes handy later in this chapter, when you simplify algebraic expressions. As another example, suppose you have this expression:

$$4x - y + 6$$

You can rearrange it in a variety of ways:

$$= 6 + 4x - y$$
$$= -y + 4x + 6$$

Because the term $4x$ has no sign, it's positive, so you can write in a plus sign as needed when rearranging terms.

REMEMBER

As long as each term's sign stays with that term, rearranging the terms in an expression has no effect on its value.

For example, suppose that $x = 2$ and $y = 3$. Here's how to evaluate the original expression and the two rearrangements:

$$\begin{aligned}
4x - y + 6 && 6 + 4x - y && -y + 4x + 6 \\
= 4(2) - 3 + 6 && = 6 + 4(2) - 3 && = -3 + 4(2) + 6 \\
= 8 - 3 + 6 && = 6 + 8 - 3 && = -3 + 8 + 6 \\
= 5 + 6 && = 14 - 3 && = 5 + 6 \\
= 11 && = 11 && = 11
\end{aligned}$$

Q. Rearrange the terms of the expression $-2b + 5a - 10 + c$ so that variable terms are all in alphabetical order and the constant is at the end.

EXAMPLE

A. **5a – 2b + c – 10.** Remember that each negative sign stays with the term it precedes.

YOUR
TURN

11 Rearrange the terms of the expression $-7k + 1 - j$ so that the variable terms are both in alphabetical order and the constant is at the end.

12 Rearrange the terms of the expression $5z + 4y - 9 - x$ so that the variable terms are all in alphabetical order and the constant is at the end.

13 Rearrange the terms of the expression $-75 + 15v - 100u - t$ so that the variable terms are all in alphabetical order and the constant is at the end.

14 Rearrange the terms of the expression $q + 3r - 10 + s - 2p$ so that the variable terms are all in alphabetical order and the constant is at the end.

Identifying the coefficient and variable

Every term in an algebraic expression has a coefficient. The *coefficient* is the signed numerical part of a term in an algebraic expression — that is, the number and the sign (+ or –) that goes with that term. For example, suppose you're working with the following algebraic expression:

$$-4x^3 + x^2 - x - 7$$

The following table shows the four terms of this expression, with each term's coefficient:

Term	Coefficient	Variable
$-4x^3$	-4	x^3
x^2	1	x^2
$-x$	-1	x
-7	-7	none

Notice that the sign associated with the term is part of the coefficient. So the coefficient of $-4x^3$ is -4.

REMEMBER

When a term appears to have no coefficient, the coefficient is actually 1. So the coefficient of x^2 is 1, and the coefficient of $-x$ is -1. And when a term is a constant (just a number), that number with its associated sign is the coefficient. So the coefficient of the term -7 is simply -7.

By the way, when the coefficient of any algebraic term is 0, the expression equals 0 no matter what the variable part looks like:

$$0x = 0 \qquad\qquad 0xyz = 0 \qquad\qquad 0x^3y^4z^{10} = 0$$

In contrast, the *variable part* of an expression is everything except the coefficient. The previous table shows the four terms of the same expression, with each term's variable part.

Q. What are the coefficients of the terms in the expression $5x - y + z - 3$?

EXAMPLE **A.** 5, -1, 1, and -3.

YOUR TURN

15 What are the coefficients of the terms in the expression $h + 4k - 3$?

16 What are the coefficients of the terms in the expression $3x - 2xy - z$?

17 What are the coefficients of the terms in the expression $-25a^3 - 60b + 10.5c - d + 300$?

18 What are the coefficients of the terms in the expression $-x^3 + 7x^2 - 9x + 1$?

Adding and Subtracting Like Terms

In algebra, you can add and subtract only *like terms* — in other words, terms in which the variable parts match exactly. In this section, I show you how to identify like terms and then you discover how to add and subtract them.

Identifying like terms

Like terms are any two algebraic terms that have the same variable part — that is, both the letters and their exponents have to be exact matches. Here are some examples:

Variable Part	Examples of Like Terms		
x	$4x$	$12x$	$99.9x$
x^2	$6x^2$	$-20x^2$	$\frac{8}{3}x^2$
y	$-y$	$1,000y$	πy
xy	$-7xy$	$800xy$	$\frac{22}{7}xy$
x^3y^3	$3x^3y^3$	$-111x^3y^3$	$3.14x^3y^3$

As you can see, in each example, the variable part in all three like terms is the same. Only the coefficient changes, and it can be any real number: positive or negative, whole number, fraction, or decimal — or even an irrational number such as π. (For more on real numbers, see Chapter 1.)

Adding and subtracting terms

Add like terms by adding their coefficients and keeping the same variable part.

REMEMBER For example, suppose you have the expression $2x + 3x$. Remember that $2x$ is just shorthand for $x + x$, and $3x$ means simply $x + x + x$. So when you add them up, you get the following:

$$= x + x + x + x + x = 5x$$

As you can see, when the variable parts of two terms are the same, you add these terms by adding their coefficients: $2x + 3x = (2 + 3)x$. The idea here is roughly similar to the idea that 2 apples + 3 apples = 5 apples.

You *cannot* add non-like terms. Here are some cases in which the variables or their exponents are different:

$$2x + 3y \qquad\qquad 2yz + 3y \qquad\qquad 2x^2 + 3x$$

In these cases, you can't simplify the expression. You're faced with a situation that's similar to 2 apples + 3 oranges. Because the units (apples and oranges) are different, you can't combine terms. (See Chapter 18 for more on how to work with units.)

Subtraction works much the same as addition. Subtract like terms by finding the difference between their coefficients and keeping the same variable part.

REMEMBER For example, suppose you have $3x - x$. Recall that $3x$ is simply shorthand for $x + x + x$. So doing this subtraction gives you the following:

$$x + x + x - x = 2x$$

No big surprises here. You simply find $(3-1)x$. This time, the idea roughly parallels the idea that $\$3 - \$1 = \$2$.

Here's another example:

$$2x - 5x$$

Again, no problem, as long as you know how to work with negative numbers (see Chapter 4 if you need details). Just find the difference between the coefficients:

$$= (2 - 5)x = -3x$$

In this case, recall that $\$2 - \$5 = -\$3$ (that is, a debt of $\$3$).

WARNING

You *cannot* subtract non-like terms. For example, you can't subtract the following:

$$7x - 4y$$

As with addition, you can't do subtraction with different variables. Think of this as trying to figure out 7 dollars − 4 pesos. Because the units in this case (dollars versus pesos) are different, you're stuck. (See Chapter 18 for more on working with units.)

EXAMPLE

Q. What is $3x + 5x$?

A. **8x.** The variable part of both terms is *x*, so you can add:

$$3x + 5x = (3+5)x = 8x$$

Q. What is $24x^3 - 7x^3$?

A. **17x³.** Subtract the coefficients:

$$24x^3 - 7x^3 = (24-7)x^3 = 17x^3$$

YOUR TURN

19 Add $4x^2y + (-9x^2y)$.

20 Add $x^3y^2 + 20x^3y^2$.

21 Subtract $-2xy^4 - 20xy^4$.

22 Subtract $-xyz - (-xyz)$.

Multiplying and Dividing Terms

REMEMBER

Unlike adding and subtracting, you can multiply non-like terms. Multiply *any* two terms by multiplying their coefficients and combining — that is, by collecting or gathering up — all the variables in each term into a single term, as I show you next.

For example, suppose you want to multiply $5x(3y)$. To get the coefficient, multiply 5×3. To get the algebraic part, combine the variables *x* and *y*:

$$= 5(3)xy = 15xy$$

Now suppose you want to multiply $2x(7x)$. Again, multiply the coefficients and collect the variables into a single term:

$$= 7(2)xx = 14xx$$

Remember that x^2 is shorthand for xx, so you can write the answer more efficiently:

$$= 14x^2$$

Here's another example. Multiply all three coefficients together and gather up the variables:

$$2x^2(3y)(4xy)$$
$$= 2(3)(4)x^2xyy$$
$$= 24x^3y^2$$

As you can see, the exponent 3 that's associated with x is just the count of how many x's appear in the problem. The same is true of the exponent 2 associated with y.

TIP

A fast way to multiply variables with exponents is to add the exponents together. For example:

$$\left(x^4y^3\right)\left(x^2y^5\right)\left(x^6y\right) = x^{12}y^9$$

In this example, I added the exponents of the x's $(4+2+6=12)$ to get the exponent of x in the expression. Similarly, I added the exponents of the y's $(3+5+1=9$ — don't forget that $y = y^1$!) to get the exponent of y in the expression.

It's customary to represent division of algebraic expressions as a fraction instead of using the division sign ($\div$). So division of algebraic terms really looks like reducing a fraction to lowest terms (see Chapter 8 for more on reducing).

To divide one algebraic term by another, follow these steps:

1. **Make a fraction of the two terms.**

   Suppose you want to divide $3xy$ by $12x^2$. Begin by turning the problem into a fraction:

   $$\frac{3xy}{12x^2}$$

2. **Cancel out factors in coefficients that are in both the numerator and the denominator.**

   In this case, you can cancel out a 3. Notice that when the coefficient in xy becomes 1, you can drop it:

   $$= \frac{xy}{4x^2}$$

3. **Cancel out any variable that's in both the numerator and the denominator.**

   You can break x^2 out as xx:

   $$= \frac{xy}{4xx}$$

Now you can clearly cancel an x in both the numerator and the denominator:

$$= \frac{y}{4x}$$

As you can see, the resulting fraction is really a reduced form of the original.

As another example, suppose you want to divide $-6x^2yz^3$ by $-8x^2y^2z$. Begin by writing the division as a fraction:

$$\frac{-6x^2yz^3}{-8x^2y^2z}$$

First, reduce the coefficients. Notice that, because both coefficients were originally negative, you can cancel out both negative signs as well:

$$= \frac{3x^2yz^3}{4x^2y^2z}$$

Now you can begin canceling variables. I do this in two steps, as before:

$$= \frac{3xxyzzz}{4xxyyz}$$

At this point, just cross out any occurrence of a variable that appears in both the numerator and the denominator:

$$= \frac{3zz}{4y}$$

$$= \frac{3z^2}{4y}$$

WARNING

You can't cancel out variables or coefficients if either the numerator or the denominator has more than one term in it. This is a very common mistake in algebra, so don't let it happen to you!

TIP

A fast way to divide is to subtract the exponents of identical variables. For each variable, subtract the exponent in the numerator minus the exponent in the denominator. When a resulting exponent is a negative number, move that variable to the denominator and remove the negative sign.

EXAMPLE

Q. What is $2x(6y)$?

A. **12xy.** To get the final coefficient, multiply the coefficients $2 \times 6 = 12$. To get the variable part of the answer, combine the variables x and y:

$$2x(6y) = 2(6)xy = 12xy$$

Q. What is $4x(-9x)$?

A. **$-36x^2$.** To get the final coefficient, multiply the coefficients $4 \times (-9) = -36$. To get the variable part of the answer, collect the variables into a single term:

$$4x(-9x) = 4(-9)xx$$

Remember that x^2 is shorthand for xx, so rewrite the answer as follows:

$$-36x^2$$

Q. What is $2x(4xy)(5y^2)$?

A. **$40x^2y^3$.** Multiply all three coefficients together and gather up the variables:

$$2x(4xy)(5y^2) = 2(4)(5)xxyy^2 = 40x^2y^3$$

In this answer, the exponent 2 that's associated with x is just a count of how many x's appear in the problem; the same is true of the exponent 3 that's associated with y.

Q. What is $(x^2y^3)(xy^5)(x^4)$?

A. **x^7y^8.** Add exponents of the three x's $(2+1+4 = 7)$ to get the exponent of x in the answer (remember that x means x^1). Add the two y exponents $(3+5 = 8)$ to get the exponent of y in the answer.

Q. What is $\dfrac{6x^3y^2}{3xy^2}$?

A. **$2x^2$.** Exponents mean repeated multiplication, so

$$\frac{6x^3y^2}{3xy^2} = \frac{6xxxyy}{3xyy}$$

Both the numerator and the denominator are divisible by 2, so reduce the coefficients just as you'd reduce a fraction:

$$= \frac{2xxxyy}{xyy}$$

Now cancel out any variables repeated in both the numerator and denominator — that is, one x and two y's both above and below the fraction bar:

$$= \frac{2xx\cancel{x}\cancel{y}\cancel{y}}{\cancel{x}\cancel{y}\cancel{y}} = 2xx$$

Rewrite this result using an exponent.

$$2x^2$$

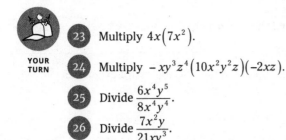

23. Multiply $4x\left(7x^2\right)$.

24. Multiply $-xy^3z^4\left(10x^2y^2z\right)(-2xz)$.

25. Divide $\dfrac{6x^4y^5}{8x^4y^4}$.

26. Divide $\dfrac{7x^2y}{21xy^3}$.

Simplifying Expressions by Combining Like Terms

Although an algebraic expression may have any number of terms, you can sometimes make it smaller and easier to work with. This process is called *simplifying* an expression. To simplify, you *combine like terms*, which means you add or subtract any terms that have identical variable parts. (To understand how to identify like algebraic terms, see the earlier section, "Knowing the Terms." The section "Adding and Subtracting Like Terms" shows you how to do the math.)

In some expressions, like terms may not all be next to each other. In this case, you may want to rearrange the terms to place all like terms together before combining them.

When rearranging terms, you *must* keep the sign (+ or −) with the term that follows it.

To finish, you usually rearrange the answer alphabetically, from highest exponent to lowest, and place any constant last. In other words, if your answer is $-5+4y^3+x^2+2x-3xy$, you'd rearrange it to read $x^2+2x-3xy+4y^3-5$. (This step doesn't change the answer, but it kind of cleans it up, so teachers love it.)

For example, suppose you're working with the following expression:

$$4x-3y+2x+y-x+2y$$

As it stands, this expression has six terms. But three terms have the variable x and the other three have the variable y. Begin by rearranging the expression so that all like terms are grouped together:

$$=4x+2x-x-3y+y+2y$$

Now you can add and subtract like terms. I do this in two steps, first for the x terms and then for the y terms:

$$=5x-3y+y+2y$$
$$=5x+0y$$
$$=5x$$

Notice that the x terms simplify to $5x$, and the y terms simplify to $0y$, which is 0, so the y terms drop out of the expression altogether.

Here's a somewhat more complicated example that has variables with exponents:

$$12x - xy - 3x^2 + 8y + 10xy + 3x^2 - 7x$$

This time, you have four different types of terms. As a first step, you can rearrange these terms so that groups of like terms are all together (I underline these four groups so you can see them clearly):

$$= \underline{12x - 7x} \ \underline{-xy + 10xy} \ \underline{-3x^2 + 3x^2} \ \underline{+8y}$$

Now combine each set of like terms:

$$= 5x + 9xy + 0x^2 + 8y$$

This time, the x^2 terms add up to 0, so they drop out of the expression altogether:

$$= 5x + 9xy + 8y$$

Q. Simplify the expression $x^2 + 2x - 7x + 1$.

A. $x^2 - 5x + 1$. The terms $2x$ and $-7x$ are like, so you can combine them:

$$x^2 + \underline{2x - 7x} + 1 = x^2 + (2 - 7)x + 1 = x^2 - 5x + 1$$

Q. Simplify the expression $4x^2 - 3x + 2 + x - 7x^2$.

A. Begin by rearranging these terms as needed so that all like terms are together:

$$\underline{4x^2 - 7x^2} \ \underline{-3x + x} + 2$$

At this point, combine the two underlined pairs of like terms:

$$(4 - 7)x^2 + (-3 + 1)x + 2 = -3x^2 - 2x + 2$$

YOUR
TURN

(27) Simplify the expression $3x^2 + 5x^2 + 2x - 8x - 1$.

(28) Simplify the expression $6x^3 - x^2 + 2 - 5x^2 - 1 + x$.

(29) Simplify the expression $2x^4 - 2x^3 + 2x^2 - x + 9 + x + 7x^2$.

(30) Simplify the expression $x^5 - x^3 + xy - 5x^3 - 1 + x^3 - xy + x$.

Removing Parentheses from an Algebraic Expression

Parentheses keep parts of an expression together as a single unit. In Chapter 4, I show you how to handle parentheses in an arithmetic expression. This skill is also useful with algebraic expressions. As you find when you begin solving algebraic equations in Chapter 23, getting rid of parentheses is often the first step toward solving a problem. In this section, I show how to handle the Big Four operations with ease.

Drop everything: Parentheses with a plus sign

When an expression contains parentheses that come right after a plus sign (+), you can just remove the parentheses. Here's an example:

$$2x + (3x - y) + 5y$$
$$= 2x + 3x - y + 5y$$

Now you can simplify the expression by combining like terms:

$$= 5x + 4y$$

When the first term inside the parentheses is negative, when you drop the parentheses, the negative sign replaces the plus sign. For example:

$$6x + (-2x + y) - 4y$$
$$= 6x - 2x + y - 4y$$
$$= 4x - 3y$$

Sign turnabout: Parentheses with a negative sign

Sometimes an expression contains parentheses that come right after a negative sign (−). In this case, change the sign of every term inside the parentheses to the opposite sign; then remove the parentheses.

Consider this example:

$$6x - (2xy - 3y) + 5xy$$

A negative sign is in front of the parentheses, so you need to change the signs of both terms in the parentheses and remove the parentheses. Notice that the term $2xy$ appears to have no sign because it's the first term inside the parentheses. This expression really means the following:

$$= 6x - (+2xy - 3y) + 5xy$$

You can see how to change the signs:

$$= 6x - 2xy + 3y + 5xy$$

At this point, you can combine the two xy terms:

$$= 6x + 3xy + 3y$$

Distribution: Parentheses with no sign

When you see nothing between a number and a set of parentheses, it means multiplication. For example:

$$2(3) = 6 \qquad 4(4) = 16 \qquad 10(15) = 150$$

This notation becomes much more common with algebraic expressions, replacing the multiplication sign ($\times$) to avoid confusion with the variable x:

$$3(4x) = 12x \qquad 4x(2x) = 8x^2 \qquad 3x(7y) = 21xy$$

REMEMBER

To remove parentheses without a sign, multiply the term outside the parentheses by every term inside the parentheses; then remove the parentheses. When you follow those steps, you're using the *distributive property*. Here's an example:

$$2(3x - 5y + 4)$$

In this case, multiply 2 by each of the three terms inside the parentheses:

$$= 2(3x) + 2(-5y) + 2(4)$$

For the moment, this expression looks more complex than the original one, but now you can get rid of all three sets of parentheses by multiplying:

$$= 6x - 10y + 8$$

As another example, suppose you have the following expression:

$$-2x(-3x + y + 6) + 2xy - 5x^2$$

Begin by multiplying $-2x$ by the three terms inside the parentheses:

$$= -2x(-3x) - 2x(y) - 2x(6) + 2xy - 5x^2$$

The expression looks worse than when you started, but you can get rid of all the parentheses by multiplying:

$$= 6x^2 - 2xy - 12x + 2xy - 5x^2$$

Now you can combine like terms:

$$= x^2 - 12x$$

In summary, here are the four possible cases.

>> **Parentheses preceded by a plus sign (+):** Just remove the parentheses. After that, you may be able to simplify the expression further by combining like terms, as I show you in the preceding section.

>> **Parentheses preceded by a negative sign (–):** Change every term inside the parentheses to the opposite sign; then remove the parentheses. After the parentheses are gone, combine like terms.

>> **Parentheses preceded by no sign (a term directly next to a set of parentheses):** Multiply the term next to the parentheses by every term inside the parentheses (make sure you include the plus or negative signs in your terms); then drop the parentheses. Simplify by combining like terms. **Remember:** To multiply identical variables, simply add the exponents. For instance, $x(x^2) = x^{1+2} = x^3$. Likewise, $-x^3(x^4) = -(x^{3+4}) = -x^7$.

>> **Two sets of adjacent parentheses:** I discuss this case in the next section.

EXAMPLE

Q. Simplify the expression $7x + (x^2 - 6x + 4) - 5$.

A. $x^2 + x - 1$. Because a plus sign precedes this set of parentheses, you can drop the parentheses.

$$7x + (x^2 - 6x + 4) - 5$$
$$= 7x + x^2 - 6x + 4 - 5$$

Now combine like terms. I do this in two steps:

$$= \underline{7x - 6x} + x^2 + \underline{4 - 5}$$

$$= x + x^2 - 1$$

Finally, rearrange the answer from highest exponent to lowest:

Q. Simplify the expression $x - 3x(x^3 - 4x^2 + 2) + 8x^4$.

A. $5x^4 + 12x^3 - 5x$. The term $-3x$ precedes this set of parentheses without a sign in between, so multiply every term inside the parentheses by $-3x$ and then drop the parentheses:

$$x - 3x(x^3 - 4x^2 + 2) + 8x^4$$
$$= x + -3x(x^3) - 3x(-4x^2) - 3x(2) + 8x^4$$
$$= x - 3x^4 + 12x^3 - 6x + 8x^4$$

Now combine like terms. I do this in two steps:

$$= x - 6x - 3x^4 + 8x^4 + 12x^3$$
$$= -5x + 5x^4 + 12x^3$$

As a final step, arrange the terms in order from highest to lowest exponent:

$$= 5x^4 + 12x^3 - 5x$$

YOUR TURN

31 Simplify the expression $3x^3 + (12x^3 - 6x) + (5 - x)$.

32 Simplify the expression $2x^4 - (-9x^2 + x) + (x + 10)$.

33 Simplify the expression $x - (x^3 - x - 5) + 3(x^2 - x)$.

34 Simplify the expression $-x^3(x^2 + x) - (x^5 - x^4)$.

FOILing: Dealing with Two Sets of Parentheses

Sometimes, expressions have two sets of parentheses next to each other without a sign between them. In that case, you need to multiply *every term* inside the first set by *every term* inside the second.

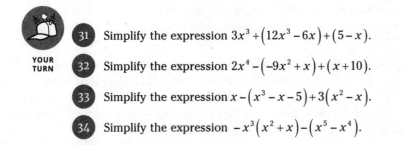

When you have two terms inside each set of parentheses, you can use a process called FOILing. This is really just the distributive property, as I show you in the following steps. The word *FOIL* is an acronym to help you make sure you multiply the correct terms. It stands for First, Outside, Inside, and Last.

TIP

Here's how the process works. In this example, you're simplifying the expression $(2x - 2)(3x - 6)$:

1. **Start out by multiplying the two *First* terms in the parentheses.**

   The first term in the first set of parentheses is $2x$, and $3x$ is the first term in the second set of parentheses: $(\underline{2x} - 2)(\underline{3x} - 6)$.

   F — Multiply the first terms: $2x(3x) = 6x^2$

2. **Multiply the two *Outside* terms.**

   The two outside terms, $2x$ and -6, are on the ends: $(\underline{2x} - 2)(3x \underline{- 6})$

   O — Multiply the outside terms: $2x(-6) = -12x$

3. **Multiply the two *Inside* terms.**

   The two inside terms are -2 and $3x$: $(2x \underline{- 2})(\underline{3x} - 6)$

   I — Multiply the inside terms: $-2(3x) = -6x$

4. **Multiply the two *Last* terms.**

The last term in the first set of parentheses is −2, and −6 is the last term in the second set: $(2x \underline{- 2})(3x \underline{- 6})$.

L — Multiply the last terms: $-2(-6) = 12$

Add these four results together to get the simplified expression:

$$(2x-2)(3x-6) = 6x^2 - 12x - 6x + 12$$

In this case, you can simplify this expression still further by combining the like terms $-12x$ and $-6x$:

$$= 6x^2 - 18x + 12$$

Notice that, during this process, you multiply every term inside one set of parentheses by every term inside the other set. FOILing just helps you keep track and make sure you've multiplied everything.

FOILing is really just an application of the distributive property, which I discuss in the section preceding this one. In other words, $(2x-2)(3x-6)$ is really the same as $2x(3x-6) - 2(3x-6)$ when distributed. Then, distributing again gives you $6x^2 - 12x - 6x + 12$.

Q. Simplify the expression $(x+4)(x-3)$.

EXAMPLE **A.** $x^2 + x - 12$. Start by multiplying the two first terms:

$(\underline{x}+4)(\underline{x}-3)$ $x \bullet x = x^2$

Next, multiply the two outside terms:

$(\underline{x}+4)(x \underline{-3})$ $x \bullet (-3) = -3x$

Now multiply the two inside terms:

$(x+\underline{4})(\underline{x}-3)$ $4 \bullet x = 4x$

Finally, multiply the two last terms:

$(x+\underline{4})(x\underline{-3})$ $4 \bullet (-3) = -12$

Add all four results together and simplify by combining like terms:

$(x+4)(x-3) = x^2 \underline{-3x + 4x} - 12 = x^2 + x - 12$

Q. Simplify the expression $x^2 - (-2x + 5)(3x - 1) + 9$.

A. $7x^2 - 17x + 14$. Begin by FOILing the parentheses. Start by multiplying the two first terms:

$(\underline{-2x}+5)(\underline{3x}-1)$ $-2x \bullet 3x = -6x^2$

Multiply the two outside terms:

$(\underline{-2x}+5)(3x\ \underline{-1})$ $-2x\bullet(-1)=2x$

Multiply the two inside terms:

$(-2x\ \underline{+5})(\ \underline{3x}\ -\ 1)$ $5\bullet 3x=15x$

Finally, multiply the two last terms:

$(-2x+\underline{5})(3x\ \underline{-1})$ $5\bullet(-1)=-5$

Add these four products together and put the result inside one set of parentheses, replacing the two sets of parentheses that were originally there:

$x^2-(-2x+5)(3x-1)+9$
$=x^2-\left(-6x^2+2x+15x-5\right)+9$

The remaining set of parentheses is preceded by a negative sign, so change the sign of every term in there to its opposite and drop the parentheses:

$=x^2+6x^2-2x-15x+5+9$

At this point, you can simplify the expression by combining like terms:

$=x^2+6x^2-2x-15x+5+9$
$=7x^2-17x+14$

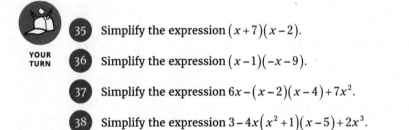

35 Simplify the expression $(x+7)(x-2)$.

36 Simplify the expression $(x-1)(-x-9)$.

37 Simplify the expression $6x-(x-2)(x-4)+7x^2$.

38 Simplify the expression $3-4x\left(x^2+1\right)(x-5)+2x^3$.

Practice Questions Answers and Explanations

(1) **28.** Substitute 3 for x in the expression:

$$x^2 + 5x + 4 = 3^2 + 5(3) + 4$$

Evaluate the expression using the order of operations. Start with the power:

$$= 9 + 5(3) + 4$$

Continue by evaluating the multiplication:

$$= 9 + 15 + 4$$

Finish up by evaluating the addition from left to right:

$$= 24 + 4 = 28$$

(2) **56.** Plug in −2 for every x in the expression:

$$5x^4 + x^3 - x^2 + 10x + 8 = 5(-2)^4 + (-2)^3 - (-2)^2 + 10(-2) + 8$$

Evaluate the expression using the order of operations. Start with the powers:

$$= 5(16) + -8 - 4 + 10(-2) + 8$$

Do the multiplication:

$$= 80 + -8 - 4 - 20 + 8$$

Finish up by evaluating the addition and subtraction from left to right:

$$= 72 - 4 + -20 + 8 = 68 - 20 + 8 = 48 + 8 = 56$$

(3) **−120.** Substitute 4 for x in the expression:

$$x(x^2 - 6)(x - 7) = 4(4^2 - 6)(4 - 7)$$

Follow the order of operations as you evaluate the expression. Starting inside the first set of parentheses, evaluate the power and then the subtraction:

$$= 4(16 - 6)(4 - 7) = 4(10)(4 - 7)$$

Find the contents of the remaining set of parentheses:

$$= 4(10)(-3)$$

Evaluate the multiplication from left to right:

$$= 40(-3) = -120$$

④ $\dfrac{(x-9)^4}{(x+4)^3} = \dfrac{81}{1,000}$ **(or 0.081).** Replace every x in the expression with a 6:

$$\dfrac{(x-9)^4}{(x+4)^3} = \dfrac{(6-9)^4}{(6+4)^3}$$

Follow the order of operations. Evaluate the contents of the set of parentheses in the numerator and then in the denominator:

$$= \dfrac{(-3)^4}{(6+4)^3} = \dfrac{(-3)^4}{10^3}$$

Continue by evaluating the powers from top to bottom:

$$= \dfrac{81}{10^3} = \dfrac{81}{1,000}$$

You can also express this fraction as the decimal 0.081.

⑤ **446.** Substitute 5 for x and 7 for y in the expression:

$$3x^2 + 5xy + 4y^2 = 3(5)^2 + 5(5)(7) + 4(7)^2$$

Evaluate using the order of operations. Start with the two powers:

$$= 3(25) + 5(5)(7) + 4(49)$$

Evaluate the multiplication from left to right:

$$= 75 + 175 + 196$$

Finally, do the addition from left to right:

$$= 250 + 196 = 446$$

⑥ **414.** Plug in −1 for each x and 9 for each y in the expression:

$$x^6y - 5xy^2 = (-1)^6(9) - 5(-1)(9)^2$$

Follow the order of operations. Start by evaluating the two powers:

$$= 1(9) - 5(-1)(81)$$

Continue by evaluating the multiplication from left to right:

$$= 9 - (-5)(81) = 9 - (-405)$$

Finally, do the subtraction:

$$= 414$$

(7) The algebraic terms are $7x^2yz, -10xy^2z, 4xyz^2, y,$ and $-z$; the constant is **2**.

(8) The six coefficients, in order, are $-2, 6, 1, -1, -8,$ and **17**.

(9) The four coefficients, in order, are $-1, -1, 1,$ and -1.

(10) $7x^2$ and $-x^2$ are like terms (the variable part is x^2); $-2x$ and $99x$ are also like terms (the variable part is x).

(11) $-j - 7k + 1$

(12) $-x + 4y + 5z - 9$

(13) $-t - 100u + 15v - 75$

(14) $-2p + q + 3r + s - 10$

(15) $1, 4,$ and -3.

(16) $3, -2,$ and -1.

(17) $-25, -60, 10.5, -1,$ and **300**.

(18) $-1, 7, -9,$ and **1**.

(19) $4x^2y + -(9x^2y) = (4 - 9)x^2y = -5x^2y$

(20) $x^3y^2 + 20x^3y^2 = (1 + 20)x^3y^2 = 21x^3y^2$

(21) $-2xy^4 - 20xy^4 = (-2 - 20)xy^4 = -22xy^4$

(22) $-xyz - (-xyz) = \left[-1 - (-1)\right]xyz = (-1 + 1)xyz = 0$

(23) **28x³.** Multiply the two coefficients to get the coefficient of the answer; then gather the variables into one term:

$$4x(7x^2) = 4(7)xx^2 = 28x^3$$

(24) **20$x^4y^5z^6$.** Multiply the coefficients $(-1 \times 10 \times -2 = 20)$ to get the coefficient of the answer. Add the x exponents $(1 + 2 + 1 = 4)$ to get the exponent of x in the answer. Add the y exponents $(3 + 2 = 5)$ to get the exponent of y in the answer. And add the z exponents $(4 + 1 + 1 = 6)$ to get the exponent of z in the answer.

(25) $\dfrac{6x^4y^5}{8x^4y^4} = \dfrac{3y}{4}$. Reduce the coefficients of the numerator and denominator just as you'd reduce a fraction:

$$\frac{6x^4y^5}{8x^4y^4} = \frac{3x^4y^5}{4x^4y^4}$$

To get the x exponent of the answer, take the x exponent in the numerator minus the x exponent in the denominator: $4 - 4 = 0$, so the x's cancel:

$$= \frac{3y^5}{4y^4}$$

To get the y exponent of the answer, take the y exponent in the numerator minus the y exponent in the denominator: $5 - 4 = 1$, so you have only y^1, or y, in the numerator:

$$= \frac{3y}{4}$$

(26) $\frac{7x^2y}{21xy^3} = \frac{x}{3y^2}$. Reduce the coefficients of the numerator and denominator just as you'd reduce a fraction:

$$\frac{7x^2y}{21xy^3} = \frac{x^2y}{3xy^3}$$

Take the x exponent in the numerator minus the x exponent in the denominator $(2 - 1 = 1)$ to get the x exponent of the answer:

$$= \frac{xy}{3y^3}$$

To get the y exponent, take the y exponent in the numerator minus the y exponent in the denominator $(1 - 3 = -2)$:

$$= \frac{xy^{-2}}{3}$$

To finish up, remove the negative sign from the y exponent and move this variable to the denominator:

$$= \frac{x}{3y^2}$$

(27) $8x^2 - 6x - 1$. Combine the following underlined like terms:

$$3x^2 + 5x^2 + 2x - 8x - 1$$
$$\underline{3x^2 + 5x^2} + \underline{2x - 8x} - 1$$
$$= 8x^2 - 6x - 1$$

(28) $6x^3 - 6x^2 + x + 1$. Rearrange the terms so like terms are next to each other:

$$6x^3 - x^2 + 2 - 5x^2 - 1 + x$$
$$= 6x^3 \underline{-x^2 - 5x^2} + x + \underline{2 - 1}$$

Now combine the underlined like terms:

$$= 6x^3 - 6x^2 + x + 1$$

29 $2x^4 - 2x^3 + 9x^2 + 9$. Put like terms next to each other:

$$2x^4 - 2x^3 + 2x^2 - x + 9 + x + 7x^2$$
$$= 2x^4 - 2x^3 + \underline{2x^2 + 7x^2} - \underline{x + x} + 9$$

Now combine the underlined like terms:

$$= 2x^4 - 2x^3 + 9x^2 + 9$$

Note that the two x terms cancel each other out.

30 $x^5 - 5x^3 + x - 1$. Rearrange the terms so like terms are next to each other:

$$x^5 - x^3 + xy - 5x^3 - 1 + x^3 - xy + x$$
$$= x^5 - \underline{x^3 - 5x^3 + x^3} + \underline{xy - xy} + x - 1$$

Now combine the underlined like terms:

$$= x^5 - 5x^3 + x - 1$$

Note that the two xy terms cancel each other out.

31 $15x^3 - 7x + 5$. A plus sign precedes both sets of parentheses, so you can drop both sets:

$$3x^3 + \left(12x^3 - 6x\right) + \left(5 - x\right)$$
$$= 3x^3 + 12x^3 - 6x + 5 - x$$

Now combine like terms:

$$= \underline{3x^3 + 12x^3} - \underline{6x - x} + 5$$
$$= 15x^3 - 7x + 5$$

32 $2x^4 + 9x^2 + 10$. A negative sign precedes the first set of parentheses, so change the sign of every term inside this set and then drop these parentheses:

$$2x^4 - \left(-9x^2 + x\right) + \left(x + 10\right)$$
$$= 2x^4 + 9x^2 - x + \left(x + 10\right)$$

A plus sign precedes the second set of parentheses, so just drop this set:

$$= 2x^4 + 9x^2 - x + x + 10$$

Now combine like terms:

$$= 2x^4 + 9x^2 + 10$$

33 $-x^3 + 3x^2 - x + 5$. A negative sign precedes the first set of parentheses, so change the sign of every term inside this set and then drop these parentheses:

$$x - \left(x^3 - x - 5\right) + 3\left(x^2 - x\right)$$
$$= x - x^3 + x + 5 + 3\left(x^2 - x\right)$$

You have no sign between the term 3 and the second set of parentheses, so multiply every term inside these parentheses by 3 and then drop the parentheses:

$$= x - x^3 + x + 5 + 3x^2 - 3x$$

Now combine like terms:

$$= \underline{x} + \underline{x} - \underline{3x} - x^3 + 5 + 3x^2$$
$$= -x - x^3 + 5 + 3x^2$$

Rearrange the answer so that the exponents are in descending order:

$$= -x^3 + 3x^2 - x + 5$$

34 $-2x^5$. You have no sign between the term $-x^3$ and the first set of parentheses, so multiply $-x^3$ by every term inside this set and then drop these parentheses:

$$-x^3\left(x^2 + x\right) - \left(x^5 - x^4\right)$$
$$= -x^5 - x^4 - \left(x^5 - x^4\right)$$

A negative sign precedes the second set of parentheses, so change every term inside this set and then drop the parentheses:

$$= -x^5 - x^4 - x^5 + x^4$$

Combine like terms, noting that the x^4 terms cancel out:

$$= \underline{-x^5} \underline{-x^5} \underline{-x^4} \underline{+x^4}$$
$$= -2x^5$$

35 $x^2 + 5x - 14$. Begin by FOILing the parentheses. Start by multiplying the two first terms:

$$\left(\underline{x} + 7\right)\left(\underline{x} - 2\right) \quad x \bullet x = x^2$$

Multiply the two outside terms:

$$\left(\underline{x} + 7\right)\left(x \underline{-2}\right) \quad x \bullet -2 = -2x$$

Multiply the two inside terms:

$$\left(x + \underline{7}\right)\left(\underline{x} - 2\right) \quad 7 \bullet x = 7x$$

Finally, multiply the two last terms:

$$(x+\underline{7})(x\underline{-2}) \quad 7\bullet-2=-14$$

Add these four products together and simplify by combining like terms:

$$x^2\underline{-2x+7x}-14=x^2+5x-14$$

(36) $-x^2-8x+9$. FOIL the parentheses. Multiply the two first terms, the two outside terms, the two inside terms, and the two last terms:

$$\begin{aligned}(\underline{x}-1)(\underline{-x}-9) \quad & x\bullet-x=-x^2\\(\underline{x}-1)(-x\underline{-9}) \quad & x\bullet-9=-9x\\(x\underline{-1})(\underline{-x}-9) \quad & -1\bullet-x=x\\(x\underline{-1})(-x\underline{-9}) \quad & -1\bullet-9=9\end{aligned}$$

Add these four products together and simplify by combining like terms:

$$-x^2\underline{-9x+x}+9=-x^2-8x+9$$

(37) $6x^2+12x-8$. Begin by FOILing the parentheses: Multiply the first, outside, inside, and last terms:

$$\begin{aligned}(\underline{x}-2)(\underline{x}-4) \quad & x\bullet x^2\\(\underline{x}-2)(x\underline{-4}) \quad & x\bullet-4=-4x\\(x\underline{-2})(\underline{x}-4) \quad & -2\bullet x=-2x\\(x\underline{-2})(x\underline{-4}) \quad & -2\bullet-4=8\end{aligned}$$

Add these four products together and put the result inside one set of parentheses, replacing the two sets of parentheses that were originally there:

$$6x-(x-2)(x-4)+7x^2=6x-\left(x^2-4x-2x+8\right)+7x^2$$

The remaining set of parentheses is preceded by a negative sign, so change the sign of every term in there to its opposite and drop the parentheses:

$$=6x-x^2+4x+2x-8+7x^2$$

Now simplify the expression by combining like terms and reordering your solution:

$$\begin{aligned}&=\underline{6x+4x+2x}-x^2+7x^2-8\\&=6x^2+12x-8\end{aligned}$$

(38) $-4x^4+22x^3-4x^2+20x+3$. Begin by FOILing the parentheses, multiplying the first, outside, inside, and last terms:

$$\left(\underline{x^2}+1\right)\left(\underline{x}-5\right) \quad x^2 \bullet x = x^3$$
$$\left(\underline{x^2}+1\right)\left(x\ \underline{-5}\right) \quad x^2 \bullet -5 = -5x^2$$
$$\left(x^2+\underline{1}\right)\left(\underline{x}-5\right) \quad 1 \bullet x = x$$
$$\left(x^2+\underline{1}\right)\left(x\ \underline{-5}\right) \quad 1 \bullet -5 = -5$$

Add these four products together and put the result inside one set of parentheses, replacing the two sets of parentheses that were originally there:

$$3 - 4x\left(x^2+1\right)\left(x-5\right) + 2x^3 = 3 - 4x\left(x^3 - 5x^2 + x - 5\right) + 2x^3$$

The remaining set of parentheses is preceded by the term $-4x$ with no sign in between, so multiply $-4x$ by every term in there and then drop the parentheses:

$$= 3 - 4x^4 + 20x^3 - 4x^2 + 20x + 2x^3$$

Now simplify the expression by combining like terms:

$$= -4x^4 + \underline{20x^3 + 2x^3} - 4x^2 + 20x + 3$$
$$= -4x^4 + 22x^3 - 4x^2 + 20x + 3$$

How did you do on the practice problems? When you're ready, show off what you know by taking the chapter quiz in the next section.

Whaddya Know? Chapter 22 Quiz

Time for a quiz! These 17 problems cover the gamut of algebra topics reviewed in this chapter. When you're done, flip to the next section for answers and complete explanations for all the questions.

1. Evaluate $3x^3 + 4x^2 - 1$ when $x = -2$.

2. Simplify: $4y + 6x^2 + 3y - 5x^2$

3. Divide and simplify: $\dfrac{8x^4y^2}{-2x^2y}$

4. Rearrange the terms in decreasing powers of the variable: $3x^3 - 2x^5 - 11 + 5x$

5. Simplify: $4x^3 - 3x^2 + 2x - 1 + x^3 - x$

6. Simplify: $5x(x+1) - 3x^2(x-2)$

7. Multiply and simplify: $(x^2 - 3)(x + 4) - 2(x^2 + 3)$

8. Simplify: $5x(-2x)$

9. Evaluate $x(x-2)^2(x+7)$ when $x = 3$.

10. Divide and simplify: $\dfrac{3x^2y^5}{6x^2y^8}$

11. Simplify: $5x - 3(5 - x) + 3$

12. Rearrange the terms in increasing values of the coefficients: $2x^2 - 3x^3 + 4x - 5x^7 + x^4$

13. Multiply and simplify: $(x + 2)(x - 7)$

14. Simplify: $-3x(x - y)$

15. Evaluate $x^2 - 2xy + 3y^2$ when $x = -1$ and $y = -2$.

16. Multiply and simplify: $(3x - 1)(4x - 2)$

17. Simplify: $3x^2(2xy)(-y^3)$

Answers to Chapter 22 Quiz

$\textcircled{1}$ **−9.** After replacing all the x's with −2, raise the powers, then multiply, and lastly add and subtract: $3(-2)^3 + 4(-2)^2 - 1 = 3(-8) + 4(4) - 1 = -24 + 16 - 1 = -9$.

$\textcircled{2}$ $x^2 + 7y$. Rearrange the terms, grouping the like-terms. Then simplify.

$$4y + 6x^2 + 3y - 5x^2 = 4y + 3y + 6x^2 - 5x^2 = (4+3)y + (6-5)x^2 = 7y + x^2$$

Place the variables in alphabetical order:

$$= x^2 + 7y$$

$\textcircled{3}$ **−4x^2y.** Rewrite with all the factors. Then reduce the fraction.

$$\frac{8x^4y^2}{-2x^2y} = \frac{2 \cdot 2 \cdot \cancel{2} \cancel{x}\cancel{x}xx\cancel{y}y}{-\cancel{2}\cancel{x}\cancel{x}\cancel{y}} = \frac{4x^2y}{-1} = -4x^2y$$

$\textcircled{4}$ **−2x^5 + 3x^3 + 5x − 11.** The variables are the x factors. When there is no variable, technically the power of x is 0: $3x^3 - 2x^5 - 11 + 5x = -2x^5 + 3x^3 + 5x - 11$.

$\textcircled{5}$ **5x^3 − 3x^2 + x − 1.** Rewrite by grouping like terms. Then simplify.

$$4x^3 - 3x^2 + 2x - 1 + x^3 - x = 4x^3 + x^3 - 3x^2 + 2x - x - 1$$
$$= (4+1)x^3 - 3x^2 + (2-1)x - 1$$
$$= 5x^3 - 3x^2 + 1x - 1$$

$\textcircled{6}$ **−3x^3 + 11x^2 + 5x.** First perform the multiplications. Then combine like terms and simplify.

$$5x(x+1) - 3x^2(x-2) = 5x^2 + 5x - 3x^3 + 6x^2$$
$$= -3x^3 + 5x^2 + 6x^2 + 5x$$
$$= -3x^3 + (5+6)x^2 + 5x$$
$$= -3x^3 + 11x^2 + 5x$$

$\textcircled{7}$ $x^3 + 2x^2 - 3x - 18$. Multiply the first term using FOIL; distribute the second product. Simplify.

$$(x^2-3)(x+4) - 2(x^2+3) = x^3 + 4x^2 - 3x - 12 - 2x^2 - 6$$
$$= x^3 + 4x^2 - 2x^2 - 3x - 12 - 6$$
$$= x^3 + (4-2)x^2 - 3x + (-12-6)$$
$$= x^3 + 2x^2 - 3x - 18$$

$\textcircled{8}$ **−10x^2.** Multiply the coefficients and variables.

$$5x(-2x) = (5)(-2)(x)(x) = -10x^2$$

$\textcircled{9}$ **30.** Replace each x with 3. Then simplify.

$$3(3-2)^2(3+7) = 3(1)^2(10) = 3(1)(10) = 30$$

10 $\dfrac{1}{2y^3}$. **Write the variable powers as products and reduce.**

$$\frac{3x^2y^5}{6x^2y^8} = \frac{3xxyyyyy}{2 \cdot 3xxyyyyyyyyy} = \frac{\cancel{3}\cancel{x}\cancel{x}\cancel{y}\cancel{y}\cancel{y}\cancel{y}\cancel{y}}{2 \cdot \cancel{3}\cancel{x}\cancel{x}\cancel{y}\cancel{y}\cancel{y}\cancel{y}\cancel{y}yyy} = \frac{1}{2y^3}$$

11 $8x - 12$. **First distribute the –3. Then combine the like terms.**

$$5x - 3(5-x) + 3 = 5x - 15 + 3x + 3$$
$$= 5x + 3x - 15 + 3$$
$$= (5+3)x + (-15+3)$$
$$= 8x - 12$$

12 $-5x^7 - 3x^3 + x^4 + 2x^2 + 4x$. **The coefficients are the numerical multipliers of the x's. Negative coefficients are smaller and come first. When there is no numerical coefficient, you assume that it is 1.**

$$2x^2 - 3x^3 + 4x - 5x^7 + x^4 = -5x^7 - 3x^3 + x^4 + 2x^2 + 4x$$

13 $x^2 - 5x - 14$. **Use FOIL:** $(x+2)(x-7) = x^2 - 7x + 2x - 14 = x^2 - 5x - 14$.

14 $-3x^2 + 3xy$. **Distribute the –3x:** $-3x(x-y) = -3x(x) - 3x(-y) = -3x^2 + 3xy$.

15 9. **Replace the x's with –1 and the y's with –2. Then simplify.**

$$(-1)^2 - 2(-1)(-2) + 3(-2)^2 = 1 - 2(-1)(-2) + 3(4)$$
$$= 1 - (4) + 12$$
$$= 9$$

16 $12x^2 - 10x + 2$. **Use FOIL.**

$$(3x-1)(4x-2) = 12x^2 - 6x - 4x + 2 = 12x^2 + (-6-4)x + 2 = 12x^2 - 10x + 2$$

17 $-6x^3y^4$. **Multiply the numerical factors. Multiply the variables by adding exponents.**

$$3x^2(2xy)(-y^3) = 3(2)(-1)(x^2)(x)(y)(y^3)$$
$$= -6(x^{2+1})(y^{1+3}) = -6x^3y^4$$

Chapter **23**

Solving Algebraic Equations

When it comes to algebra, solving equations is the main event.

Solving an algebraic equation means finding out what number the variable (usually *x*) stands for. Not surprisingly, this process is called *solving for x,* and when you know how to do it, your confidence — not to mention your grades in your algebra class — will soar through the roof.

This chapter is all about solving for *x*. First, I show you a few informal methods to solve for *x* when an equation isn't too difficult. Then I show you how to solve more difficult equations by thinking of them as a balance scale.

The balance scale method is really the heart of algebra (yes, algebra has a heart, after all!). When you understand this simple idea, you're ready to solve more complicated equations, using all the tools I show you in Chapter 22, such as simplifying expressions and removing parentheses. You find out how to extend these skills to algebraic equations. Finally, I show you how cross-multiplying (see Chapter 10) can make solving algebraic equations with fractions a piece of cake.

By the end of this chapter, you'll have a solid grasp of a bunch of ways to solve equations for the elusive and mysterious x.

Understanding Algebraic Equations

An algebraic equation is an equation that includes at least one variable — that is, a letter (such as x) that stands for a number. *Solving* an algebraic equation means finding out what number x stands for.

In this section, I show you the basics of how a variable like x works its way into an equation in the first place. Then I show you a few quick ways to *solve for x* when an equation isn't too difficult.

Using x in equations

As you discover in Chapter 6, an *equation* is a mathematical statement that contains an equals sign. For example, here's a perfectly good equation:

$$7 \times 9 = 63$$

At its heart, a variable (such as x) is nothing more than a placeholder for a number. You're probably used to equations that use other placeholders: One number is purposely left as a blank or replaced by an underline or a question mark, and you're supposed to fill it in. Usually, this number comes after the equals sign. For example:

$$8 + 2 = \qquad\qquad 12 - 3 = \underline{\quad} \qquad\qquad 14 \div 7 = ?$$

As soon as you're comfortable with addition, subtraction, or whatever, you can switch the equation around a bit:

$$9 + \underline{\quad} = 14 \qquad\qquad ? \times 6 = 18 \qquad\qquad 100 - \square = 70$$

When you stop using underlines, question marks, and boxes and start using variables such as x to stand for the part of the equation you want to figure out, bingo! You have an algebra problem:

$$4 + 1 = x \qquad\qquad x - 13 = 30 \qquad\qquad 12 \div x = 3$$

Choosing among four ways to solve algebraic equations

You don't need to call an exterminator just to kill a bug. Similarly, algebra is strong stuff, and you don't always need it to solve an algebraic equation.

Generally, you have four ways to solve algebraic equations such as the ones I introduce earlier in this chapter. In this section, I introduce them in order of difficulty.

Eyeballing easy equations

You can solve easy problems just by looking at them. This is called solving equations by *inspection*. For example:

$$5 + x = 6$$

When you look at this problem, you can probably see that $x = 1$. When a problem is this easy and you can see the answer, you don't need to go to any particular trouble to solve it.

Rearranging slightly harder equations

When you can't see an answer just by looking at a problem, sometimes rearranging the problem helps to turn it into one that you can solve using a Big Four operation. For example:

$$6x = 96$$

You can rearrange this problem using inverse operations, as I show you in Chapter 5, changing multiplication to division:

$$x = \frac{96}{6}$$

Now solve the problem by division to find that $x = 16$.

Guessing and checking equations

You can solve some equations by guessing an answer and then checking to see whether you're right. For example, suppose you want to solve the following equation:

$$3x + 7 = 19$$

To find out what x equals, start by guessing that $x = 2$. Now check to see whether you're right by substituting 2 for x in the equation:

$$3(2) + 7 = 13 \quad \text{WRONG! (13 is less than 19.)}$$
$$3(5) + 7 = 22 \quad \text{WRONG! (22 is greater than 19.)}$$
$$3(4) + 7 = 19 \quad \text{RIGHT!}$$

With only three guesses, you found that $x = 4$.

Applying algebra to more difficult equations

When an algebraic equation gets hard enough, you find that looking at it and rearranging it just isn't enough to solve it. For example:

$$11x - 13 = 9x + 3$$

You probably can't tell what x equals just by looking at this problem. You also can't solve it just by rearranging it, using an inverse operation. And guessing and checking would be very tedious. Here's where algebra comes into play.

Algebra is especially useful because you can follow mathematical rules to find your answer. Throughout the rest of this chapter, I show you how to use the rules of algebra to turn tough problems like this one into problems that you can solve.

Q. In the equation $x + 3 = 10$, what's the value of x?

A. **$x = 7$.** You can solve this problem through simple inspection. Because $7 + 3 = 10$, $x = 7$.

Q. Solve the equation $7x = 224$ for x.

A. **$x = 32$.** Turn the problem around using the inverse of multiplication, which is division:

$7 \cdot x = 224$ means $224 \div 7 = x$.

Now you can solve the problem easily using long division:

$224 \div 7 = 32$

Q. Find the value of x in the equation $8x - 20 = 108$.

A. **$x = 16$.** Guess what you think the answer may be. For example, perhaps it's $x = 10$:

$$8(10) - 20 = 80 - 20 = 60$$

Because 60 is less than 108, that guess was too low. Try a higher number — say, $x = 20$:

$$8(20) - 20 = 160 - 20 = 140$$

Because 140 is greater than 108, this guess was too high. But now you know that the right answer is between 10 and 20. So now try $x = 15$:

$$8(15) - 20 = 120 - 20 = 100$$

The number 100 is only a little less than 108, so this guess was only a little too low. Try $x = 16$:

$$8(16) - 20 = 128 - 20 = 108$$

The result 108 is correct, so $x = 16$.

Q. Solve for x: $8x^2 - x + x^2 + 4x - 9x^2 = 18$.

A. $x = 6$. Rearrange the expression on the left side of the equation so that all like terms are next to each other:

$$8x^2 + x^2 - 9x^2 - x + 4x = 18$$

Combine like terms:

$$3x = 18$$

Notice that the x^2 terms cancel each other out. Because $3(6) = 18$, you know that $x = 6$.

YOUR TURN

1 Solve for x in each case just by looking at the equation.

    (a) $x + 5 = 13$

    (b) $18 - x = 12$

    (c) $4x = 44$

    (d) $\dfrac{30}{x} = 3$

2 Use the correct inverse operation to rewrite and solve each problem.

    (a) $x + 41 = 97$

    (b) $100 - x = 58$

    (c) $13x = 273$

    (d) $\dfrac{238}{x} = 17$

3 Find the value of x in each equation by guessing and checking.

    (a) $19x + 22 = 136$

    (b) $12x - 17 = 151$

    (c) $19x - 8 = 600$

    (d) $x^2 + 3 = 292$

4 Simplify the equation and then solve for x using any method you like:

    (a) $x^5 - 16 + x + 20 - x^5 = 24$

    (b) $5xy + x - 2xy + 27 - 3xy = 73$

    (c) $6x - 3 + x^2 - x + 8 - 5x = 30$

    (d) $-3 + x^3 + 4 + x - x^3 - 1 = 2xy + 7 - x - 2xy + x$

The Balancing Act: Solving for x

As I show you in the preceding section, some problems are too complicated to find out what the variable (usually x) equals just by eyeballing it or rearranging it. For these problems, you need a reliable method for getting the right answer. I call this method the *balance scale*.

The balance scale allows you to *solve for x* — that is, find the number that x stands for — in a step-by-step process that always works. In this section, I show you how to use the balance scale method to solve algebraic equations.

Striking a balance

REMEMBER

The equals sign in any equation means that both sides balance. To keep that equals sign, you have to maintain that balance. In other words, whatever you do to one side of an equation, you have to do to the other.

For example, here's a balanced equation:

$$\underset{\Delta}{1+2=3}$$

If you add 1 to one side of the equation, the scale goes out of balance.

$$\underset{\Delta}{1+2+1 \neq 3}$$

But if you add 1 to *both* sides of the equation, the scale stays balanced:

$$\underset{\Delta}{1+2+1=3+1}$$

You can add any number to the equation, as long as you do it to both sides. And in math, *any number* means x:

$$1+2+x=3+x$$

Remember that x is the same wherever it appears in a single equation or problem.

This idea of changing both sides of an equation equally isn't limited to addition. You can just as easily subtract an x, or even multiply or divide by x, as long as you do the same to both sides of the equation:

$$\text{Subtract: } 1+2-x=3-x$$
$$\text{Multiply: } (1+2)x=3x$$
$$\text{Divide: } \frac{1+2}{x}=\frac{3}{x} \quad \text{(when } x \neq 0)$$

Using the balance scale to isolate x

The simple idea of balance is at the heart of algebra, and it enables you to find out what x is in many equations. When you solve an algebraic equation, the goal is to *isolate x* — that is, to get x alone on one side of the equation and some number on the other side. In algebraic equations of middling difficulty, this is a three-step process:

1. **Get all constants (non-x terms) on one side of the equation.**

2. **Get all x-terms on the other side of the equation.**

3. **Divide to isolate x.**

For example, take a look at the following problem:

$$11x - 13 = 9x + 3$$

As you follow the steps, notice how I keep the equation balanced at each step:

1. **Get all the constants on one side of the equation by adding 13 to both sides of the equation:**

$$
\begin{array}{l}
11x - 13 = 9x + 3 \\
\underline{+13 = +13} \\
11x = 9x + 16
\end{array}
$$

Because you've obeyed the rules of the balance scale, you know that this new equation is also correct. Now the only non-x term (16) is on the right side of the equation.

2. **Get all the x-terms on the other side by subtracting 9x from both sides of the equation:**

$$
\begin{array}{l}
11x = 9x + 16 \\
\underline{-9x -9x} \\
2x = 16
\end{array}
$$

Again, the balance is preserved, so the new equation is correct.

3. **Divide by 2 to isolate x:**

$$\frac{2x}{2} = \frac{16}{2}$$
$$x = 8$$

To check this answer, you can simply substitute 8 for x in the original equation:

$$
\begin{aligned}
11(8) - 13 &= 9(8) + 3 \\
88 - 13 &= 72 + 3 \\
75 &= 75 \checkmark
\end{aligned}
$$

This checks out, so 8 is the correct value of x.

Q. Use the balance scale method to find the value of x in the equation $5x - 6 = 3x + 8$.

EXAMPLE **A.** $x = 7$. To get all constants on the right side of the equation, add 6 to both sides, which causes the -6 to cancel out on the left side of the equation:

$$
\begin{array}{l}
5x - 6 = 3x + 8 \\
\underline{+6 + 6} \\
5x = 3x + 14
\end{array}
$$

The right side of the equation still contains a $3x$. To get all x terms on the left side of the equation, subtract $3x$ from both sides:

$$5x = 3x + 14$$
$$\underline{-3x \quad -3x}$$
$$2x = \qquad 14$$

Divide by 2 to isolate x:

$$\frac{2x}{2} = \frac{14}{2}$$
$$x = 7$$

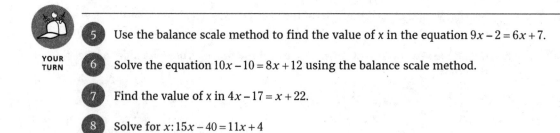

YOUR TURN

5 Use the balance scale method to find the value of x in the equation $9x - 2 = 6x + 7$.

6 Solve the equation $10x - 10 = 8x + 12$ using the balance scale method.

7 Find the value of x in $4x - 17 = x + 22$.

8 Solve for x: $15x - 40 = 11x + 4$

Rearranging Equations and Isolating x

When you understand how algebra works like a balance scale, as I show you in the preceding section, you can begin to solve more-difficult algebraic equations. The basic strategy is always the same: Changing both sides of the equation equally at every step, try to isolate x on one side of the equation.

In this section, I show you how to put your skills from Chapter 22 to work solving equations. First, I show you how rearranging the terms in an expression is similar to rearranging them in an algebraic equation. Next, I show you how removing parentheses from an equation can help you solve it. Finally, you discover how cross-multiplication is useful for solving algebraic equations with fractions.

Rearranging terms on one side of an equation

Rearranging terms becomes all-important when working with equations. For example, suppose you're working with this equation:

$$5x - 4 = 2x + 2$$

When you think about it, this equation is really two expressions connected with an equals sign. And of course, that's true of *every* equation. That's why everything you find out about expressions in Chapter 22 is useful for solving equations. For example, you can rearrange the terms on one side of an equation. So here's another way to write the same equation:

$$-4 + 5x = 2x + 2$$

And here's a third way:

$$-4 + 5x = 2 + 2x$$

This flexibility to rearrange terms comes in handy when you're solving equations.

Moving terms to the other side of the equals sign

Earlier in this chapter, I show you how an equation is similar to a balance scale. For example, take a look at Figure 23-1.

To keep the scale balanced, if you add or remove anything on one side, you must do the same on the other side. For example:

$$\begin{array}{r} 2x - 3 = 11 \\ \underline{-2x \qquad -2x} \\ -3 = 11 - 2x \end{array}$$

Now take a look at these two versions of this equation side by side:

$$2x - 3 = 11 \quad -3 = 11 - 2x$$

In the first version, the term $2x$ is on the left side of the equals sign. In the second, the term $-2x$ is on the right side. This example illustrates an important rule.

When you move any term in an expression to the other side of the equals sign, change its sign (from plus to negative or from negative to plus).

REMEMBER

As another example, suppose you're working with this equation:

$$4x - 2 = 3x + 1$$

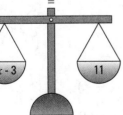

FIGURE 23-1:
Showing how an equation is similar to a balance scale.

2x - 3   11

You have x's on both sides of the equation, so say you want to move the $3x$. When you move the term $3x$ from the right side to the left side, you have to change its sign from plus to negative (technically, you're subtracting $3x$ from both sides of the equation).

$$4x - 2 - 3x = 1$$

After that, you can simplify the expression on the left side of the equation by combining like terms:

$$x - 2 = 1$$

At this point, you can probably see that $x = 3$ because $3 - 2 = 1$. But just to be sure, move the -2 term to the right side and change its sign:

$$x = 1 + 2$$
$$x = 3$$

To check this result, substitute a 3 wherever x appears in the original equation:

$$4x - 2 = 3x + 1$$
$$4(3) - 2 = 3(3) + 1$$
$$12 - 2 = 9 + 1$$
$$10 = 10$$

As you can see, moving terms from one side of an equation to the other can be a big help when you're solving equations.

Q. Find the value of x in the equation $7x - 6 = 4x + 9$.

EXAMPLE **A.** $x = 5$. Rearrange the terms of the equation so that the x terms are on one side and the constants are on the other. I do this in two steps:

$$7x - 6 = 4x + 9$$
$$7x = 4x + 9 + 6$$
$$7x - 4x = 9 + 6$$

Combine like terms on both sides of the equation:

$$3x = 15$$

Divide by 3 to isolate x.

$$\frac{3x}{3} = \frac{15}{3}$$
$$x = 5$$

Q. Find the value of x in the equation $3 - (7x - 13) = 5(3 - x) - x$.

A. $x = 1$. Before you can begin rearranging terms, remove the parentheses on both sides of the equation. On the left side, the parentheses are preceded by a negative sign, so change the sign of every term and remove the parentheses:

$$3 - 7x + 13 = 5(3 - x) - x$$

On the right side, no sign comes between the 5 and the parentheses, so multiply every term inside the parentheses by 5 and remove the parentheses:

$$3 - 7x + 13 = 15 - 5x - x$$

Now you can solve the equation in three steps. Put the x terms on one side and the constants on the other, remembering to switch the signs as needed:

$$7x = 15 - 5x - x - 3 - 13$$
$$-7x + 5x + x = 15 - 3 - 13$$

Combine like terms on both sides of the equation:

$$-x = -1$$

Divide by -1 to isolate x:

$$\frac{-x}{-1} = \frac{-1}{-1}$$
$$x = 1$$

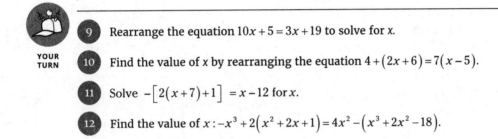

YOUR TURN

9. Rearrange the equation $10x + 5 = 3x + 19$ to solve for x.

10. Find the value of x by rearranging the equation $4 + (2x + 6) = 7(x - 5)$.

11. Solve $-\left[2(x + 7) + 1\right] = x - 12$ for x.

12. Find the value of x: $-x^3 + 2(x^2 + 2x + 1) = 4x^2 - (x^3 + 2x^2 - 18)$.

Removing parentheses from equations

Chapter 22 gives you a treasure trove of tricks for simplifying expressions, and they come in handy when you're solving equations. One key skill from that chapter is removing parentheses from expressions. This tactic is also indispensable when you're solving equations.

For example, suppose you have the following equation:

$$5x + (6x - 15) = 30 - (x - 7) + 8$$

Your mission is to get all the x terms on one side of the equation and all the constants on the other. As the equation stands, however, x terms and constants are "locked together" inside parentheses. In other words, you can't isolate the x terms from the constants. So before you can isolate terms, you need to remove the parentheses from the equation.

Recall that an equation is really just two expressions connected by an equals sign. So you can start working with the expression on the left side. In this expression, the parentheses begin with a plus sign (+), so you can just remove them:

$$5x + 6x - 15 = 30 - (x - 7) + 8$$

Now move on to the expression on the right side. This time, the parentheses come right after a negative sign (−). To remove them, change the sign of both terms inside the parentheses: x becomes $-x$, and -7 becomes 7:

$$5x + 6x - 15 = 30 - x + 7 + 8$$

Bravo! Now you can isolate x terms to your heart's content. Move the $-x$ from the right side to the left, changing it to x:

$$5x + 6x - 15 + x = 30 + 7 + 8$$

Next, move -15 from the left side to the right, changing it to 15:

$$5x + 6x + x = 30 + 7 + 8 + 15$$

Now combine like terms on both sides of the equation:

$$12x = 30 + 7 + 8 + 15$$
$$12x = 60$$

Finally, get rid of the coefficient 12 by dividing:

$$\frac{12x}{12} = \frac{60}{12}$$
$$x = 5$$

As usual, you can check your answer by substituting 5 into the original equation wherever x appears:

$$5x + (6x - 15) = 30 - (x - 7) + 8$$
$$5(5) + [6(5) - 15] = 30 - (5 - 7) + 8$$
$$25 + (30 - 15) = 30 - (-2) + 8$$
$$25 + 15 = 30 + 2 + 8$$
$$40 = 40$$

Here's one more example:

$$11 + 3(-3x + 1) = 25 - (7x - 3) - 12$$

As in the preceding example, start out by removing both sets of parentheses. This time, however, on the left side of the equation, you have no sign between 3 and $(-3x + 1)$. But again, you can put your skills from Chapter 22 to use. To remove the parentheses, multiply 3 by both terms inside the parentheses:

$$11 - 9x + 3 = 25 - (7x - 3) - 12$$

On the right side, the parentheses begin with a negative sign, so remove the parentheses by changing all the signs inside the parentheses:

$$11 - 9x + 3 = 25 - 7x + 3 - 12$$

Now you're ready to isolate the x terms. I do this in one step, but take as many steps as you want:

$$-9x + 7x = 25 + 3 - 12 - 11 - 3$$

At this point, you can combine like terms:

$$-2x = 2$$

To finish, divide both sides by -2:

$$x = -1$$

Copy this example, and work through it a few times with the book closed.

Q. What is the value of x in the equation $x - (7x + 2) = -3(4x - 5) - 13 + 4x$?

EXAMPLE **A.** 2. To begin, remove parentheses:

$$x - (7x + 2) = -3(4x - 5) - 13 + 4x$$
$$x - 7x - 2 = -12x + 15 - 13 + 4x$$

Combine like terms on each side of the equation:

$$-6x - 2 = -8x + 2$$

Isolate x:

$$2x - 2 = 2$$
$$2x = 4$$
$$x = 2$$

13 Solve for x: $5x + (-17 - 7x) = 3(x + 9) + 1$

YOUR
TURN
14 What is the value of x in this equation: $-7(10 - x) + 115 = 2(x + 60) + 30$?

15 Solve: $6x - 3(1 + x) + (5 - x) = 4(x + 1) + 7 - (x - 3)$

16 Find x: $4x - (3x + (x - 1)) = -(x + (6x + 31 + 5(2 - x)) - 9x)$

Cross-multiplying

In algebra, cross-multiplication helps to simplify equations by removing unwanted fractions (and, honestly, when are fractions ever wanted?). As I discuss in Chapter 10, you can use cross-multiplication to find out whether two fractions are equal. You can use this same idea to solve algebra equations with fractions, like this one:

$$\frac{x}{2x-2} = \frac{2x+3}{4x}$$

This equation looks hairy. You can't do the division or cancel anything out because the fraction on the left has two terms in the denominator, and the fraction on the right has two terms in the numerator (see Chapter 22 for info on dividing algebraic terms). However, an important piece of information that you have is that the fraction equals the fraction. So if you cross-multiply these two fractions, you get two results that are also equal:

$$x(4x) = (2x+3)(2x-2)$$

At this point, you have something you know how to work with. The left side is easy:

$$4x^2 = (2x+3)(2x-2)$$

The right side requires a bit of FOILing (flip to Chapter 22 for details):

$$4x^2 = 4x^2 - 4x + 6x - 6$$

Now all the parentheses are gone, so you can isolate the x terms. Because most of these terms are already on the right side of the equation, isolate them on that side:

$$6 = 4x^2 - 4x + 6x - 4x^2$$

Combining like terms gives you a pleasant surprise:

$$6 = 2x$$

The two x^2 terms cancel each other out. You may be able to eyeball the correct answer, but here's how to finish:

$$\frac{6}{2} = \frac{2x}{2}$$
$$3 = x$$

To check your answer, substitute 3 back into the original equation:

$$\frac{x}{2x-2} = \frac{2x+3}{4x}$$
$$\frac{3}{2(3)-2} = \frac{2(3)+3}{4(3)}$$
$$\frac{3}{6-2} = \frac{6+3}{12}$$
$$\frac{3}{4} = \frac{3}{4}$$

So the answer $x = 3$ is correct.

EXAMPLE

Q. Use cross-multiplication to solve the equation $\frac{2x}{3} = x - 3$.

A. $x = 9$. Cross-multiply to get rid of the fraction in this equation. To do this, turn the right side of the equation into a fraction by inserting a denominator of 1:

$$\frac{2x}{3} = \frac{x-3}{1}$$

Now cross-multiply:

$$2x(1) = 3(x-3)$$

Remove the parentheses from both sides (as I show you in Chapter 22):

$$2x = 3x - 9$$

At this point, you can rearrange the equation and solve for x, as I show you earlier in this chapter:

$$2x - 3x = -9$$
$$-x = -9$$
$$\frac{-x}{-1} = \frac{-9}{-1}$$
$$x = 9$$

Q. Use cross-multiplication to solve the equation $\frac{2x+1}{x+1} = \frac{6x}{3x+1}$.

A. $x = 1$. In some cases, after you cross-multiply, you may need to FOIL one or both sides of the resulting equation. First, cross-multiply to get rid of the fraction bar in this equation:

$$(2x+1)(3x+1) = 6x(x+1)$$

Now remove the parentheses on the left side of the equation by FOILing (as I show you in Chapter 22):

$$6x^2 + 2x + 3x + 1 = 6x(x+1)$$

To remove the parentheses from the right side, multiply $6x$ by every term inside the parentheses, and then drop the parentheses:

$$6x^2 + 2x + 3x + 1 = 6x^2 + 6x$$

At this point, you can rearrange the equation and solve for x:

$$1 = 6x^2 + 6x - 6x^2 - 2x - 3x$$

Notice that the two x^2 terms cancel each other out:

$$1 = 6x - 2x - 3x$$
$$1 = x$$

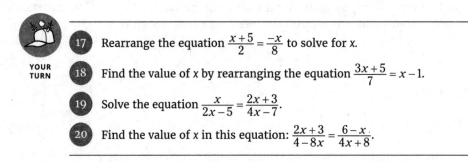

17. Rearrange the equation $\frac{x+5}{2} = \frac{-x}{8}$ to solve for x.

18. Find the value of x by rearranging the equation $\frac{3x+5}{7} = x - 1$.

19. Solve the equation $\frac{x}{2x-5} = \frac{2x+3}{4x-7}$.

20. Find the value of x in this equation: $\frac{2x+3}{4-8x} = \frac{6-x}{4x+8}$.

Practice Questions Answers and Explanations

1

(a) $x = 8$, because $8 + 5 = 13$.

(b) $x = 6$, because $18 - 6 = 12$.

(c) $x = 11$, because $4(11) = 44$.

(d) $x = 10$, because $\dfrac{30}{10} = 3$.

2

(a) $x = 56$. Change the addition to subtraction: $x + 41 = 97$ is the same as $97 - 41$, so $x = 56$.

(b) $x = 42$. Change the subtraction to addition: $100 - x = 58$ means the same thing as $100 - 58 = x$, so $x = 42$.

(c) $x = 21$. Change the multiplication to division: $13x = 273$ is equivalent to $\dfrac{273}{13} = x$, so $x = 21$.

(a) $x = 14$. Switch around the division: $\dfrac{238}{x} = 17$ means $\dfrac{238}{17} = x$, so $x = 14$.

3

(a) $x = 6$. First, try $x = 10$:

$$19(10) + 22 = 190 + 22 = 212$$

212 is greater than 136, so this guess is too high. Try $x = 5$:

$$19(5) + 22 = 95 + 22 = 117$$

117 is only a little less than 136, so this guess is a little too low. Try $x = 6$:

$$19(6) + 22 = 114 + 22 = 136$$

136 is correct, so $x = 6$.

(b) $x = 14$. First, try $x = 10$:

$$12(10) - 17 = 120 - 17 = 103$$

103 is less than 151, so this guess is too low. Try $x = 20$:

$$12(20) - 17 = 240 - 17 = 223$$

223 is greater than 151, so this guess is too high. Therefore, x is between 10 and 20. Try $x = 15$:

$$12(15) - 17 = 180 - 17 = 163$$

163 is a little greater than 151, so this guess is a little too high. Try $x = 14$:

$$12(14) - 17 = 168 - 17 = 151$$

151 is correct, so $x = 14$.

(c) $x = 32$. First, try $x = 10$:

$$19(10) - 8 = 190 - 8 = 182$$

182 is much less than 600, so this guess is much too low. Try $x = 30$:

$$19(30) - 8 = 570 - 8 = 562$$

562 is still less than 600, so this guess is still too low. Try $x = 35$:

$$19(35) - 8 = 665 - 8 = 657$$

657 is greater than 600, so this guess is too high. Therefore, x is between 30 and 35. Try $x = 32$:

$$19(32) - 8 = 608 - 8 = 600$$

600 is correct, so $x = 32$.

(d) **$x = 17$.** First, try $x = 10$:

$$10^2 + 3 = 100 + 3 = 103$$

103 is less than 292, so this guess is too low. Try $x = 20$:

$$20^2 + 3 = 400 + 3 = 403$$

403 is greater than 292, so this guess is too high. Therefore, x is between 10 and 20. Try $x = 15$:

$$15^2 + 3 = 225 + 3 = 228$$

228 is less than 292, so this guess is too low. Therefore, x is between 15 and 20. Try $x = 17$:

$$17^2 + 3 = 289 + 3 = 292$$

292 is correct, so $x = 17$.

④

(a) **$x = 20$.** Rearrange the expression on the left side of the equation so that all like terms are next to each other:

$$\underline{x^5 - x^5} + x + \underline{20 - 16} = 24$$

Combine like terms:

$$x + 4 = 24$$

Notice that the two x^5 terms cancel each other out. Because $20 + 4 = 24$, you know that $x = 20$.

(b) **$x = 46$.** Rearrange the expression on the left side of the equation:

$$\underline{5xy - 2xy - 3xy} + x + 27 = 73$$

Combine like terms:

$$x + 27 = 73$$

Notice that the three xy terms cancel each other out. Because $x + 27 = 73$ means $73 - 27 = x$, you know that $x = 46$.

(c) **$x = 5$.** Rearrange the expression on the left side of the equation so that all like terms are adjacent:

$$\underline{6x - x - 5x} + x^2 + \underline{8 - 3} = 30$$

Combine like terms:

$$x^2 + 5 = 30$$

Notice that the three x terms cancel each other out. Try $x = 10$:

$$10^2 + 5 = 100 + 5 = 105$$

105 is greater than 30, so this guess is too high. Therefore, x is between 0 and 10. Try $x = 5$:

$$5^2 + 5 = 25 + 5 = 30$$

This result is correct, so $x = 5$.

(d) **$x = 7$.** Rearrange the expression on the left side of the equation:

$$\underline{-3 + 4 - 1} + \underline{x^3 - x^3} + x = 2xy + 7 - x - 2xy + x$$

Combine like terms:

$$x = 2xy + 7 - x - 2xy + x$$

Notice that the three constant terms cancel each other out, and so do the two x^3 terms. Now rearrange the expression on the right side of the equation:

$$x = 2xy - 2xy + 7 - x + x$$

Combine like terms:

$$x = 7$$

Notice that the two xy terms cancel each other out, and so do the two x terms. Therefore, $x = 7$.

⑤ **$x = 3$.** To get all constants on the right side of the equation, add 2 to both sides:

$$
\begin{array}{rl}
9x - 2 &= 6x + 7 \\
\underline{+2} & \underline{+2} \\
9x &= 6x + 9
\end{array}
$$

To get all x terms on the left side, subtract $6x$ from both sides:

$$
\begin{array}{rl}
9x &= 6x + 9 \\
\underline{-6x} & \underline{-6x} \\
3x &= 9
\end{array}
$$

Divide by 3 to isolate x:

$$\frac{3x}{3} = \frac{9}{3}$$
$$x = 3$$

(6) $x = 11$. Move all constants on the right side of the equation by adding 10 to both sides:

$$10x - 10 = 8x + 12$$
$$\underline{\quad +10 \qquad +10 \quad}$$
$$10x \qquad 8x + 22$$

To get all x terms on the left side, subtract $8x$ from both sides:

$$10x = 8x + 22$$
$$\underline{-8x - 8x \qquad}$$
$$2x = \qquad 22$$

Divide by 2 to isolate x:

$$\frac{2x}{2} = \frac{22}{2}$$
$$x = 11$$

(7) $x = 13$. Add 17 to both sides to get all constants on the right side of the equation:

$$4x - 17 = x + 22$$
$$\underline{\quad +17 \qquad +17 \quad}$$
$$4x \qquad = x + 39$$

Subtract x from both sides to get all x terms on the left side:

$$4x = +39$$
$$\underline{-x \quad -x \quad}$$
$$3x = \quad 39$$

Divide by 3 to isolate x:

$$\frac{3x}{3} = \frac{39}{3}$$
$$x = 13$$

(8) $x = 11$. To get all constants on the right side of the equation, add 40 to both sides:

$$15x - 40 = 11x + 4$$
$$\underline{\quad +40 \qquad +40 \quad}$$
$$15x \qquad = 11x + 44$$

To get all x terms on the left side, subtract $11x$ from both sides:

$$15x = 11x + 44$$

$$\underline{-11x \quad -11x}$$

$$4x = \qquad 44$$

Divide by 4 to isolate x:

$$\frac{4x}{4} = \frac{44}{4}$$

$$x = 11$$

⑨ **$x = 2$.** Rearrange the terms of the equation so that the x terms are on one side and the constants are on the other. I do this in two steps:

$$10x + 5 = 3x + 19$$

$$10x = 3x + 19 - 5$$

$$10x - 3x = 19 - 5$$

Combine like terms on both sides:

$$7x = 14$$

Divide by 7 to isolate x:

$$\frac{7x}{7} = \frac{14}{7}$$

$$x = 2$$

⑩ **$x = 9$.** Before you can begin rearranging terms, remove the parentheses on both sides of the equation. On the left side, the parentheses are preceded by a plus sign, so just drop them:

$$4 + (2x + 6) = 7(x - 5)$$

$$4 + 2x + 6 = 7(x - 5)$$

On the right side, no sign comes between the number 7 and the parentheses, so multiply 7 by every term inside the parentheses and then drop the parentheses:

$$4 + 2x + 6 = 7x - 35$$

Now you can solve for x by rearranging the terms of the equation. Group the x terms on one side and the constants on the other. I do this in two steps:

$$4 + 6 = 7x - 35 - 2x$$

$$4 + 6 + 35 = 7x - 2x$$

Combine like terms on both sides:

$$45 = 5x$$

Divide by 5 to isolate x:

$$\frac{45}{5} = \frac{5x}{5}$$
$$9 = x$$

(11) $x = -1$. Before you can begin rearranging terms, remove the parentheses on the left side of the equation. Start with the inner parentheses, multiplying 2 by every term inside that set:

$$-\left[2(x+7)+1\right] = x-12$$
$$-\left[2x+14+1\right] = x-12$$

Next, remove the remaining parentheses, switching the sign of every term within that set:

$$-2x-14-1 = x-12$$

Now you can solve for x by rearranging the terms of the equation:

$$-2x-14-1+12 = x$$
$$-14-1+12 = x+2x$$

Combine like terms on both sides:

$$-3 = 3x$$

Divide by 3 to isolate x:

$$\frac{-3}{3} = \frac{3x}{3}$$
$$-1 = x$$

(12) $x = 4$. Before you can begin rearranging terms, multiply the terms in the left-hand parentheses by 2 and remove the parentheses on both sides of the equation:

$$-x^3 + 2\left(x^2 + 2x + 1\right) = 4x^2 - \left(x^3 + 2x^2 - 18\right)$$
$$-x^3 + 2x^2 + 4x + 2 = 4x^2 - \left(x^3 + 2x^2 - 18\right)$$
$$-x^3 + 2x^2 + 4x + 2 = 4x^2 - x^3 - 2x^2 + 18$$

Rearrange the terms of the equation:

$$-x^3 + 2x^2 + 4x + 2 - 4x^2 + x^3 + 2x^2 = 18$$
$$-x^3 + 2x^2 + 4x - 4x^2 + x^3 + 2x^2 = 18 - 2$$

Combine like terms on both sides (notice that the x^3 and x^2 terms all cancel out):

$$4x = 16$$

Divide by 4 to isolate x:

$$\frac{4x}{4} = \frac{16}{4}$$
$$x = 4$$

(13) $x = -9$. Begin by removing parentheses:

$$5x + (-17 - 7x) = 3(x + 9) + 1$$
$$5x - 17 - 7x = 3x + 27 + 1$$

Combine like terms:

$$-17 - 2x = 3x + 28$$

Isolate x:

$$-17 = 5x + 28$$
$$-45 = 5x$$
$$-9 = x$$

(14) $x = 21$. Begin by removing parentheses:

$$-7(10 - x) + 115 = 2(x + 60) + 30$$
$$-70 + 7x + 115 = 2x + 120 + 30$$

Combine like terms:

$$7x + 45 = 2x + 150$$

Isolate x:

$$5x + 45 = 150$$
$$5x = 105$$
$$x = 21$$

(15) $x = -12$. Begin by removing parentheses:

$$6x - 3(1 + x) + (5 - x) = 4(x + 1) + 7 - (x - 3)$$
$$6x - 3 - 3x + 5 - x = 4x + 4 + 7 - x + 3$$

Combine like terms:

$$2x + 2 = 3x + 14$$

Isolate x:

$$2 = x + 14$$
$$-12 = x$$

(16) **$x = 6$.** Begin by removing parentheses, starting on the inside and working your way out:

$$4x - \left(3x + \left(x - 1\right)\right) = -\left(x + \left(6x + 31 + 5\left(2 - x\right)\right) - 9x\right)$$
$$4x - \left(3x + x - 1\right) = -\left(x + \left(6x + 31 + 10 - 5x\right) - 9x\right)$$
$$4x - 3x - x + 1 = -\left(x + 6x + 31 + 10 - 5x - 9x\right)$$
$$4x - 3x - x + 1 = -x - 6x - 31 - 10 + 5x + 9x$$

Combine like terms:

$$1 = 7x - 41$$

Isolate x:

$$42 = 7x$$
$$6 = x$$

(17) **$x = -4$.** Remove the fraction from the equation by cross-multiplying:

$$\frac{x + 5}{2} = \frac{-x}{8}$$
$$8\left(x + 5\right) = -2x$$

Multiply to remove the parentheses from the left side of the equation:

$$8x + 40 = -2x$$

At this point, you can solve for x:

$$40 = -2x - 8x$$
$$40 = -10x$$
$$\frac{40}{-10} = \frac{-10x}{-10}$$
$$-4 = x$$

(18) **$x = 3$.** Change the right side of the equation to a fraction by attaching a denominator of 1. Remove the fraction bar from the equation by cross-multiplying:

$$\frac{3x + 5}{7} = \frac{x - 1}{1}$$
$$3x + 5 = 7\left(x - 1\right)$$

Multiply 7 by each term inside the parentheses to remove the parentheses from the right side of the equation:

$$3x + 5 = 7x - 7$$

Now solve for x:

$$5 = 7x - 7 - 3x$$
$$5 + 7 = 7x - 3x$$
$$12 = 4x$$
$$3 = x$$

(19) **$x = 5$.** Remove the fractions from the equation by cross-multiplying:

$$\frac{x}{2x - 5} = \frac{2x + 3}{4x - 7}$$
$$x(4x - 7) = (2x + 3)(2x - 5)$$

Remove the parentheses from the left side of the equation by multiplying through by x; remove parentheses from the right side of the equation by FOILing:

$$4x^2 - 7x = 4x^2 - 10x + 6x - 15$$

Rearrange the equation:

$$4x^2 - 7x - 4x^2 + 10x - 6x = -15$$

Notice that the two x^2 terms cancel each other out:

$$-7x + 10x - 6x = -15$$
$$-3x = -15$$
$$\frac{-3x}{-3} = \frac{-15}{-3}$$
$$x = 5$$

(20) **$x = 0$.** Remove the fractions from the equation by cross-multiplying:

$$\frac{2x + 3}{4 - 8x} = \frac{6 - x}{4x + 8}$$

$$(2x + 3)(4x + 8) = (6 - x)(4 - 8x)$$

FOIL both sides of the equation to remove the parentheses:

$$8x^2 + 16x + 12x + 24 = 24 - 48x - 4x + 8x^2$$

At this point, rearrange terms so you can solve for *x*:

$$8x^2 + 16x + 12x + 24 + 48x + 4x - 8x^2 = 24$$
$$8x^2 + 16x + 12x + 48x + 4x - 8x^2 = 24 - 24$$

Notice that the x^2 terms and the constant terms drop out of the equation:

$$16x + 12x + 48x + 4x = 0$$
$$80x = 0$$
$$\frac{80x}{80} = \frac{0}{80}$$
$$x = 0$$

When you're ready, try out your skills in the next section, where the chapter quiz awaits.

Whaddya Know? Chapter 23 Quiz

Ready for a math workout? Then try out these 12 questions, which bring together the equation-solving strategies you discovered in this chapter. When you're done, the section that follows includes complete answers and explanations for every question.

Solve each equation for x.

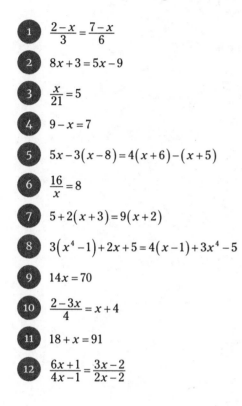

1 $\dfrac{2-x}{3} = \dfrac{7-x}{6}$

2 $8x + 3 = 5x - 9$

3 $\dfrac{x}{21} = 5$

4 $9 - x = 7$

5 $5x - 3(x - 8) = 4(x + 6) - (x + 5)$

6 $\dfrac{16}{x} = 8$

7 $5 + 2(x + 3) = 9(x + 2)$

8 $3(x^4 - 1) + 2x + 5 = 4(x - 1) + 3x^4 - 5$

9 $14x = 70$

10 $\dfrac{2 - 3x}{4} = x + 4$

11 $18 + x = 91$

12 $\dfrac{6x + 1}{4x - 1} = \dfrac{3x - 2}{2x - 2}$

Answers to Chapter 23 Quiz

(1) $x = -3$. First cross-multiply and distribute the numbers.

$$\frac{2-x}{3} = \frac{7-x}{6}$$
$$6(2-x) = 3(7-x)$$
$$12 - 6x = 21 - 3x$$

Add 6x to each side and subtract 21 from each side.

$$
\begin{array}{rcccl}
12 & - & 6x & = & 21 - 3x \\
& + & 6x & & + 6x \\
\hline
12 & & & = & 21 + 3x
\end{array}
\quad \text{and then} \quad
\begin{array}{rcl}
12 & = & 21 + 3x \\
-21 & & -21 \\
\hline
-9 & = & 3x
\end{array}
$$

Divide each side by 3.

$$\frac{-9}{3} = \frac{3x}{3}$$
$$-3 = x$$

(2) $x = -4$. Subtract 5x and 3 from each side.

$$
\begin{array}{rcccl}
8x & + & 3 & = & 5x - 9 \\
-5x & & & & -5x \\
\hline
3x & + & 3 & = & -9
\end{array}
\quad \text{and} \quad
\begin{array}{rcl}
3x + 3 & = & -9 \\
-3 & & -3 \\
\hline
3x & = & -12
\end{array}
$$

Divide each side by 3.

$$\frac{3x}{3} = \frac{-12}{3}$$
$$x = -4$$

(3) $x = 105$. Multiply each side of the equation by 21.

$$\frac{x}{21} = 5$$
$$21 \cdot \frac{x}{21} = 5 \cdot 21$$
$$x = 105$$

4 $x = 2$. Subtract 9 from each side, and then multiply each side by −1.

$$\begin{array}{r} 9 \; - \; x \; = \; 7 \\ \underline{-9 \hspace{2.8cm} -9} \\ - \; x \; = \; -2 \end{array} \quad \text{and then} \quad \begin{array}{c} -1(-x) = -1(-2) \\ x = 2 \end{array}$$

5 $x = 5$. First distribute the numbers and the negative over the terms in the parentheses.

$$5x - 3(x - 8) = 4(x + 6) - (x + 5)$$
$$5x - 3x + 24 = 4x + 24 - x - 5$$

Combine like terms on each side of the equation.

$$(5 - 3)x + 24 = (4 - 1)x + 24 - 5$$
$$2x + 24 = 3x + 19$$

Subtract $2x$ and 19 from each side of the equation.

$$\begin{array}{r} 2x \; + \; 24 \; = \; 3x \; + \; 19 \\ \underline{-2x \hspace{1.8cm} -2x \hspace{1.5cm}} \\ 24 \; = \; x \; + \; 19 \end{array} \quad \text{and} \quad \begin{array}{r} 24 \; = \; x \; + \; 19 \\ \underline{-19 \hspace{2.4cm} -19} \\ 5 \; = \; x \end{array}$$

6 $x = 2$. Multiply each side of the equation by x. Then divide by 8.

$$\cancel{x} \cdot \frac{16}{\cancel{x}} = 8 \cdot x$$
$$16 = 8x$$
$$\frac{16}{8} = \frac{\cancel{8}x}{\cancel{8}}$$
$$2 = x$$

7 $x = -1$. Distribute the numbers over the parentheses, and then combine like terms on the left side.

$$5 + 2(x + 3) = 9(x + 2)$$
$$5 + 2x + 6 = 9x + 18$$
$$2x + 11 = 9x + 18$$

Subtract $2x$ and 18 from each side of the equation.

$$\begin{array}{r} 2x \; + \; 11 \; = \; 9x \; + \; 18 \\ \underline{-2x \hspace{2.0cm} -2x \hspace{1.5cm}} \\ 11 \; = \; 7x \; + \; 18 \end{array} \quad \text{and then} \quad \begin{array}{r} 11 \; = \; 7x \; + \; 18 \\ \underline{-18 \hspace{2.2cm} -18} \\ -7 \; = \; 7x \end{array}$$

Divide each side of the equation by 7.

$$\frac{-7}{7} = \frac{\cancel{7}x}{\cancel{7}}$$
$$-1 = x$$

8 $x = \dfrac{11}{2}$. Distribute the numbers over the parentheses, and then combine like terms on both sides and rearrange the terms in decreasing powers of the variable.

$$3(x^4 - 1) + 2x + 5 = 4(x - 1) + 3x^4 - 5$$
$$3x^4 - 3 + 2x + 5 = 4x - 4 + 3x^4 - 5$$
$$3x^4 + 2x + 2 = 3x^4 + 4x - 9$$

Subtract $3x^4$ from each side of the equation.

$$3x^4 - 3x^4 + 2x + 2 = 3x^4 - 3x^4 + 4x - 9$$
$$2x + 2 = 4x - 9$$

Subtract $2x$ from each side and add 9 to each side.

$$
\begin{array}{rcrcr}
2x & + \ 2 & = & 4x & - \ 9 \\
-2x & & & -2x & \\
\hline
 & 2 & = & 2x & - \ 9
\end{array}
\quad \text{and then} \quad
\begin{array}{rcrcr}
2 & = & 2x & - \ 9 \\
+9 & & & + \ 9 \\
\hline
11 & = & 2x &
\end{array}
$$

Divide each side of the equation by 2.

$$\frac{11}{2} = \frac{\cancel{2}x}{\cancel{2}}$$
$$\frac{11}{2} = x$$

9 $x = 5$. Divide each side of the equation by 14.

$$\frac{\cancel{14}x}{\cancel{14}} = \frac{70}{14}$$
$$x = 5$$

10 $x = -2$. First multiply each side of the equation by 4 and distribute the 4 on the right.

$$\cancel{4} \cdot \frac{2 - 3x}{\cancel{4}} = 4(x + 4)$$
$$2 - 3x = 4x + 16$$

Add 3x to each side and subtract 16 from each side.

$$
\begin{array}{rcl}
2 \;-\; 3x &=& 4x \;+\; 16 \\
\underline{\;+\; 3x} & & \underline{+3x} \\
2 &=& 7x \;+\; 16
\end{array}
\quad \text{and then} \quad
\begin{array}{rcl}
2 &=& 7x \;+\; 16 \\
\underline{-16} & & \underline{\;-\;16} \\
-14 &=& 7x
\end{array}
$$

Now divide each side of the equation by 7.

$$
\frac{-14}{7} = \frac{\cancel{7}x}{\cancel{7}}
$$
$$
-2 = x
$$

11 $x = 73$. Subtract 18 from each side of the equation.

$$
\begin{array}{rcl}
18 \;+\; x &=& 91 \\
\underline{-18} & & \underline{-18} \\
x &=& 73
\end{array}
$$

12 $x = 4$. First cross-multiply. Then FOIL for the products on each side of the equation.

$$
\frac{6x+1}{4x-1} = \frac{3x-2}{2x-2}
$$
$$
(6x+1)(2x-2) = (3x-2)(4x-1)
$$
$$
12x^2 - 12x + 2x - 2 = 12x^2 - 3x - 8x + 2
$$
$$
12x^2 - 10x - 2 = 12x^2 - 11x + 2
$$

Subtract $12x^2$ from each side of the equation.

$$
12x^2 - 12x^2 - 10x - 2 = 12x^2 - 12x^2 - 11x + 2
$$
$$
-10x - 2 = -11x + 2
$$

Add 11x to each side and add 2 to each side.

$$
\begin{array}{rcl}
-10x \;-\; 2 &=& -11x \;+\; 2 \\
\underline{11x} & & \underline{11x} \\
x \;-\; 2 &=& 2
\end{array}
\quad \text{and then} \quad
\begin{array}{rcl}
x \;-\; 2 &=& 2 \\
\underline{+2} & & \underline{+2} \\
x & =& 4
\end{array}
$$

Chapter **24**

Tackling Algebra Word Problems

Word problems that require algebra are among the toughest problems that students face — and the most common. Teachers just love algebra word problems because they bring together a lot of what you know, such as solving algebra equations (see Chapter 23) and turning words into numbers (see Chapters 7 and 15). And standardized tests virtually always include these types of problems.

In this chapter, I show you a five-step method for using algebra to solve word problems. Then I give you a bunch of examples that take you through all five steps.

Along the way, I give you some important tips that can make solving word problems easier. First, I show you how to choose a variable that makes your equation as simple as possible. Next, I give you practice organizing information from the problem into a chart. By the end of this chapter, you'll have a solid understanding of how to solve a wide variety of algebra word problems.

Solving Algebra Word Problems in Five Steps

Everything from Chapter 23 comes into play when you use algebra to solve word problems, so if you feel a little shaky on solving algebraic equations, flip back to that chapter for some review.

Throughout this section, I use the following word problem as an example:

> In three days, Alexandra sold a total of 31 tickets to her school play. On Tuesday, she sold twice as many tickets as on Wednesday. And on Thursday, she sold exactly 7 tickets. How many tickets did Alexandra sell on each day, Tuesday through Thursday?

Organizing the information in an algebra word problem by using a chart or picture is usually helpful. Here's what I came up with:

Tuesday:	Twice as many as on Wednesday
Wednesday:	?
Thursday:	7
Total:	31

At this point, all the information is in the chart, but the answer still may not be jumping out at you. In this section, I outline a step-by-step method that enables you to solve this problem — and much harder ones as well.

Here are the five steps for solving most algebra word problems:

1. **Declare a variable.**
2. **Set up the equation.**
3. **Solve the equation.**
4. **Answer the question that the problem asks.**
5. **Check your answer.**

Declaring a variable

As you know from Chapter 22, a variable is a letter that stands for a number. Most of the time, you don't find the variable x (or any other variable, for that matter) in a word problem. That omission doesn't mean you don't need algebra to solve the problem. It just means that you're going to have to put x into the problem yourself and decide what it stands for.

 When you *declare a variable*, you say what that variable means in the problem you're solving.

REMEMBER Here are some examples of variable declarations:

> Let m = the number of dead mice that the cat dragged into the house.
>
> Let p = the number of times Marianne's husband promised to take out the garbage.
>
> Let c = the number of complaints Arnold received after he painted his garage door purple.

In each case, you take a variable (m, p, or c) and give it a meaning by attaching it to a number.

Notice that the earlier chart for the sample problem has a big question mark next to *Wednesday*. This question mark stands for *some number*, so you may want to declare a variable that stands for this number. Here's how you do it:

Let w = the number of tickets that Alexandra sold on Wednesday.

Whenever possible, choose a variable with the same initial as what the variable stands for. This practice makes remembering what the variable means a lot easier, which will help you later in the problem.

For the rest of the problem, every time you see the variable w, keep in mind that it stands for the number of tickets that Alexandra sold on Wednesday.

Setting up the equation

After you have a variable to work with, you can go through the problem again and find other ways to use this variable. For example, Alexandra sold twice as many tickets on Tuesday as on Wednesday, so she sold $2w$ tickets on Tuesday. Now you have a lot more information to fill in on the chart:

Tuesday:	Twice as many as on Wednesday	$2w$
Wednesday:	?	w
Thursday:	7	7
Total:	31	31

You know that the total number of tickets, or the sum of the tickets she sold on Tuesday, Wednesday, and Thursday, is 31. With the chart filled in like that, you're ready to set up an equation to solve the problem:

$$2w + w + 7 = 31$$

Solving the equation

After you set up an equation, you can use the tricks from Chapter 23 to solve the equation for w. Here's the equation one more time:

$$2w + w + 7 = 31$$

For starters, remember that $2w$ really means $w + w$. So on the left, you know you really have $w + w + w$, or $3w$; you can simplify the equation a little bit, as follows:

$$3w + 7 = 31$$

The goal at this point is to try to get all the terms with w on one side of the equation and all the terms without w on the other side. So on the left side of the equation, you want to get rid of the 7. The inverse of addition is subtraction, so subtract 7 from both sides:

$$\begin{array}{r} 3w + 7 = 31 \\ \underline{-7 \quad -7} \\ 3w \quad\; = 24 \end{array}$$

You now want to isolate w on the left side of the equation. To do this, you have to undo the multiplication by 3, so divide both sides by 3:

$$\frac{3w}{3} = \frac{24}{3}$$
$$w = 8$$

Answering the question

You may be tempted to think that, after you've solved the equation, you're done. But you still have a bit more work to do. Look back at the problem, and you see that it asks you this question:

How many tickets did Alexandra sell on each day, Tuesday through Thursday?

At this point, you have some information that can help you solve the problem. The problem tells you that Alexandra sold 7 tickets on Thursday. And because $w = 8$, you now know that she sold 8 tickets on Wednesday. And on Tuesday, she sold twice as many on Wednesday, so she sold 16. So Alexandra sold 16 tickets on Tuesday, 8 on Wednesday, and 7 on Thursday.

Checking your work

To check your work, compare your answer to the problem, line by line, to make sure every statement in the problem is true:

In three days, Alexandra sold a total of 31 tickets to her school play.

That part is correct because $16 + 8 + 7 = 31$.

On Tuesday, she sold twice as many tickets as on Wednesday.

Correct, because she sold 16 tickets on Tuesday and 8 on Wednesday.

And on Thursday, she sold exactly 7 tickets.

Yep, that's right, too, so you're good to go.

Q. Three sisters spent the day at the pool swimming laps. Tory swam 25 more laps than Sara, who swam twice as many laps as Petra. If together Tory and Sara swam 185 laps, how many laps did Petra swim?

(a) Let p equal the number of laps that Petra swam. How many laps did Sara and Tory swim in terms of p?

(b) Set up an equation and solve it for p.

(c) Answer the question.

(d) Check your work.

A.

(a) Sara $= 2p$; Tory $= 2p + 25$.

(b) $p = 40$

$$2p + 2p + 25 = 185$$
$$4p + 25 = 185$$
$$4p = 160$$
$$p = 40$$

(c) Petra swam 40 laps.

(d) If Petra swam 40 laps, Sarah swam 80 laps $(40 \times 2 = 80)$, and Tory swam 105 laps $(80 + 25 = 105)$, so Sarah and Tory swam 185 laps $(80 + 105 = 185)$.

YOUR TURN

1 Annette and Benjamin both sold books of raffle tickets for their school play. Benjamin sold four fewer than twice as many books as Annette, and together they sold 29 books. How many books did Annette sell?

(a) Let a equal the number of books that Annette sold. How many did Benjamin sell, in terms of a?

(b) Set up an equation and solve it for a.

(c) Answer the question.

(d) Check your work.

2 A museum houses a collection of works by artist Maxine Brioche. The collection includes 8 more sculptures than mobiles and three times as many canvasses as mobiles, for a total of 38 pieces. How many mobiles does the collection boast?

(a) Let m equal the number of mobiles in the collection. How many sculptures and canvasses does the collection include, in terms of m?

(b) Set up an equation and solve it for m.

(c) Answer the question.

(d) Check your work.

3 Three friends are playing pinochle. Roy has half as many points as Keith, and Keith has 210 points fewer than Brad. If together, the three men have a total of 960 points, how many points does Brad have?

(a) Let r = the number of points that Roy has. How many points do Keith and Brad have, in terms of r?

(b) Set up an equation and solve it for r.

(c) Answer the question.

(d) Check your work.

4 Alison read a book for school in three days. On the second day, she read three times as many pages as she had read on the first day, and then on the third day she finished the book by reading 130 fewer pages than she had read in the first two days combined. If the book has a total of 358 pages, how many pages did she read of it on the third day?

(a) Let f = the number of pages that Alison read on the first day. How many pages did she read on the second and third days, in terms of f?

(b) Set up an equation and solve it for f.

(c) Answer the question.

(d) Check your work.

Choosing Your Variable Wisely

REMEMBER

Declaring a variable is simple, as I show you earlier in this chapter, but you can make the rest of your work a lot easier when you know how to choose your variable wisely. Whenever possible, choose a variable so that the equation you have to solve has no fractions, which are much more difficult to work with than whole numbers.

For example, suppose you're trying to solve this problem:

> Irina has three times as many clients as Toby. If they have 52 clients altogether, how many clients does each person have?

The key sentence in the problem is "Irina has *three times as many* clients as Toby." It's significant because it indicates a relationship between Irina and Toby that's based on either *multiplication or division*. And to avoid fractions, you want to avoid division wherever possible.

TIP

Whenever you see a sentence that indicates you need to use either multiplication or division, choose your variable to represent the *smaller* number. In this case, Toby has fewer clients than Irina, so choosing t as your variable is the smart move.

Suppose you begin by declaring your variable as follows:

> Let t = the number of clients that Toby has.

Then, using that variable, you can make this chart:

Irina	$3t$
Toby	t

No fraction! To solve this problem, set up this equation:

Irina + Toby = 52

Plug in the values from the chart:

$3t + t = 52$

Now you can solve the problem easily, using what I show you in Chapter 23:

$4t = 52$
$t = 13$

Toby has 13 clients, so Irina has 39. Checking this result — which I recommend highly earlier in this chapter! — you find that $13 + 39 = 52$.

Now suppose that, instead, you take the opposite route and decide to declare a variable as follows:

Let i = the number of clients that Irina has.

Given that variable, you have to represent Toby's clients using the fraction $\frac{i}{3}$, which leads to the same answer but a *lot* more work.

EXAMPLE

Q. Abby lives twice as many miles from school as Kaitlin, and Kaitlin lives 7 miles farther from school than Giada does. If Abby lives 20 miles farther from the school than Giada, how far does Kaitlin live from it?

(a) Declare a variable using the initial of one of the girls, so that the resulting equation will not have fractions.

(b) Represent the values of the other two girls in terms of this variable.

(c) Set up the equation and solve it.

(d) Answer the question.

(e) Check your work.

A.

(a) Let g = the number of miles that Giada lives from school.

(b) Kaitlin $= g + 7$; Abby $= 2(g + 7) = 2g + 14$.

(c) $g = 6$

$2g + 14 = g + 20$
$g + 14 = 20$
$g = 6$

(d) Kaitlin lives 13 miles from school $(6 + 7 = 13)$.

(e) If Giada lives 6 miles from school, Kaitlin lives 13 miles from school $(6 + 7 = 13)$, and Abby lives 26 miles from school $(13 \times 2 = 26)$, so Abby lives 20 miles farther from school than Giada does $(26 = 6 + 20)$.

5 Boris earned $165 less than Carlos, and Carlos earned twice as much as Tran. If Boris and Tran earned a total of $1,425, how much did Carlos earn?

(a) Declare a variable using the initial of one of the boys, so that the resulting equation will not have fractions.

(b) Represent the amounts that the other two boys earned in terms of this variable.

(c) Set up the equation and solve it.

(d) Answer the question.

(e) Check your work.

6 Darryl needs to buy a jacket, a pair of shoes, and a pair of pants for a job interview. The pair of shoes he wants costs $65 less than the jacket, and the price of the pants is one-fifth that of the jacket. If the total cost of the three items is $254, what will Darryl pay for the shoes?

(a) Declare a variable using the initial of one of the items, so that the resulting equation will not have fractions.

(b) Represent the values of the other two items in terms of this variable.

(c) Set up the equation and solve it.

(d) Answer the question.

(e) Check your work.

7 Evan, Jack, and Nate are all playing a game in which the winning score is 100. Jack's score is one more than twice Nate's score, and one-third Evan's score. If Evan has 73 more points than Nate, how many points does Jack need to win the game?

(a) Declare a variable using the initial of one of the boys, so that the resulting equation will not have fractions.

(b) Represent the other two boys' scores in terms of this variable.

(c) Set up the equation and solve it.

(d) Answer the question.

(e) Check your work.

8 Evelyn participated in a bicycle race with three parts: a flat leg, an uphill leg, and a downhill leg. The flat leg took her 20 minutes less than twice the length of the uphill leg, and the uphill leg took her 20 minutes less than twice the length of the downhill leg. If her total time was 3 hours and 34 minutes, how long did Evelyn take to complete the flat leg of the race?

(a) Declare a variable using the initial of one of the legs of the race, so that the resulting equation will not have fractions.

(b) Represent the values of the other two legs in terms of this variable.

(c) Set up the equation and solve it.

(d) Answer the question.

(e) Check your work.

Solving More-Complex Algebraic Problems

Algebra word problems become more complex when the number of people or things you need to find out increases. In this section, the complexity increases to four and then five people. When you're done, you should feel comfortable solving algebra word problems of significant difficulty.

Charting four people

As in the previous section, a chart can help you organize information so you don't get confused. Here's a problem that involves four people:

> Alison, Jeremy, Liz, and Raymond participated in a canned goods drive at work. Liz donated three times as many cans as Jeremy, Alison donated twice as many as Jeremy, and Raymond donated 7 more than Liz. Together the two women donated two more cans than the two men. How many cans did the four people donate altogether?

The first step, as always, is declaring a variable. Remember that, to avoid fractions, you want to declare a variable based on the person who brought in the fewest cans. Liz donated more cans than Jeremy, and so did Alison. Furthermore, Raymond donated more cans than Liz. So because Jeremy donated the fewest cans, declare your variable as follows:

Let j = the number of cans that Jeremy donated.

Now you can set up your chart as follows:

Jeremy	j
Liz	$3j$
Alison	$2j$
Raymond	$\text{Liz} + 7 = 3j + 7$

This setup looks good because, as expected, there are no fractional amounts in the chart. The next sentence tells you that the women donated two more cans than the men, so make a word problem, as I show you in Chapter 7:

$$\text{Liz} + \text{Alison} = \text{Jeremy} + \text{Raymond} + 2$$

You can now substitute into this equation as follows:

$$3j + 2j = j + 3j + 7 + 2$$

With your equation set up, you're ready to solve. First, isolate the algebraic terms:

$$3j + 2j - j - 3j = 7 + 2$$

Combine like terms:

$$j = 9$$

Almost without effort, you've solved the equation, so you know that Jeremy donated 9 cans. With this information, you can go back to the chart, plug in 9 for j, and find out how many cans the other people donated: Liz donated 27, Alison donated 18, and Raymond donated 34. Finally, you can add up these numbers to conclude that the four people donated 88 cans altogether.

To check the numbers, read through the problem and make sure they work at every point in the story. For example, together Liz and Alison donated 45 cans, and Jeremy and Raymond donated 43, so the women really did donate 2 more cans than the men.

Crossing the finish line with five people

Here's one final example, the most difficult in this chapter, in which you have five people to work with.

> Five friends are keeping track of how many miles they run. So far this month, Mina has run 12 miles, Suzanne has run 3 more miles than Jake, and Kyle has run twice as far as Victor. But tomorrow, after they all complete a 5-mile run, Jake will have run as far as Mina and Victor combined, and the whole group will have run 174 miles. How far has each person run so far?

The most important point to notice in this problem is that there are two sets of numbers: the miles that all five people have run up to *today* and their mileage including *tomorrow*. And each person's mileage tomorrow will be 5 miles greater than their mileage today. Here's how to set up a chart:

	Today	Tomorrow $(\text{Today} + 5)$
Jake		
Kyle		
Mina		
Suzanne		
Victor		

With this chart, you're off to a good start to solve this problem. Next, look for that statement early in the problem that connects two people by either multiplication or division. Here it is:

> Kyle has run *twice as far* as Victor.

Because Victor has run fewer miles than Kyle, declare your variable as follows:

> Let v = the number of miles that Victor has run up to *today*.

Notice that I added the word *today* to the declaration to be very clear that I'm talking about Victor's miles *before* the 5-mile run tomorrow.

At this point, you can begin filling in the chart:

	Today	Tomorrow $(\text{Today}+5)$
Jake		
Kyle	$2v$	$2v+5$
Mina	12	17
Suzanne		
Victor	v	$v+5$

As you can see, I left out the information about Jake and Suzanne because I can't represent it using the variable v. I've also begun to fill in the *Tomorrow* column by adding 5 to my numbers in the *Today* column.

Now I can move on to the next statement in the problem:

But tomorrow, Jake will have run as far as Mina and Victor combined.

I can use this to fill in Jake's information:

	Today	Tomorrow $(\text{Today}+5)$
Jake	$17+v$	$17+v+5$
Kyle	$2v$	$2v+5$
Mina	12	17
Suzanne		
Victor	v	$v+5$

In this case, I first filled in Jake's *tomorrow* distance ($17 + v + 5$) and then subtracted 5 to find out his *today* distance. Now I can use the information that today Suzanne has run 3 more miles than Jake:

	Today	Tomorrow $(\text{Today}+5)$
Jake	$17+v$	$17+v+5$
Kyle	$2v$	$2v+5$
Mina	12	17
Suzanne	$17+v+3$	$17+v+8$
Victor	v	$v+5$

With the chart filled in like this, you can begin to set up your equation. First, set up a word equation, as follows:

$$\text{Jake tomorrow} + \text{Kyle tomorrow} + \text{Mina tomorrow} + \text{Suzanne tomorrow} +$$
$$\text{Victor tomorrow} = 174$$

Now just substitute information from the chart into this word equation to set up your equation:

$$17 + v + 5 + 2v + 5 + 17 + 17 + v + 8 + v + 5 = 174$$

As always, begin solving by isolating the algebraic terms:

$$v + 2v + v + v = 174 - 17 - 5 - 5 - 17 - 17 - 8 - 5$$

Next, combine like terms:

$$5v = 100$$

Finally, to get rid of the coefficient in the term $5v$, divide both sides by 5:

$$\frac{5v}{5} = \frac{100}{5}$$
$$v = 20$$

You now know that Victor's total distance up to *today* is 20 miles. With this information, you substitute 20 for v and fill in the chart, as follows:

	Today	Tomorrow$\left(\text{Today} + 5\right)$
Jake	37	42
Kyle	40	45
Mina	12	17
Suzanne	40	45
Victor	20	25

The *Today* column contains the answers to the question the problem asks. To check this solution, make sure that every statement in the problem is true. For example, tomorrow the five people will have run a total of 174 miles because

$$42 + 45 + 17 + 45 + 25 = 174$$

Copy down this problem, close the book, and work through it for practice.

EXAMPLE

Q. Ray and Donna both receive $10.00 allowance each week from their parents. This week, before receiving their allowances, Ray had three times as much money as his sister. After receiving their allowances, he had $13 more than twice as much as her. How much money did they have together after receiving their allowances?

(a) Declare a variable to represent the amount that one of the siblings had before receiving their allowance.

(b) Make a table to organize the amounts the siblings had before and after receiving their allowances, and fill it in.

(c) Set up an equation and solve it.

(d) Answer the question.

(e) Check your work.

A.

(a) Let d = the amount that Donna had before receiving her allowance

(b)

	Before Allowances	After Allowances
Donna	d	$d+10$
Ray	$3d$	$3d+10$

(c) $d = 23$

$$2(d+10)+13 = 3d+10$$
$$2d+20+13 = 3d+10$$
$$2d+33 = 3d+10$$
$$33 = d+10$$
$$23 = d$$

(d) Together, Donna and Ray had $112 after receiving their allowances:
$(23+10+(23\times3)+10 = 33+69+10 = 112)$.

(e) Before receiving their allowances, Donna had $23 and Ray had $69 $(23\times3=69)$, so they had a total of $92. Afterwards, they had a total of $20 more, so they had $112 total.

YOUR TURN

9 A college class that has 73 students includes twice as many freshmen as sophomores, seven fewer juniors than sophomores, and three times as many juniors as seniors. How many freshmen are in the class?

(a) Declare a variable to represent the number in the smallest group of students.

(b) Write expressions to represent the number of students in the other three groups.

(c) Set up and solve this equation.

(d) Answer the question.

(e) Check your work.

10 Anthony is on the football team at school, and his brother Dylan is on the wrestling team. At the start of this school year, Anthony's weight was 110 percent of Dylan's. But then on instructions from their coaches, Anthony gained 8 pounds and Dylan lost 5 pounds. By the end of the school year, Anthony's weight was 120 percent of Dylan's. What did the two boys weigh at the end of the school year?

(a) Declare a variable to represent the weight of one of the boys at the start of the school year.

(b) Make a table to organize the amounts the boys weighed at the start and end of the year, and fill it in.

(c) Set up an equation and solve it.

(d) Answer the question.

(e) Check your work.

11 Three friends received bonuses at work. Artie received twice as much money as Barry, and Elaine received $150 more than Barry. They all placed their windfalls in the same mutual fund, which increased by 30 percent over the next year. The result was that Artie ended up with $260 more than Elaine. How much money did Barry make on his mutual fund investment?

(a) Declare a variable to represent the amount that one employee received as a bonus.

(b) Make a table to organize the amounts of the employees' bonuses and the results of their investments, and fill it in.

(c) Set up an equation and solve it.

(d) Answer the question.

(e) Check your work.

Practice Questions Answers and Explanations

①

(a) Benjamin $= 2a - 4$.

(b) $a = 11$

$$a + 2a - 4 = 29$$
$$3a - 4 = 29$$
$$3a = 33$$
$$a = 11$$

(c) Annette sold 11 books.

(d) If Annette sold 11 books, Benjamin sold 18 books $(2 \times 11 - 4 = 18)$, and $11 + 18 = 29$.

②

(a) Sculptures $= m + 8$; canvasses $= 3m$.

(b) $m = 6$

$$m + m + 8 + 3m = 38$$
$$5m + 8 = 38$$
$$5m = 30$$
$$m = 6$$

(c) The collection includes 6 mobiles.

(d) If the collection has 6 mobiles, it has 14 sculptures $(6 + 8 = 14)$, and it has 18 canvasses $(6 \times 3 = 18)$, so it has a total of 38 pieces $(6 + 14 + 18 = 38)$.

③

(a) Keith $= 2r$; Brad $= 2r + 210$.

(b) $r = 150$

$$r + 2r + 2r + 210 = 960$$
$$5r + 210 = 960$$
$$5r = 750$$
$$r = 150$$

(c) Brad has 510 points $(2 \times 150 + 210 = 300 + 210 = 510)$.

(d) If Roy has 150 points, Keith has 300 points $(150 \times 2 = 300)$, and Brad has 510 points $(300 + 210 = 500)$, so altogether, the three men have 960 points $(150 + 300 + 510 = 960)$.

④

(a) Second day $= 3f$; third day $= f + 3f - 130 = 4f - 130$

(b) $f = 61$

$$f + 3f + 4f - 130 = 358$$
$$8f - 130 = 358$$
$$8f = 488$$
$$f = 61$$

(c) Alison read 114 pages on the third day $(4 \times 61 - 130 = 244 - 130 = 114)$.

(d) If Alison read 61 pages on the first day, she read 183 pages on the second day $(61 \times 3 = 183)$, and 114 pages on the third day $(61 + 183 - 130 = 114)$, so she read a total of 358 pages $(61 + 183 + 114 = 358)$.

5

(a) Let t = the number of dollars that Tran earned.

(b) Carlos $= 2t$; Boris $= 2t - 165$.

(c) $t = 530$

$$t + 2t - 165 = 1,425$$
$$3t - 165 = 1,425$$
$$3t = 1,590$$
$$t = 530$$

(d) Carlos earned $1,060$ $(530 \times 2 = 1,060)$.

(e) If Tran earned 530 and Carlos earned $1,060$, Boris earned 895 $(1,060 - 165 = 895)$, so Boris and Tran earned a total of $1,425$ $(895 + 530 = 1,425)$.

6

(a) Let p = the price of the pants.

(b) Jacket $= 5p$; shoes $= 5p - 65$.

(c) $p = 29$

$$p + 5p + 5p - 65 = 254$$
$$11p - 65 = 254$$
$$11p = 319$$
$$p = 29$$

(d) Darryl will pay 80 for the shoes $(29 \times 5 - 65 = 145 - 65 = 80)$.

(e) If Darryl pays 29 for the pants, and 80 for the shoes, he pays 145 for the jacket $(29 \times 5 = 145)$, so he pays a total of 254 $(29 + 80 + 145 = 254)$.

7

(a) Let n = Nate's score.

(b) Jack $= 2n + 1$; Evan $= 3(2n + 1) = 6n + 3$.

(c) $n = 14$

$$6n + 3 = n + 73$$
$$5n + 3 = 73$$
$$5n = 70$$
$$n = 14$$

(d) Jack needs 71 points to win the game. $($Jack's score is $14 \times 2 + 1 = 29$, and $100 - 29 = 71.)$

(e) If Nate has 14 points and Jack has 29 points, then Evan has 87 points $(14 \times 6 + 3 = 84 + 3 = 87)$, so Evan has 73 more points than Nate $(87 - 14 = 73)$.

(8)

(a) Let d = the length of time Evelyn took for the downhill leg.

(b) Uphill $= 2d - 20$; flat $= 2(2d - 20) - 20 = 4d - 40 - 20 = 4d - 60$.

(c) $d = 42$

$$d + 2d - 20 + 4d - 60 = 214 \quad (3 \text{ hours and } 34 \text{ minutes} = 214 \text{ minutes.})$$
$$7d - 80 = 214$$
$$7d = 294$$
$$d = 42$$

(d) Evelyn took 1 hour and 48 minutes to complete the flat leg of the race
$\left(42 \times 4 - 60 = 160 - 68 = 108 \text{ minutes} = 1 \text{ hour and } 48 \text{ minutes} \right)$.

(e) If Evelyn took 42 minutes to complete the downhill leg and 108 minutes to complete the flat leg, she took 64 minutes to complete the uphill leg $\left(42 \times 2 - 20 = 84 - 20 = 64 \right)$, so she took a total of 3 hours and 34 minutes to complete the race $\left(42 + 64 + 108 = 214 \text{ minutes} = 3 \text{ hours and } 34 \text{ minutes} \right)$.

(9)

(a) Let s = the number of seniors. There are more freshmen than sophomores, more sophomores than juniors, and more juniors than seniors, so the smallest group is seniors.

(b) Juniors $= 3s$; sophomores $= 3s + 7$; freshmen $= 2(3s + 7)$.

(c) $s + 3s + (3s + 7) + 2(3s + 7) = 73$. The four groups added together total 73 students.

$$s = 4$$
$$s = 3s + (3s + 7) + 2(3s + 7) = 73$$
$$s + 3s + 3s + 7 + 6s + 14 = 73$$
$$13s + 21 = 73$$
$$13s = 52$$
$$s = 4$$

(d) 38. There are 4 seniors, and there are three times as many junior as seniors, so there are 12 juniors. There are 7 more sophomores than juniors, so there are 19 sophomores. Finally, there are twice as many freshmen as sophomores, so there are 38 freshmen.

(e) If there are 4 seniors, there are 12 juniors, 19 sophomores, and 38 freshmen, which equals 73 students.

(10)

(a) Let d = Dylan's weight at the start of the school year.

(b)

	Start of year	End of year
Dylan	d	$d - 5$
Anthony	$1.1d$	$1.1d + 8$

(c) $d = 140$

$$1.2(d-5) = 1.1d + 8$$
$$1.2d - 6 = 1.1d + 8$$
$$0.1d - 6 = 8$$
$$0.1d = 14$$
$$d = 140$$

(d) At the end of the year, Dylan weighed 135 pounds and Anthony weighed 162 $(140 - 5 = 135; 140 \times 1.1 + 8 = 154 + 8 = 162)$.

(e) If Dylan weighed 140 pounds at the start of the year, Anthony weighed 154 pounds $(140 \times 14 = 154)$. Afterwards, Dylan weighed 135 pounds and Anthony weighed 162 pounds, so Anthony weighed 120% of Dylan's weight $(135 \times 1.2 = 162)$.

(a) Let b equal the amount that Barry received as a bonus.

(b)

	Bonus	After Investment
Artie	$2b$	$1.3(2b) = 2.6b$
Barry	b	$1.3b$
Elaine	$b + 150$	$1.3(b + 150) = 1.3b + 195$

(c) $b = 350$
$$2.6b = 1.3b + 195 + 260$$
$$2.6b = 1.3b + 455$$
$$1.3b = 455$$
$$b = 350$$

(d) Barry made $105 on his investment $(30\% \text{ of } 350 = \$105)$.

(e) If Barry received a $350 bonus, then Artie received $700 $(350 \times 2 = 700)$ and Elaine received $500 $(350 + 150 = 500)$. Each earned 30% on their investment, so Artie earned $210, bringing his total up to $910, and Elaine earned $150, bringing her total up to $650. Thus, Artie ended up with $260 more than Elaine $(910 - 260 = 650)$.

Try out the chapter quiz in the next section for another chance to test and strengthen your word problem-solving skills.

Whaddya Know? Chapter 24 Quiz

The ten questions that follow provide you with another opportunity to practice solving algebra word problems. When you're finished, check out the solutions and explanations in the section that follows.

 Alexa and Freddie collected $355 for the annual holiday fundraiser. If Alexa collected $115 more than Freddie, then how much did each contribute?

 My recipe calls for twice as many apples as bananas and 3 more pears than bananas. If I need a total of 19 pieces of fruit, then how many of each do I need?

 George worked twice as many volunteer hours in March as he did in February. If the total number of hours was 93, then how many volunteer hours did he work in each month?

 Helen is 3 years more than 10 times the age of her niece. If the sum of their ages is 36, then how old are they?

5 In one 24-hour day, Jon spends twice as much time sleeping as he does on the phone and three-fourths as much time on the computer as he spends sleeping. He spends the other 6 hours exercising, eating, and reading. How many more hours does he spend on the computer than on the phone?

 The animal shelter has 16 more cats than dogs and 21 fewer rabbits than cats. If they have a total of 242 of these animals, then how many of each are at the shelter?

7 Abe is 4 years younger than Maeve. In 5 years the sum of their ages will be 44. How old are they now?

8 There used to be twice as many girls as boys in the class. When 6 new girls were added and 1 boy left, there were 3 times as many girls as boys. How many of each are there in the class right now?

9 Adam has twice as many pennies as Ben. Carly has three times as many pennies as Adam. Don has four more pennies than Carly. If they have a total of 79 pennies, then how many do each of them have?

10 Pants cost $5 more than a shirt. And, when you add $7 in tax, the total comes to $72. How much do the items of clothing cost?

Answers to Chapter 24 Quiz

1 **Alexa \$235, Freddie \$120.** Let f represent the amount that Freddie collected. Then Alexa collected $f + 115$. The sum of their contributions is written as $f + (f + 115) = 355$. Solve the equation for f.

$$f + (f + 115) = 355$$
$$2f + 115 = 355$$
$$2f = 240$$
$$f = 120$$

So Freddie collected \$120 and Alexa collected $\$120 + \$115 = \$235$.

2 **4 bananas, 8 apples, 7 pears.** Let b represent the number of bananas. Then $2b$ is the number of apples, and $b + 3$ is the number of pears. The total number of pieces of fruit is 19, written as $b + 2b + b + 3 = 19$. Solve for b.

$$b + 2b + b + 3 = 19$$
$$4b + 3 = 19$$
$$4b = 16$$
$$b = 4$$

So there are 4 bananas, 8 apples, and 7 pears.

3 **February 31, March 62.** Let f represent the number of volunteer hours he worked in February. The $2f$ is the number he worked in March. The total number of hours is 93, so $2f + f = 93$. Solve for f.

$$3f = 93$$
$$f = 31$$

He worked 31 hours in February and 62 hours in March.

4 **Niece 3, Helen 33.** Let n represent the niece's age. Then Helen's age is $3 + 10n$. The sum of their ages is 36, so $n + 3 + 10n = 36$. Solve for n.

$$n + 3 + 10n = 36$$
$$11n + 3 = 36$$
$$11n = 33$$
$$n = 3$$

So Helen's niece is 3 and Helen is $3 + 10(3) = 33$.

(5) **Phone 4, sleep 8, computer 6.** Let p represent the number of hours he spent on the phone. Then $2p$ is the amount of time he spent sleeping, and $\frac{3}{4}(2p) = \frac{3}{2}p$ is the amount of time he spent on the computer. Add the 6 hours, and you have $p + 2p + \frac{3}{2}p + 6 = 24$. Solve for p.

$$p + 2p + \frac{3}{2}p + 6 = 24$$
$$\frac{9}{2}p + 6 = 24$$
$$\frac{9}{2}p = 18$$
$$p = 18 \cdot \frac{2}{9} = 4$$

Four hours on the phone means twice that or 8 hours sleeping. And $\frac{3}{4}$ of 8 is 6, so that's 6 hours on the computer.

(6) **Dogs 77, cats 93, rabbits 72.** Let d represent the number of dogs. Then $d + 16$ is the number of cats and $d + 16 - 21 = d - 5$ is the number of rabbits. The total of 242 animals represented by $d + d + 16 + d - 5 = 242$. Solve for d.

$$d + d + 16 + d - 5 = 242$$
$$3d + 11 = 242$$
$$3d = 231$$
$$d = 77$$

There are 77 dogs, $77 + 16 = 93$ cats, and $77 - 5 = 72$ rabbits.

(7) **Maeve 19, Abe 15.** Let m represent Maeve's age. Then $m - 4$ is Abe's age.

	Age Now	Age in 5 Years
Maeve	m	$m + 5$
Abe	$m - 4$	$m - 4 + 5 = m + 1$
Sum		44

Writing the equation for the sum of their ages in 5 years and solving for m:

$$m + 5 + m + 1 = 44$$
$$2m + 6 = 44$$
$$2m = 38$$
$$m = 19$$

Maeve is 19 now, and Abe is $19 - 4 = 15$. In 5 years they'll be 24 and 20, and the sum of those ages is 44.

Boys 8, girls 24. Let b represent the number of boys there used to be; this means there used to be $2b$ girls.

	Former Number	Changes
Boys	b	$b-1$
Girls	$2b$	$2b+6$

After the changes, the number of girls was three times the number of boys. This is represented with $2b+6 = 3(b-1)$. Solve for b.

$$2b+6 = 3(b-1)$$
$$2b+6 = 3b-3$$
$$9 = b$$

If there were 9 boys before the 1 boy left, then there are 8 boys now.

If there were 2(9) or 18 girls before the 6 were added, then there are 24 girls now.

24 is 3 times 8.

⑨ **Adam 10, Ben 5, Carly 30, Don 34.** Let b represent the number of pennies that Ben has. Then Adam has $2b$ pennies. If Carly has three times what Adam has, then she has $3(2b)$ pennies. And Don has 4 more than Carly, so he has $3(2b)+4$ pennies. Altogether they have 79 pennies, which you write as $b+2b+3(2b)+3(2b)+4 = 79$. Solve for b.

$$b+2b+3(2b)+3(2b)+4 = 79$$
$$b+2b+6b+6b+4 = 79$$
$$15b+4 = 79$$
$$15b = 75$$
$$b = 5$$

So Ben has 5 pennies, Adam has twice that or 10 pennies, Carly has three times Adam's or 30 pennies, and Don has 4 more than Carly or 34 pennies.

⑩ **Shirt \$30, pants \$35.** Let the cost of a shirt be represented by h. Then the pants cost $h+5$. Add in the tax, and the total cost of the clothing is $h+h+5+7 = 72$. Solve for h.

$$h+h+5+7 = 72$$
$$2h+12 = 72$$
$$2h = 60$$
$$h = 30$$

The shirt costs \$30, and the pants cost \$35.

- » **Graphing equations on the *xy*-plane**

- » **Understanding the most basic linear equation, $y = x$**

- » **Working with the slope-intercept form equation, $y = mx + b$**

- » **Identifying and calculating the slope of a line**

- » **Using the slope and the intercept to graph linear equations**

Chapter **25**

Graphing Algebraic Equations

G raphing is an important part of algebra, providing a way to visualize the connections between numbers.

In this chapter, I introduce you to the *xy*-plane, where most graphing in algebra happens. I show you how to identify and plot points using coordinate pairs (x, y), and how to graph equations by plotting points.

Then, you discover how to work with the slope-intercept form of the linear equation, $y = mx + b$. You work with the slope m and the y-intercept b to graph equations. I show you two ways to find the slope of a line. Finally, you use the slope and the intercept to graph linear equations.

Graphing on the xy-Plane

The *xy-plane* is a mathematical construction defined by a pair of axes, where every point is labeled by a unique pair of points of the form (x, y). Figure 25-1 shows the xy-plane.

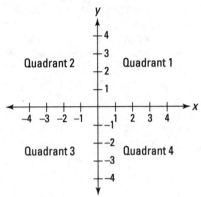

FIGURE 25-1:
The *xy*-plane.

Understanding the axes, the origin, and the quadrants

The *x*-axis and *y*-axis, collectively known as the *axes*, form the basis of the xy-plane. Essentially, these are a pair of number lines set at right angles, as shown in Figure 25-2. The arrows indicate that each axis extends infinitely in both directions. Thus, the xy-plane also extends infinitely in every direction.

These two axes divide the xy-plane into four regions called *quadrants* that extend infinitely outward from the axes.

The intersection point of the two axes is called the *origin*.

Plotting coordinates on the xy-plane

Every point on the xy-plane is labeled uniquely as a pair of *coordinate points*, or *coordinates*, of the form (x, y). The coordinates of the origin are $(0, 0)$.

Starting from the origin, you can find the coordinates of any point by counting first on the *x*-axis, and then on the *y*-axis. This is called *plotting a point*.

Figure 25-2 shows you how to plot four points on the xy-plane.

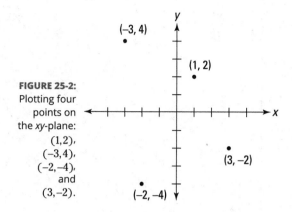

FIGURE 25-2:
Plotting four
points on
the *xy*-plane:
(1,2),
(−3,4),
(−2,−4),
and
(3,−2).

Graphing equations on the xy-plane

You can graph an equation on the *xy*-plane, point by point, by plugging in a few values for *x* and finding the resulting value for *y* in each case. For example, to plot the equation $y = 2x - 3$, plug in *x*-values of 0, 1, 2, and 3 into the equation and solve for *y*:

$$y = 2(0) - 3 \qquad y = 2(1) - 3 \qquad y = 2(2) - 3 \qquad y = 2(3) - 3$$
$$y = 0 - 3 \qquad y = 2 - 3 \qquad y = 4 - 3 \qquad y = 6 - 3$$
$$y = -3 \qquad y = -1 \qquad y = 1 \qquad y = 3$$

Thus, the graph of this equation includes the four points (0,−3), (1,−1), (2,1), and (3,3). The result is the line shown in Figure 25-3.

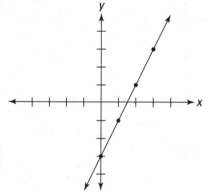

FIGURE 25-3:
Graphing
the equation
$y = 2x - 3$.

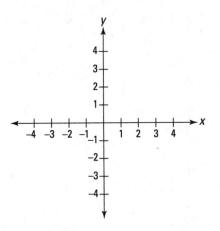

 Q. Plot the following four points on the *xy*-plane shown here.

    (a) (3,4)

    (b) (4,−1)

EXAMPLE

    (c) (−1,2)

    (d) (−2,−3)

A.

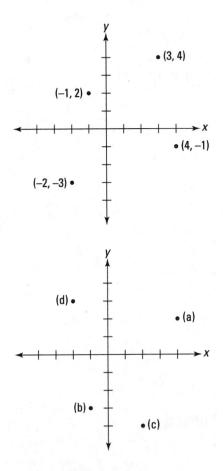

Q. Identify the *xy*-coordinates of the four points shown here.

A.

   (a) (4, 2)

   (b) (−1, −3)

   (c) (2, −4)

   (d) (−2, 3)

Q. For the equation $y = -2x + 4$, calculate the value of *y* for each of the following values of *x*. Then write the corresponding *xy*-coordinate for each result.

   (a) When $x = 0$, $y =$ ____.

   (b) When $x = 1$, $y =$ ____.

   (c) When $x = 2$, $y =$ ____.

   (d) When $x = 3$, $y =$ ____.

A.

   (a) 4; (0, 4): $y = -2(0) + 4 = 0 + 4 = 4$.

   (b) 2; (1, 2): $y = -2(1) + 4 = -2 + 4 = 2$.

   (c) 0; (2, 0): $y = -2(2) + 4 = -4 + 4 = 0$.

   (d) −2; (3, −2): $y = -2(3) + 4 = -6 + 4 = -2$.

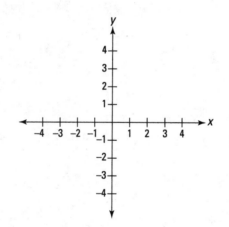

Q. Plot each the four *xy*-coordinates you found in the previous question on the *xy*-plane shown here, then draw the line that connects them.

A.

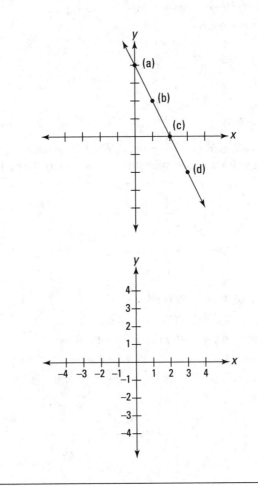

1 Plot the following four points on the *xy*-plane shown here.

(a) $(2,3)$

(b) $(-4,1)$

(c) $(-3,-2)$

(d) $(4,-3)$

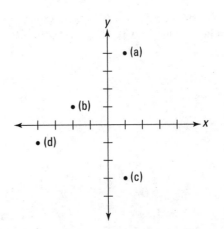

Identify the *xy*-coordinates of the four points shown here.

 For the equation $y = -x + 3$, calculate the value of *y* for each of the following values of *x*. Then write the corresponding *xy*-coordinate for each result.

(a) When $x = 0$, $y =$ ____.

(b) When $x = 1$, $y =$ ____.

(c) When $x = 2$, $y =$ ____.

(d) When $x = 3$, $y =$ ____.

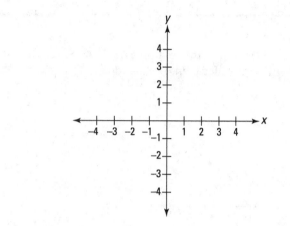

 Plot each the four *xy*-coordinates you found in the previous question on the *xy*-plane shown here, then draw the line that connects them.

Understanding Linear Equations

In the previous section, you discover how to graph lines on the *xy*-plane. When an equation results in a line, it's called a *linear equation*. In this section, you expand your understanding of linear equations.

Knowing the most basic linear equation

The most basic linear equation is $y = x$. To see why this equation results in a line when graphed on the *xy*-plane, you can plot some points on the graph. For example, if $x = 0$, then $y = 0$, so plot the point (0,0) on the graph. Similarly, if $x = 1$, then $y = 1$, so you can also plot the point (1,1). And if $x = 2$, then $y = 2$, so plot the point (2,2).

In Table 25-1, I provide a list of five *x*-values, the *y*-value that each produces, and its resulting point on the *xy*-graph.

Table 25-1   Plotting the Points for the Linear Equation $y = x$.

x	y	(x, y)
-2	-2	(-2,-2)
-1	-1	(-1,-1)
0	0	(0,0)
1	1	(1,1)
2	2	(2,2)

In Figure 25-4, I show you the resulting graph. As you can see, this graph is a line, which is why the equation $y = x$ is called a linear equation.

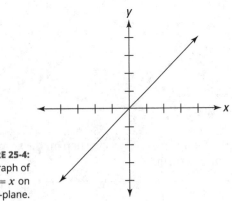

FIGURE 25-4:
The graph of
$y = x$ on
the xy-plane.

Changing the slope (m)

With one small change to the most basic linear equation, $y = x$, you can change its graph. This change occurs when you change this equation to $y = mx$, and then change the value of m.

For example, in Figure 25-5, you can see the results when you change m to both $\frac{1}{2}$ and 2. As you can see, the graph of the line $y = \frac{1}{2}x$ slopes upward in the positive direction only half as quickly as $y = x$ does. Similarly, $y = 2x$ slopes upward twice as steeply.

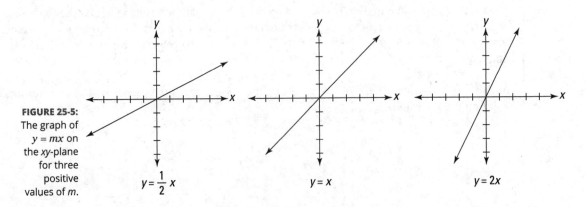

FIGURE 25-5:
The graph of
$y = mx$ on
the xy-plane
for three
positive
values of m.

$y = \dfrac{1}{2}x$ $y = x$ $y = 2x$

In contrast, Figure 25-6 shows what happens when the m value is negative. When m is -1, the resulting equation $y = -x$ creates a line graph that slopes downward rather than upward as you move in the positive direction. When m is $-\frac{1}{2}$, the equation becomes $y = -\frac{1}{2}x$, and the graph slopes downward half as sharply. And when m is -2, the equation changes to $y = -2x$, and the resulting graph slopes down twice as fast.

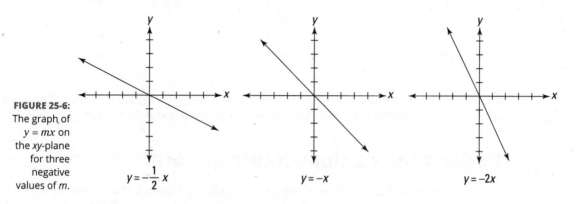

FIGURE 25-6:
The graph of $y = mx$ on the xy-plane for three negative values of m.

$y = -\frac{1}{2}x$ $y = -x$ $y = -2x$

As you can see from these examples, the value m in the equation $y = mx$ directly affects the direction and steepness of the slope in the resulting graph. For this reason, the value m is called the *slope* of the line.

Changing the y-intercept (b)

When you make another change to the linear equation, $y = x$, you can move its graph up or down. This change occurs when you change this equation to $y = x + b$, and then change the value of b.

For example, changing the value of b to 1, 2, or 3 moves the graph up 1, 2, or 3 units. As you can see in Figure 25-7, the resulting line always crosses the y-axis at the b-value.

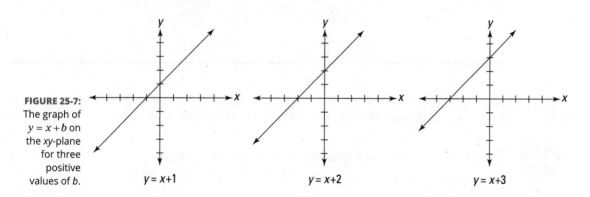

FIGURE 25-7:
The graph of $y = x + b$ on the xy-plane for three positive values of b.

$y = x+1$ $y = x+2$ $y = x+3$

In a similar way, if you change the value of b to -1, -2, or -3, the graph moves down 1, 2, or 3 units. Figure 25-8 shows you that in each case, the resulting line again crosses the y-axis at the b-value.

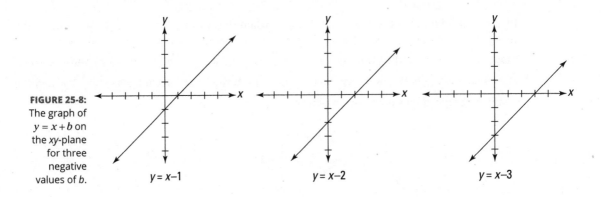

FIGURE 25-8:
The graph of
$y = x + b$ on
the xy-plane
for three
negative
values of b.

$y = x-1$　　　　　　$y = x-2$　　　　　　$y = x-3$

As you can see, the b-value in the equation $y = x + b$ determines where the resulting line on the xy-plane intersects with the y-axis. For this reason, b is called the y-intercept.

Understanding slope-intercept form

When you combine the basic linear equation, $y = x$, with a slope m and a y-intercept b, you've got all the pieces necessary to understand linear equations. The result is the *slope-intercept form* of a linear equation:

$$y = mx + b$$

The values m and b change the slope and y-intercept of the line $y = x$. To see how this works, check out Table 25-2, where I present four linear equations.

Table 25-2　The Slope and y-Intercept of Three Linear Equations

Equation	Slope	y-Intercept
$y = 3x - 4$	3	−4
$y = \frac{1}{4}x + 1$	$\frac{1}{4}$	1
$y = -2x + 3$	−2	3
$y = -\frac{1}{3}x - 1$	$-\frac{1}{3}$	−1

In Figure 25-9, you see the graphs of four equations on the xy-plane. In each case, the slope and y-intercept changes the basic equation $y = x$ in predictable ways.

Q. What is the slope and y-intercept of the following equations?

　(a)　$y = 5x + 6$

　(b)　$y = -\frac{3}{4}x - 2$

　(c)　$y = x - \frac{2}{5}$

　(d)　$y = -x + 100$

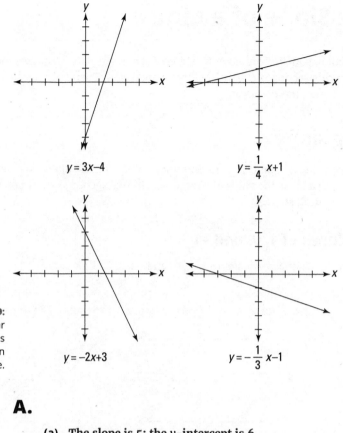

$y = 3x-4$

$y = \frac{1}{4}x+1$

$y = -2x+3$

$y = -\frac{1}{3}x-1$

FIGURE 25-9:
Four linear
equations
graphed on
the *xy*-plane.

A.

    (a) The slope is 5; the *y*-intercept is 6.

    (b) The slope is $-\frac{3}{4}$; the *y*-intercept is -2.

    (c) The slope is 1; the *y*-intercept is $-\frac{2}{5}$.

    (d) The slope is -1; the *y*-intercept is 100.

5 What is the slope and *y*-intercept of the following equations?

YOUR TURN

    (a) $y = -3x + 2$

    (b) $y = \frac{2}{3}x - 8$

    (c) $y = 100x - 60$

    (d) $y = -\frac{7}{8}x + 14$

6 What is the equation for a line with each of the following attributes?

    (a) The slope is 5, and the y-intercept is -6.

    (b) The slope is -1, and the y-intercept is $\frac{7}{10}$.

    (c) The slope is $\frac{2}{9}$, and the y-intercept is $-\frac{1}{3}$.

    (d) The slope is $-\frac{3}{5}$, and the y-intercept is 1.

Measuring the Slope of a Line

In the previous section, you discover how the slope m of a line $y = mx + b$ determines the direction of that line. In this section, I clarify how the slope works. Then I show you a few ways to find the slope of a line on the xy-plane.

Estimating slope

The *slope* of a line on the xy-graph is a measurement of steepness. Slope is always measured from left to right — that is, as the value of x increases. In this section, I show you how to begin visualizing and estimating slope.

Identifying slopes of 1, 0, and –1

Figure 25-10 shows three lines with slopes of 1, 0, and –1.

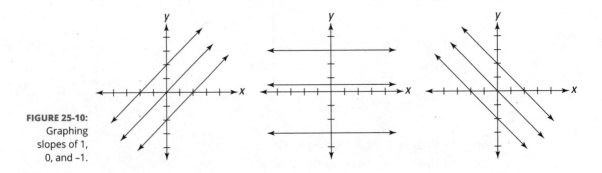

FIGURE 25-10:
Graphing slopes of 1, 0, and –1.

As you can see, a line with a slope of 1 *rises* at the same rate that it moves to the right. And a line with a slope of 0 is completely flat. Finally, a line with a slope of –1 *falls* at the same rate that it moves to the right.

Distinguishing positive and negative slope

When the slope of a line is positive, that line rises as it proceeds from left to right — that is, as x increases. In contrast, a line with a negative slope falls as x increases.

Figure 25-11 shows two xy-graphs with a variety of lines. All of the lines shown in the graph on the left have positive slopes, and all of those on the right have negative slopes.

Identifying vertical lines as undefined

When a line on the xy-graph is vertical, that line is considered to be infinitely steep. Speaking mathematically, a vertical line has an *undefined slope* — that is, its slope cannot be stated as a number, so it doesn't exist.

FIGURE 25-11:
Graphing
positive and
negative
slopes.

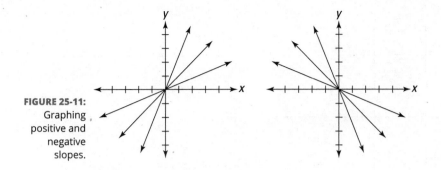

Be careful to remember the difference between vertical lines that have an undefined slope and horizontal lines that have a slope of 0.

Q. Decide whether the slope of each line is positive, negative, zero, or undefined.

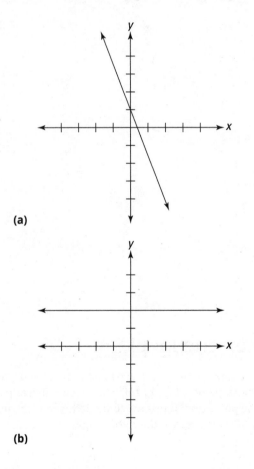

(a)

(b)

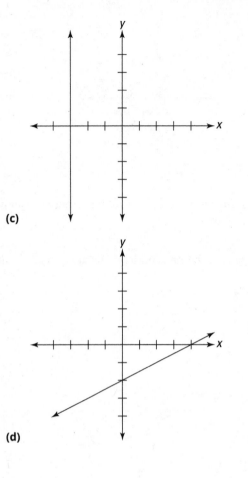

(c)

(d)

A.

(a) **Negative slope.**

(b) **Zero slope.**

(c) **Undefined slope.**

(d) **Positive slope.**

Eyeballing slope on the xy-plane

You can often find the slope of a line on the *xy*-plane by looking at it and using the mnemonic **up/down and over.** To find the slope of a line in this way, pick out two points that the line intersects. Then, starting at the point that's furthest to the left, count the number of steps **up or down** and then the number of units **over** (to the right) required to reach the second point.

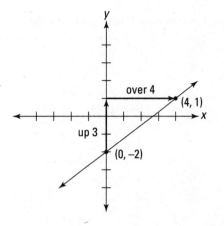

For example, the line in the graph shown here passes through the two points $(0,-2)$ and $(4,1)$. Starting at the point on the left, which is $(0,-2)$, the path **up 3, over 4** takes you to the point $(4,1)$. Now, interpret this path as follows:

up	3	over	4
+	3	/	4

Thus, the path **up 3, over 4** translates to $+\frac{3}{4}$ — that is, a slope of $\frac{3}{4}$.

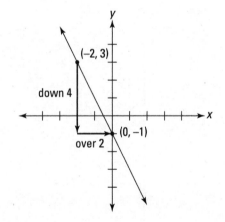

As another example, this time the line passes through the two points $(-2,3)$ and $(0,-1)$. Starting at $(-2,3)$, the path **down 4, over 2** takes you to the point $(0,-1)$. In this case, interpret the path in this way:

down	4	over	2
–	4	/	2

So this time, the path **down 4, over 2** translates to $-\frac{4}{2} = -2$, for a slope of -2.

Q. Find the slope of each line by counting **up/down** _____, **over** _____.

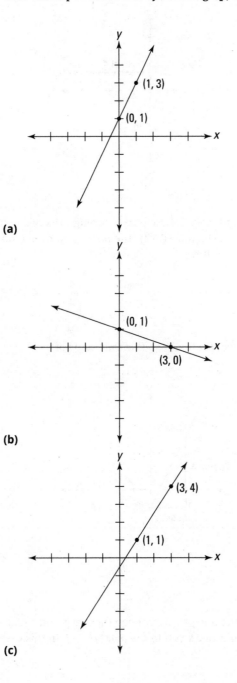

(a)

(b)

(c)

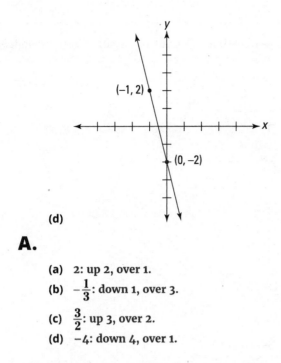

(d)

A.

(a) 2: up 2, over 1.

(b) $-\frac{1}{3}$: down 1, over 3.

(c) $\frac{3}{2}$: up 3, over 2.

(d) −4: down 4, over 1.

Using the two-point slope formula

In the previous section, you used two points on a graph to find the slope of a line. Even if you don't have a graph to work with, you can still find the slope of any line that passes through two points (x_1, y_1) and (x_2, y_2) using the *two-point slope formula*:

$$m = \frac{y_2 - y_1}{x_2 - x_1}$$

For example, to find the slope of a line on the *xy*-plane that passes through the points $(3,8)$ and $(-5,2)$, plug these values into the formula as follows:

$$m = \frac{2 - 8}{-5 - 3}$$

Now, simplify:

$$= \frac{-6}{-8} = \frac{3}{4}$$

So the slope of this line is $\frac{3}{4}$.

Q. What is the slope of the line that passes through the points $(-1,7)$ and $(3,-3)$?

A. −5. Plug these values into the formula and simplify:

EXAMPLE

$$m = \frac{y_2 - y_1}{x_2 - x_1} = \frac{-3 - 7}{3 - (-1)} = \frac{-10}{4} = -\frac{5}{2}$$

YOUR TURN

7 Figure out whether the slope of each line is positive, negative, zero, or undefined.

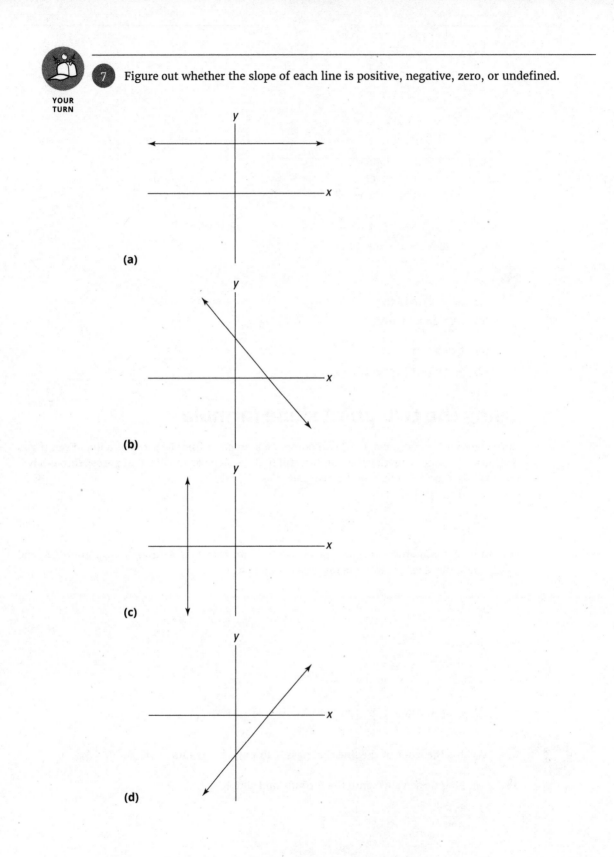

(a)

(b)

(c)

(d)

8 Find the slope of each line by counting **up/down** _____, **over** _____.

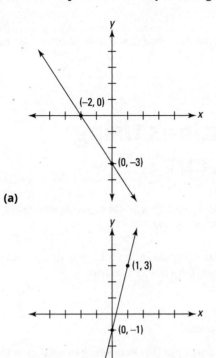

(a)

(b)

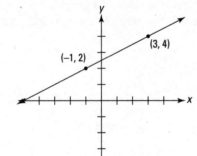

(c)

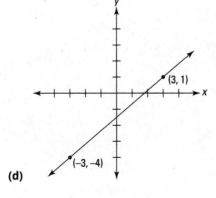

(d)

9 What is the slope of a line on the xy-plane that passes through the points $(1,7)$ and $(5,-11)$?

 10 What is the slope of a line on the xy-plane that passes through the points $(-2,-9)$ and $(10,7)$?

Graphing Linear Equations Using the Slope and y-intercept

Earlier in this chapter, I show you how to graph a linear equation by calculating x- and y-values and then plotting these points on the xy-plane.

A more elegant way to graph a linear equation in the form $y = mx + b$ is to use slope m and the y-intercept b to find a pair of points on the line. To do this:

1. **Plot the y-intercept.**

2. **Use the slope to plot a second point.**

3. **Connect these two points with a straight line extending in both directions. This line is the graph of the equation.**

For example, suppose you want to graph the line $y = 3x - 1$. Begin by plotting the y-intercept $(0,-1)$. Now, starting from this point, use the slope of 3 to move **up 3, over 1** to the point $(1,2)$. The line that connects these two points is the graph of the equation, as shown here:

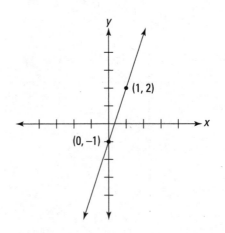

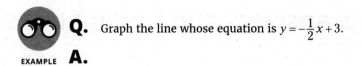 **Q.** Graph the line whose equation is $y = -\frac{1}{2}x + 3$.

EXAMPLE **A.**

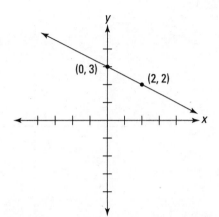

Begin by plotting the y-intercept $(0,3)$. Now, starting from this point, use the slope of $-\frac{1}{2}$ to move **down 1, over 2** to the point $(2,2)$. The line that connects these two points is the graph of the equation, as shown here:

YOUR TURN

⓫ Graph the linear equation $y = -2x - 1$.

⓬ Graph the line of the equation $y = \frac{3}{4}x - 2$.

Practice Questions Answers and Explanations

1

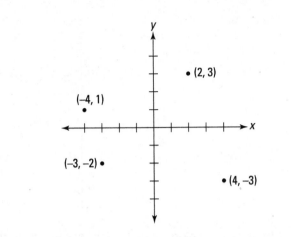

2

    (a) $(1,4)$

    (b) $(-2,1)$

    (c) $(1,-3)$

    (d) $(-4,-1)$

3

    (a) $3; (0,3). y = 0 + 3 = 3.$

    (b) $2; (1,2). y = -1 + 3 = 2.$

    (c) $1; (2,1). y = -2 + 3 = 1.$

    (d) $0; (3,0). y = -3 + 3 = 0.$

4

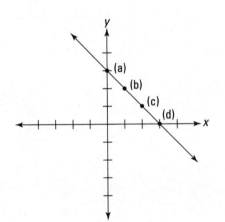

(5)

   (a) The slope is -3; the y-intercept is 2.

   (b) The slope is $\frac{2}{3}$; the y-intercept is -8.

   (c) The slope is 100; the y-intercept is -60.

   (d) The slope is $-\frac{7}{8}$; the y-intercept is 14.

(6)

   (a) $y = 5x - 6$

   (b) $y = -x + \frac{7}{10}$

   (c) $y = \frac{2}{9}x - \frac{1}{3}$

   (d) $y = -\frac{3}{5}x + 1$

(7)

   (a) Zero slope.

   (b) Negative slope.

   (c) Undefined slope.

   (d) Positive slope.

(8)

   (a) $-\frac{3}{2}$: down 3, over 2.

   (b) 4: up 4, over 1.

   (c) $\frac{1}{2}$: up 2, over 4.

   (d) $\frac{5}{6}$: up 5, over 6.

(9) $-\frac{9}{2}$. Plug these values into the formula and simplify:

$$m = \frac{y_2 - y_1}{x_2 - x_1} = \frac{-11 - 7}{5 - 1} = \frac{-18}{4} = -\frac{9}{2}$$

(10) $\frac{4}{3}$. Plug these values into the formula and simplify:

$$m = \frac{y_2 - y_1}{x_2 - x_1} = \frac{7 - (-9)}{10 - (-2)} = \frac{7 + 9}{10 + 2} = \frac{16}{12} = \frac{4}{3}$$

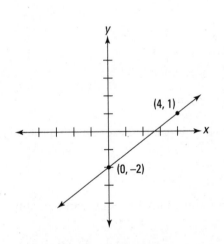

Begin by plotting the y-intercept $(0,-1)$. Now, starting from this point, use the slope of -2 to move down 2, over 1 to the point $(1,-3)$. The line that connects these two points is the graph of the equation, as shown in the figure above.

12

Begin by plotting the y-intercept $(0,-2)$. Now, starting from this point, use the slope of $\frac{3}{4}$ to move up 3, over 4 to the point $(4,1)$. The line that connects these two points is the graph of the equation, as shown in the figure above.

When you're ready, the chapter quiz in the next section provides more practice on the graphing skills you attain in this chapter.

Whaddya Know? Chapter 25 Quiz

Ready for a quiz? The 24 questions in this section will test the skills you learn in this chapter. When you're done, check out the section that follows for answers and explanations.

Questions 1–4: Refer to the figure shown here to identify the coordinates of the points.

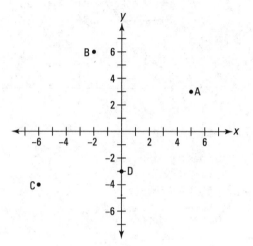

1. What are the coordinates of point A?

2. What are the coordinates of point B?

3. What are the coordinates of point C?

4. What are the coordinates of point D?

5. What is the slope of the line shown in the figure?

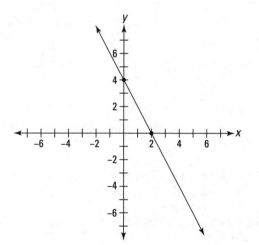

Questions 6–8: Refer to this figure to identify the line described.

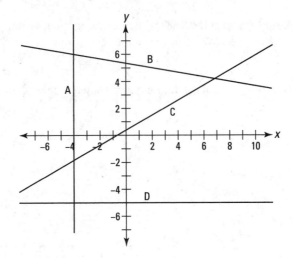

 6 Which line has a positive slope?

7 Which line has a 0 slope?

8 Which line has no slope?

9 Write the equation, in slope–intercept form, of the line shown in the figure.

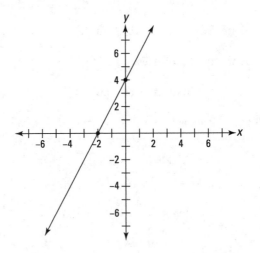

 10 Write the equation of the line, in slope–intercept form, that goes through the points $(1,9)$ and $(4,9)$.

11 Given the equation of the line $y = 3x - 4$, what are the coordinates of the point on the line where it crosses the y-axis?

 12 Write the equation of the line, in slope–intercept form, that has a slope of 7 and a y-intercept of $(0,-2)$.

13 What is the slope of the line that goes through the points $(4,-2)$ and $(3,0)$?

14 Given the equation of the line $y = -9x$, what are the coordinates of the y-intercept?

15 Write the equation of the line, in slope–intercept form, that goes through the points $(0,-2)$ and $(4,6)$.

16 Given the equation of the line $y = -\frac{1}{2}x + \frac{1}{3}$, what are the coordinates of the y-intercept?

17 Write the equation of the line, in slope–intercept form, that has a y-intercept of 4 and slope of 0.

18 Given the equation of the line $y = -x + 3$, what is the y-value when the x-value is –1?

19 Given the equations of the two lines $y = -3x + 4$ and $y = -4x + 3$, which line is steeper?

20 Given the line A shown in the figure, which line has a different slope than line A?

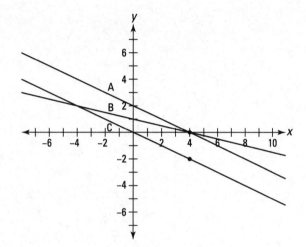

21 Given the line shown in the figure, what is the slope of the line?

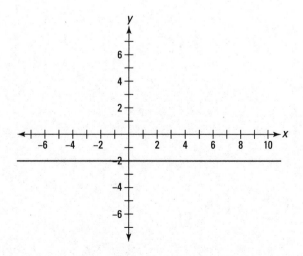

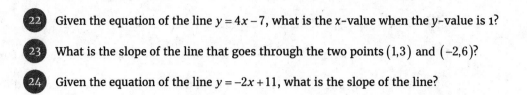

22 Given the equation of the line $y = 4x - 7$, what is the x-value when the y-value is 1?

23 What is the slope of the line that goes through the two points $(1,3)$ and $(-2,6)$?

24 Given the equation of the line $y = -2x + 11$, what is the slope of the line?

Answers to Chapter 25 Quiz

1. $(5,3)$. Point A is in the first quadrant, where both the x and y coordinates are positive.

2. $(-2,6)$. Point B is in the second quadrant, where the x-coordinate is negative, because it's on the left side of the y-axis.

3. $(-6,-4)$. Points in the third quadrant are to the left of the y-axis and below the x-axis, so their coordinates are both negative.

4. $(0,-3)$. This point is on the y-axis, so the x-coordinate is 0.

5. -2. Starting at the y-intercept of $(0,4)$, you count 4 units down and then 2 units to the right to get to the x-intercept. Making the "4 units down" a negative number, the slope is $\frac{-4}{2} = -2$.

6. **C.** Starting at any point on the line, you move upward and then to the right, making the slope a positive number.

7. **D.** Moving from one point to another on the line, you never move upward or downward, so the slope is 0.

8. **A.** Choosing any point on the line and then moving to another, you never move left or right, so the x-value in the slope is 0. You can't divide by 0, so there's no possible slope value.

9. $y = 2x + 4$. The y-intercept is $(0,4)$, so the b value in the equation is 4. The line moves upward 4 units when you move 2 units to the right, so the slope, the m value, is 2.

10. $y = 9$. The slope of this line is 0, so it's a horizontal line with a y-intercept of $(0,9)$. Using the slope-intercept form, you have $y = 0 \cdot x + 9$, which is written as just $y = 9$.

11. $(0,-4)$. The y-intercept is found in the b value, which is -4.

12. $y = 7x - 2$. The value of m is 7 and b is -2.

13. -2. Use the slope formula: $m = \frac{0-(-2)}{3-4} = \frac{2}{-1} = -2$.

14. $(0,0)$. There is no b value showing in the equation, so $b = 0$, and the coordinates of the point are (0,0).

15. $y = 2x - 2$. First, find the slope using the coordinates of the points, which is $m = \frac{6-(-2)}{4-0} = \frac{8}{4} = 2$. The value of b is found in the y-intercept, (0,-2). So you have $y = 2x - 2$.

16. $\left(0, \frac{1}{3}\right)$. The b-value is $\frac{1}{3}$. Use this as the y-coordinate of the point.

17. $y = 4$. If b equals 4 and m equals 0, then you have $y = 0 \cdot x + 4$, which is written $y = 4$.

18. **4.** Replace the x in the equation with -1 and solve for y: $y = -(-1) + 3$ simplifies to $y = 1 + 3 = 4$. So the y-value is 4.

19. $y = -4x + 3$. You can graph the lines to check this out, but, even though -3 is closer to 0 on the number line, the number -4 has the greater absolute value and creates the steeper line.

20. **B.** Line B crosses line A, so it isn't parallel to A and has a different slope.

21 **0.** The line is horizontal, so its slope is 0. Take any two points on the line and the slope formula will confirm this. For example, using the points $(1,-2)$ *and* $(4,-2)$, you have

$$m = \frac{-2-(-2)}{4-1} = \frac{-2+2}{3} = \frac{0}{3} = 0.$$

22 **2.** Replace the y in the equation with 1, and you have $1 = 4x - 7$. Solve for x by adding 7 to each side and then dividing by 4: $1 + 7 = 4x - 7 + 7$ simplifies to $8 = 4x$. Dividing each side by 4, $2 = x$.

23 **-1.** Using the slope formula, $m = \frac{6-3}{-2-1} = \frac{3}{-3} = -1.$

24 **-2.** The equation is in the form $y = mx + b$, so the slope is the coefficient of x, which is -2.

Index

associative property, 59–62

axes, 554

B

balance scale method, 503–506

base numbers

  changing, 332–334

  defined, 66

basic unit, 369

borrowing, subtracting mixed numbers with, 222–224

box, 407, 409–410

braces ({}), 86, 452

brackets ([]), 86

C

calculating

  discounts, 317–318

  long division, 22–24

  probability, 440–441

  slope of a line, 564–572

  stacked addition, 18–19

  stacked multiplication, 21–22

  stacked subtraction, 19–21

canceling units of measurement, 376

capacity, 366

cardinality, of sets, 453

carrying, adding mixed numbers with, 217–220

center, of circles, 405

changing

  base numbers, 332–334

  between decimals and fractions, 262–270

  between English and metric systems, 372–377

  fractions to decimals, 267–270

  between improper fractions and mixed numbers, 174–178

  percentages to/from decimals, 287–288

  percentages to/from fractions, 287–290

  slope (m), 560–561

  to/from percentages, 287

  units of measurement, 375–377

  y-intercept (b), 561–562

Cheat Sheet (website), 3

choosing variables, 536–538

circles, 399, 405–406

circumference, of circles, 405

coefficients

  defined, 470

  identifying, 472–473

combining like terms, 479–480

commutative property, 56–59, 471–472

comparing fractions, with cross-multiplication, 178–179

complement, of a set, 457–458

composite numbers

  about, 10–11

  identifying, 136–138

cone, 408–409, 410

conversion factors, 375–377

converting

  base numbers, 332–334

  between decimals and fractions, 262–270

  between English and metric systems, 372–377

  fractions to decimals, 267–270

  between improper fractions and mixed numbers, 174–178

  percentages to/from decimals, 287–288

  percentages to/from fractions, 287–290

  slope (m), 560–561

  to/from percentages, 287

  units of measurement, 375–377

  y-intercept (b), 561–562

coordinate points, plotting, 554–555

counting by numbers, 9

counting numbers, 13

counting zeros, 348–350

cross-multiplication

  about, 512–514

  comparing fractions with, 178–179

cube roots, 330–331, 407, 408

division *(continued)*
  of fractions, 192–195
  inverse operations and, 56
  of like terms, 475–479
  of mixed numbers, 214–215
  of powers of ten, 350–351
  with scientific notation, 355–356
  of units, 364–365
division sign, as an operator, 18
divisor, 257
dodecahedron, 408
doesn't equal (≠), 63

E

eight (8), divisible by, 134–135
elements, of sets, 452–454
eleven (11), divisible by, 132–134
ellipsis (. . .), 452
empty set, 454
English system, 365–369, 372–373, 377–380
equal sets, 453
equals sign (=), 504, 507–509
equations
  about, 78
  algebraic
    about, 499–500
    balanced, 504
    choosing a method for solving, 501–503
    cross-multiplying, 512–514
    example questions and answers, 502–503, 505–506, 508–509, 511, 513–514
    graphing (*See* graphing)
    isolating x, 504–514
    moving terms, 507–509
    practice questions answers and explanations, 515–524
    quiz questions and answers, 525–529
    rearranging, 506–514
    removing parentheses from, 509–511
    solving for x, 503–506
    using x in equations, 500

combined with evaluation and expressions, 79
defined, 77
graphing on x-y plane, 555–559
linear
  about, 559–560
  changing slope (m), 560–561
  changing y-intercept (b), 561–562
  graphing using slope and y-intercept, 572–573
  slope-intercept form, 562–563
setting up, 533
solving percent problems with, 295–298
turning word problems into word, 105–109
equilateral triangle, 397–398
equivalent sets, 453
estimating
  distance, 373
  in English system, 372–377
  in metric system, 372–377
  slope, 564–566
  speed, 373
  temperature, 374
  value, 32–33
  volume, 373–374
  weight, 374
evaluation
  about, 79
  of algebraic expressions, 467–469
  combined with equations and expressions, 79
  defined, 77
even numbers, 8–9
Example icon, 2
example questions and answers
  algebra word problems, 535, 537, 542–543
  algebraic equations, 502–503, 505–506, 508–509, 511, 513–514
  algebraic expressions, 469, 470, 472, 473, 475, 477–478, 480, 483–484, 485–486
  decimals, 251, 255, 257, 261, 263, 266, 270
  divisibility, 129–130, 133, 135, 136, 137–138

mixed-operator expressions, 82–83

nested parentheses, 86–87

parentheses and powers, 85–86

powers, 83–87

practice questions answers and
  explanations, 89–97

prioritizing parentheses, 84–85

quiz questions and answers, 98–101

Three E's, 78–79

order of precedence. *See* order of operations

origin, 554

P

parallel lines, 394

parallelogram, 398

parentheses (())

  with minus signs, 481–482

  for multiplication, 17

  nested, 86–87

  with no sign, 482–485

  operations and, 59–62

  with plus signs, 481

  powers and, 85–86

  prioritizing, 84–85

  removing

    from algebraic expressions, 481–484

    from equations, 509–511

PEMDAS, 80. *See also* order of operations

percentages

  about, 285–286

  converting to/from

    about, 287

    decimals, 287–288

    fractions, 287–290

  example questions and answers, 287, 288,
    289, 290, 292, 293, 294, 298

  greater than 100%, 286

  multiplying in word problems, 313–315

  percent increases/decreases in word
    problems, 316–318

practice questions answers and
  explanations, 299–302

quiz questions and answers, 303–306

relationship with statistics and
  probability, 434

solving

  percent problems with equations, 295–298

  problems, 290–294

types of problems, 294–295

in word problems, 309–310

pi (π) symbol, 405

place value

  about, 28–30

  identifying for decimals, 246–247

placeholders, 248–249

plane geometry, 394–396

plotting a point, 554–555

plotting coordinates, 554–555

plus sign (+)

  as an operator, 16

  parentheses with, 481

points, in geometry, 394

polygons

  about, 396, 397

  large, 399

  quadrilaterals, 398–399

  triangles, 397–398

polyhedrons, 407–408

positive slope, 564

powers and roots

  about, 83–84, 329

  changing base numbers, 332–334

  cubic numbers, 330–331

  cubic roots, 330–331

  example questions and answers, 332,
    333–334, 335, 336–337, 338–339, 340

  exponents of 0, 334–335

  finding

    powers of fractions, 333

    powers of negative numbers, 332–333

About the Author

Mark Zegarelli is a teacher, tutor, and co-founder of Simple Step Learning (www.simple steplearning.com), a web-based educational resource. He is the author of numerous *For Dummies* books, including *Basic Math and Pre-Algebra For Dummies, SAT Math For Dummies,* and *Calculus II For Dummies.* He holds degrees in both English and math from Rutgers University.

Mark currently works online with parents of preschool and school-age children, teaching kids surprisingly advanced math in a fun way, and showing their parents how to support their learning.

Contact Mark at www.markzegarelli.com.

Dedication

光耀永远。

Author's Acknowledgments

This is my eleventh *For Dummies* book and, again, I'm pleased and proud to have shared the journey with so many talented, helpful people who call me to my best. Many thanks to Lindsay Lefevere, Michelle Hacker, Christina Guthrie, Marylouise Wiack, Amy Nicklin, and Mohammed Zafar Ali.

Thank you to Mary Jane Sterling for supplying the quizzes that end each chapter. And thank you again to Chris Mark for helping to edit the final manuscript.

And a special round of applause for everybody at Castro Coffee Company for brewing that all-important first cup of the day!

Publisher's Acknowledgments

Executive Editor: Lindsay Sandman Lefevere

Project Manager and Development Editor:
  Christina N. Guthrie

Managing Editor: Michelle Hacker

Copy Editor: Marylouise Wiack

Technical Editor: Amy Nicklin

Production Editor: Mohammed Zafar Ali

Cover Image: © Adobe Stock/ lukbar

Leverage the power

Dummies is the global leader in the reference category and one of the most trusted and highly regarded brands in the world. No longer just focused on books, customers now have access to the dummies content they need in the format they want. Together we'll craft a solution that engages your customers, stands out from the competition, and helps you meet your goals.

Advertising & Sponsorships

Connect with an engaged audience on a powerful multimedia site, and position your message alongside expert how-to content. Dummies.com is a one-stop shop for free, online information and know-how curated by a team of experts.

- Targeted ads
- Video
- Email Marketing
- Microsites
- Sweepstakes sponsorship

20 **MILLION** PAGE VIEWS EVERY SINGLE MONTH

15 MILLION UNIQUE VISITORS PER MONTH

43% OF ALL VISITORS ACCESS THE SITE VIA THEIR MOBILE DEVICES

700,000 NEWSLETTER SUBSCRIPTION TO THE INBOXES OF

300,000 UNIQUE INDIVIDUALS EVERY WEEK

# of dummies

Custom Publishing

Reach a global audience in any language by creating a solution that will differentiate you from competitors, amplify your message, and encourage customers to make a buying decision.

- Apps
- Books
- eBooks
- Video
- Audio
- Webinars

Brand Licensing & Content

Leverage the strength of the world's most popular reference brand to reach new audiences and channels of distribution.

For more information, visit dummies.com/biz

PERSONAL ENRICHMENT

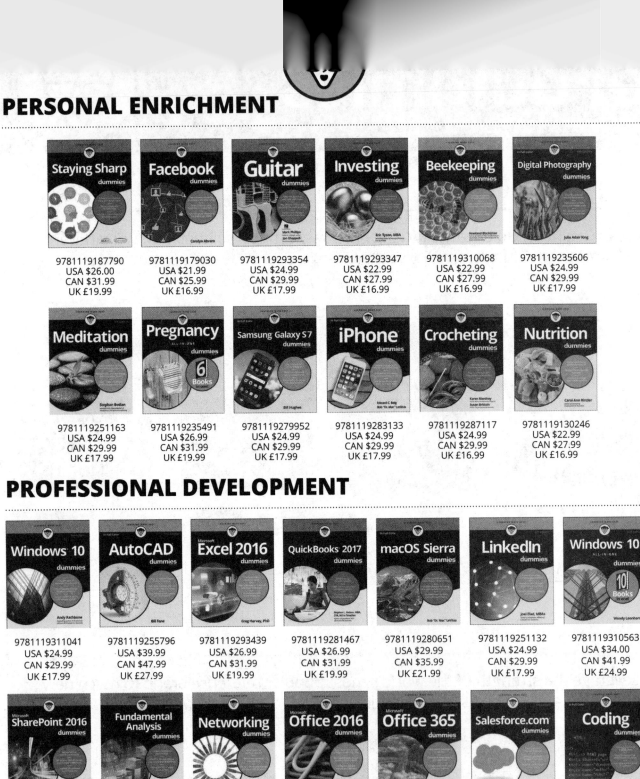

Staying Sharp dummies
9781119187790
USA $26.00
CAN $31.99
UK £19.99

Facebook dummies
Carolyn Abram
9781119179030
USA $21.99
CAN $25.99
UK £16.99

Guitar dummies
Mark Phillips
Jon Chappell
9781119293354
USA $24.99
CAN $29.99
UK £17.99

Investing dummies
Eric Tyson, MBA
9781119293347
USA $22.99
CAN $27.99
UK £16.99

Beekeeping dummies
Howland Blackiston
9781119310068
USA $22.99
CAN $27.99
UK £16.99

Digital Photography dummies
Julie Adair King
9781119235606
USA $24.99
CAN $29.99
UK £17.99

Meditation dummies
Stephan Bodian
9781119251163
USA $24.99
CAN $29.99
UK £17.99

Pregnancy ALL-IN-ONE dummies
9781119235491
USA $26.99
CAN $31.99
UK £19.99

Samsung Galaxy S7 dummies
Bill Hughes
9781119279952
USA $24.99
CAN $29.99
UK £17.99

iPhone dummies
Edward C. Baig
Bob "Dr. Mac" LeVitus
9781119283133
USA $24.99
CAN $29.99
UK £17.99

Crocheting dummies
Karen Manthey
Susan Brittain
9781119287117
USA $24.99
CAN $29.99
UK £16.99

Nutrition dummies
Carol Ann Rinzler
9781119130246
USA $22.99
CAN $27.99
UK £16.99

PROFESSIONAL DEVELOPMENT

Windows 10 dummies
Andy Rathbone
9781119311041
USA $24.99
CAN $29.99
UK £17.99

AutoCAD dummies
Bill Fane
9781119255796
USA $39.99
CAN $47.99
UK £27.99

Excel 2016 dummies
Greg Harvey, PhD
9781119293439
USA $26.99
CAN $31.99
UK £19.99

QuickBooks 2017 dummies
Stephen L. Nelson, MBA, CPA, MS in Taxation
9781119281467
USA $26.99
CAN $31.99
UK £19.99

macOS Sierra dummies
Bob "Dr. Mac" LeVitus
9781119280651
USA $29.99
CAN $35.99
UK £21.99

LinkedIn dummies
Joel Elad, MBAs
9781119251132
USA $24.99
CAN $29.99
UK £17.99

Windows 10 ALL-IN-ONE dummies
Woody Leonhard
9781119310563
USA $34.00
CAN $41.99
UK £24.99

SharePoint 2016 dummies
Rosemarie Withee
Ken Withee
9781119181705
USA $29.99
CAN $35.99
UK £21.99

Fundamental Analysis dummies
Matt Krantz
9781119263593
USA $26.99
CAN $31.99
UK £19.99

Networking dummies
Doug Lowe
9781119257769
USA $29.99
CAN $35.99
UK £21.99

Office 2016 dummies
Wallace Wang
9781119293477
USA $26.99
CAN $31.99
UK £19.99

Office 365 dummies
Rosemarie Withee
Ken Withee
Jennifer Reed
9781119265313
USA $24.99
CAN $29.99
UK £17.99

Salesforce.com dummies
Liz Kao
Jon Paz
9781119239314
USA $29.99
CAN $35.99
UK £21.99

Coding dummies
Nikhil Abraham
9781119293323
USA $29.99
CAN $35.99
UK £21.99

Learning Made Easy

ACADEMIC

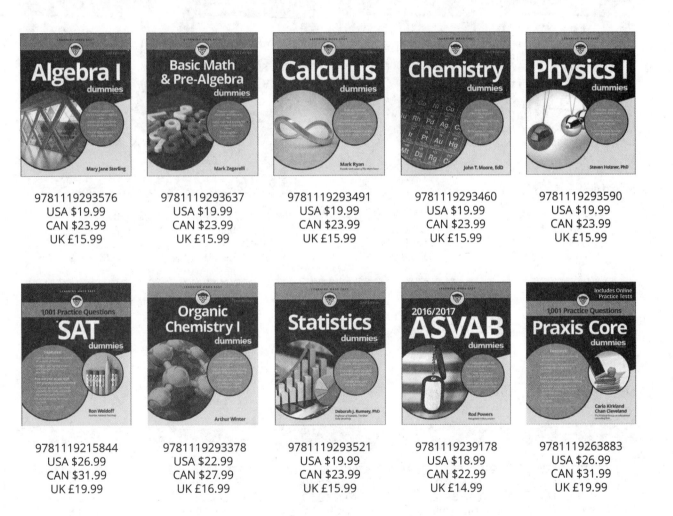

Algebra I dummies Mary Jane Sterling	**Basic Math & Pre-Algebra** dummies Mark Zegarelli	**Calculus** dummies Mark Ryan	**Chemistry** dummies John T. Moore, EdD	**Physics I** dummies Steven Holzner, PhD
9781119293576 USA $19.99 CAN $23.99 UK £15.99	9781119293637 USA $19.99 CAN $23.99 UK £15.99	9781119293491 USA $19.99 CAN $23.99 UK £15.99	9781119293460 USA $19.99 CAN $23.99 UK £15.99	9781119293590 USA $19.99 CAN $23.99 UK £15.99
SAT dummies 1,001 Practice Questions Ron Woldoff	**Organic Chemistry I** dummies Arthur Winter	**Statistics** dummies Deborah J. Rumsey, PhD	**2016/2017 ASVAB** dummies Rod Powers	**Praxis Core** dummies 1,001 Practice Questions Carla Kirkland, Chan Cleveland
9781119215844 USA $26.99 CAN $31.99 UK £19.99	9781119293378 USA $22.99 CAN $27.99 UK £16.99	9781119293521 USA $19.99 CAN $23.99 UK £15.99	9781119239178 USA $18.99 CAN $22.99 UK £14.99	9781119263883 USA $26.99 CAN $31.99 UK £19.99

Available Everywhere Books Are Sold

dummies.com

Small books for big imaginations

GETTING STARTED WITH Coding
Get Creative with Code!
Camille McCue, PhD
9781119177173
USA $9.99
CAN $9.99
UK £8.99

MODDING Minecraft
Build Your Own Minecraft Mods!
Sarah Guthals, PhD
Stephen Foster, PhD
Lindsay Handley, PhD
9781119177272
USA $9.99
CAN $9.99
UK £8.99

MAKING YouTube VIDEOS
Star in Your Own Video!
Nick Willoughby
9781119177241
USA $9.99
CAN $9.99
UK £8.99

DESIGNING Digital Games
Create Games with Scratch!
Derek Breen
9781119177210
USA $9.99
CAN $9.99
UK £8.99

GETTING STARTED WITH Raspberry Pi
Program Your Raspberry Pi!
Richard Wentk
9781119262657
USA $9.99
CAN $9.99
UK £6.99

EXPERIMENTING WITH Science
Think, Test, and Learn!
9781119291336
USA $9.99
CAN $9.99
UK £6.99

CREATING Digital Animations
Animate Stories with Scratch!
Derek Breen
9781119233527
USA $9.99
CAN $9.99
UK £6.99

GETTING STARTED WITH Engineering
Think Like an Engineer!
9781119291220
USA $9.99
CAN $9.99
UK £6.99

WRITING Computer Code
Learn the Language of Computers!
Chris Minnick and Eva Holland
9781119177302
USA $9.99
CAN $9.99
UK £8.99

Unleash Their Creativity

dummies.com

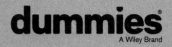

dummies
A Wiley Brand